Control Systems Design

Vladimir Zakian (Editor)

Control Systems Design

A New Framework

With 89 Figures

Vladimir Zakian, PhD
12 Cote Green Road
Marple Bridge
Stockport
SK6 5EH
UK

British Library Cataloguing in Publication Data
Control systems design : a new framework
   1. Automatic control
   I. Zakian, Vladimir
   629.8
ISBN 1852339136

Library of Congress Cataloging-in-Publication Data
Control systems design : a new framework / [edited by] Vladimir Zakian.
      p. cm.
   Includes bibliographical references and index.
   ISBN 1-85233-913-6
   1. Automatic control.   I. Zakian, Vladimir.
   TJ213.C5728   2005
   629.8—dc22
                                                                          2004051252

Apart from any fair dealing for the purposes of research or private study, or criticism or review, as permitted under the Copyright, Designs and Patents Act 1988, this publication may only be reproduced, stored or transmitted, in any form or by any means, with the prior permission in writing of the publishers, or in the case of reprographic reproduction in accordance with the terms of licences issued by the Copyright Licensing Agency. Enquiries concerning reproduction outside those terms should be sent to the publishers.

ISBN 1-85233-913-6
Springer Science+Business Media
springeronline.com

© Springer-Verlag London Limited 2005

The use of registered names, trademarks, etc. in this publication does not imply, even in the absence of a specific statement, that such names are exempt from the relevant laws and regulations and therefore free for general use.

The publisher makes no representation, express or implied, with regard to the accuracy of the information contained in this book and cannot accept any legal responsibility or liability for any errors or omissions that may be made.

Typesetting: Camera ready by editor
Printed in the United States of America
69/3830-543210   Printed on acid-free paper   SPIN 11318958

Preface

In recent decades, a new framework for the design of control systems has emerged. Its development was prompted in the 1960s by two factors. First, was the arrival of interactive computing facilities, which opened new avenues for design, relying more on numerical methods. In this way, routine computational tasks, which are a significant part of design, are left to the computer, thus allowing the designer to focus on the formulation of the design problem, which requires creative skills. This has led to a major shift in the field of design, with more emphasis placed on general principles for the formulation of design problems. Second, is a perceived disparity, concerning the aims of control, between conventional control theory and a more precise and generally understood meaning of control. Conventional design theory requires the system to have a good margin of stability. In the new framework, control is achieved when the errors and other controlled variables are kept within respective specified tolerances, whatever the disturbances to the system. This criterion reflects more accurately the nature of the classical control problem and is the criterion used in some industries. In many situations, involving what are called critical systems, some controlled variables must not exceed their prescribed tolerances and this means that there is no satisfactory alternative to the explicit use of the criterion that requires certain outputs to be bounded by their tolerances.

The conventional framework for the design of control systems was formed during the five-year period ending in 1945. During that period, known ideas from the fields of electrical circuit theory and servomechanisms, with feedback theory as their common ground, were merged to form a practical approach to design that was eventually adopted by the control community. This framework has been greatly elaborated and generalised, and now contains many design methods that are essentially equivalent. However, it remains, in essence, a practical approach to design and it gained acceptance because of its wide range of practical applications.

An essential characteristic of the new framework is that, like the conventional framework, its design methods are useable in practice. This characteristic has resulted from the way the new framework has been developed, by an evolutionary process. Starting with a minimum of theory, new design approaches, making use of numerical methods, were evolved and these were tested on challenging design problems. Successful tests led to new design theories and methods that were, in turn, tested and so on. Throughout this process, theory has been kept within the level needed to fulfil the aims of design. At present, not only can designs satisfying the conventional criterion be achieved but also critical systems, which are beyond the scope of the conventional framework, can be designed with equal facility.

The new framework has involved the development of aspects of theory that have no exact counterpart in the conventional framework. These are concerned with general principles of design and comprise the principle of inequalities and the principle of matching. These principles facilitate an accurate and realistic formulation of the design problem and form part of the foundation of the new framework, upon which a superstructure of design methods has been built.

The time has come to bring together, into a book, the components of the new framework that have been scattered in the research literature. This is accomplished in the chapters of the book, where the authors aim to explain, revise, correct or expand the ideas and results contained in published papers. Some of the material has not been published before. The book is addressed to those who use or develop practical methods for the design of control systems and to those concerned with the theory underlying such methods.

Although the new framework is still growing, its conceptual basis is now sufficiently coherent to allow further systematic development and it contains sufficient methods to permit a wide range of practical applications. However, because it is relatively new, it offers the researcher unsolved problems, some of which are highlighted in the book.

The principal aim of the book is to present the new framework. A secondary aim is to show how the conventional framework can be made more effective by the use of the method of inequalities. The method involves the use of numerical processes together with two principles of design, one of which is based on the classical concept of stability, and the other is the principle of inequalities. This secondary aim recognises that the conventional framework is likely to have, for the foreseeable future, a continuing role in the field of control, despite the fact that its range of applications is more restricted than that of the new framework.

It is hoped that the book will provide, students and researchers in universities and practitioners in industry, ready access to this field. It is also intended to be a source book from which other, perhaps more integrated or specifically oriented, books on this subject can be written.

Each chapter of the book is almost as self-contained as a paper in a journal. References to the literature mainly indicate the primary sources of the material. However, the chapters were written in accordance with an overall plan for the book. Cross-references to other chapters indicate where certain topics are dealt with more fully or may indicate the sources new material. The chapters are grouped into four parts and their order is intended to give a logical rather than a chronological sequence for the development of the subject, with Part I providing some introductions to the material in other parts. Nonetheless, the reader might find it more appropriate to choose another sequence. A reader interested in basic principles might start with Part I, while a reader interested in applications might start with the chapters in Part IV, which contain applications and case studies of the basic principles and methods developed in other parts, showing how various challenging practical problems can be formulated and solved. These problems are of two kinds, those that are formulated in terms of the conventional criterion of design and those, involving critical systems, which can be formulated and solved within the new framework. Parts II and III contain the essential computational and numerical methods required to put the framework into practice.

V. Zakian
February 2005

Acknowledgements The editor acknowledges the cooperation of all the contributors, in the preparation of this book. Thanks are due to James Whidborne and Toshiyuki Satoh for formatting the book to the publishers specifications.

Contents

List of Contributors xii

Part I. Basic Principles

1 Foundation of Control Systems Design
Vladimir Zakian 3
1.1 Need for New Foundation 3
1.2 The Principle of Inequalities 19
1.3 The Principle of Matching 32
1.4 A Class of Linear Couples 43
1.5 Well-constructed Environment-system Models 48
1.6 The Method of Inequalities 61
1.7 The Node Array Method for Solving Inequalities 68
References 92

Part II. Computational Methods (with Numerical Examples)

2 Matching Conditions for Transient Inputs
Paul Geoffrey Lane 97
2.1 Introduction 97
2.2 Finiteness of Peak Output 98
2.3 Evaluation of Peak Output 99
2.4 Example 106
2.5 Miscellaneous Results 111
2.6 Conclusion 115
References 119

3 Matching to Environment Generating Persistent Disturbances
Toshiyuki Satoh 121
3.1 Introduction 121
3.2 Preliminaries 123
3.3 Computation of Peak Output via Convex Optimisation 124
3.4 Algorithm for Computing Peak Output 135

3.5	Numerical Example	135
3.6	Conclusions	143
References		144

4 LMI-based Design

Takahiko Ono 145

4.1	Introduction	145
4.2	Preliminary	146
4.3	Problem Formulation	149
4.4	Controller Design via LMI	150
4.5	Numerical Example	161
4.6	Conclusion	163
References		164

5 Design of a Sampled-data Control System

Takahiko Ono 165

5.1	Introduction	165
5.2	Design for SISO Systems	167
5.3	Design for MIMO Systems	184
5.4	Design Example	186
5.5	Conclusion	189
References		189

Part III. Search Methods (with Numerical Tests)

6 A Numerical Evaluation of the Node Array Method

Toshiyuki Satoh 193

6.1	Introduction	193
6.2	Detection of Stuck Local Search	194
6.3	Special Test Problems	195
6.4	Test Results	208
6.5	Effect of Stopping Rule 3	210
6.6	Conclusions	211
6.A	Appendix — Moving Boundaries Process with the Rosenbrock Trial Generator	211
References		215

7 A Simulated Annealing Inequalities Solver

James F Whidborne 219

7.1	Introduction	219
7.2	The Metropolis Algorithm	220
7.3	A Simulated Annealing Inequalities Solver	221
7.4	Numerical Test Problems	224
7.5	Control Design Benchmark Problems	225
7.6	Conclusions	228

7.A Appendix – the Objective Functions 228
References 229

8 Multi-objective Genetic Algorithms for the Method of Inequalities

Tung-Kuan Liu and Tadashi Ishihara 231
8.1 Introduction 231
8.2 Auxiliary Vector Index 232
8.3 Genetic Inequalities Solver 238
8.4 Numerical Test Problems 243
8.5 Control Design Benchmark Problems 245
8.6 Conclusions 246
References 247

Part IV. Case Studies

9 Design of Multivariable Industrial Control Systems by the Method of Inequalities

Oluwafemi Taiwo 251
9.1 Introduction 251
9.2 Application of the Method of Inequalities to Distillation Columns 255
9.3 Design of Multivariable Controllers for an Advanced Turbofan Engine by the Method of Inequalities 269
9.4 Improvement of Turbo-alternator Response by the Method of Inequalities 277
References 281

10 Multi-objective Control using the Principle of Inequalities

G P Liu 287
10.1 Introduction 287
10.2 Multi-objective Optimal-tuning PID Control 288
10.3 Multi-objective Robust Eigenstructure Assignment 294
10.4 Multi-objective Critical Control 302
References 308

11 A MoI Based on $\mathcal{H}^\infty$ Theory — with a Case Study

James F Whidborne 311
11.1 Introduction 311
11.2 Preliminaries 313
11.3 A Two Degree-of-freedom $\mathcal{H}^\infty$ Method 314
11.4 A MoI for the Two Degree-of-freedom Formulation 319
11.5 Example — Distillation Column Controller Design 320
11.6 Conclusions 325
References 325

12 Critical Control of the Suspension for a Maglev Transport System
James F Whidborne 327
12.1 Introduction 327
12.2 Theory 329
12.3 Model 330
12.4 Design Specifications 332
12.5 Performance for Control System Design 332
12.6 Design using the MoI 333
12.7 Conclusions 337
References 337

13 Critical Control of Building under Seismic Disturbance
Suchin Arunsawatwong 339
13.1 Introduction 339
13.2 Computation of Performance Measure 341
13.3 Model of Building 344
13.4 Design Formulation 347
13.5 Numerical Results 349
13.6 Discussion and Conclusions 352
References 353

14 Design of a Hard Disk Drive System
Takahiko Ono 355
14.1 Introduction 355
14.2 HDD Systems Design 356
14.3 Performance Evaluation 362
14.4 Conclusions 367
References 367

15 Two Studies of Robust Matching
Oluwafemi Taiwo 369
15.1 Introduction 369
15.2 Robust Matching and Vague Systems 371
15.3 Robust Matching for Plants with Recycle 374
15.4 Robust Matching for the Brake Control of a Heavy-duty Truck
     – a Critical System 380
References 385

Index 387

List of Contributors

Suchin Arunsawatwong
Department of Electrical Engineering
Faculty of Engineering
Chulalongkorn University
Bangkok 10330
Thailand
e-mail: suchin.a@chula.ac.th

Tadashi Ishihara
Faculty of Science and Technology
Fukushima University
Kanayagawa 1, Fukushima, 960-1296
Japan
e-mail:
ishihara@educ.fukushima-u.ac.jp

Paul Geoffrey Lane
Federal Agricultural Research
Centre (FAL)
Institute for Technology and
Biosystems Engineering,
Bundesallee 50
38116 Braunschweig,
Germany
e-mail: Paul.Lane@fal.de

Guoping Liu
School of Electronics
University of Glamorgan
Pontypridd CF37 1GY
United Kingdom
e-mail: gpliu@glam.ac.uk

Tung-Kuan Liu
Department of Mechanical Automation Engineering
National Kaohsiung First University
of Science and Technology
2 Juoyue Road, Nantz District
Kaohsiung 811
Taiwan, R.O.C.
e-mail: tkliu@ccms.nkfust.edu.tw

Takahiko Ono
Department of Information Machines
and Interfaces
Graduate School of Information
Sciences
Hiroshima City University
3-4-1 Ozuka-Higashi
Asa-Minami-Ku, Hiroshima,
731-3194
Japan
e-mail:
ono@cntl.im.hiroshima-cu.ac.jp

Toshiyuki Satoh
Department of Machine Intelligence
and Systems Engineering
Faculty of Systems Science and
Technology
Akita Prefectural University,
84-4, Ebinokuchi, Tsuchiya
Honjoh, Akita, 015-0055
Japan
e-mail: tsatoh@akita-pu.ac.jp

Oluwafemi Taiwo
Chemical Engineering Department
Obafemi Awolowo University
Ile-Ife
Nigeria
e-mail: ftaiwo@oauife.edu.ng

James F. Whidborne
Department of Aerospace Sciences
Cranfield University
Bedfordshire MK43 0AL
United Kingdom
e-mail:
j.f.whidborne@cranfield.ac.uk

Vladimir Zakian
12 Cote Green Road
Marple Bridge, Stockport SK6 5EH
United Kingdom
e-mail: vzakian@onetel.com

Part I

Basic Principles

1 Foundation of Control Systems Design

Vladimir Zakian

Abstract. The need for a new conceptual foundation for the design of control systems is explained. A new foundation is proposed, comprising a definition of control and three principles of systems design: the principle of inequalities, the principle of matching and the principle of uniform stability. A design theory is built on this foundation. The theory brings into sharper focus, hitherto elusive but central, concepts of tolerance to disturbances and over-design. It also gives ways of characterising a good design. The theory is shown to be the basis of design methods that can cope with, important and commonly occurring, design problems involving critical systems and other problems, where strict bounds on responses are required. The method of inequalities, that can be used to design such systems, is discussed.

1.1 Need for New Foundation

Following a brief examination of the foundation of the conventional framework for the design of control systems, the need for a new framework is identified. The components of the foundation of this new framework are outlined in this section and developed in some detail in the rest of the chapter.

1.1.1 The Conventional Foundation for Design

Although control mechanisms have been known since antiquity, two well known papers, Maxwell's (1868) and Nyquist's (1932), have been influential in forming the foundation of what is now the mainstream theory for the design of control systems, with its remarkable successes and, as will be seen, some significant limitations. These papers introduced two key ideas, respectively called stability and sensitivity, which constitute the foundation of the conventional framework for control systems design. The two ideas are here reviewed informally and briefly, in the context of more recent developments that concern the foundation of control systems design. Accordingly attention is, in this section, focused on the meaning of control and not on the means (that is to say, the design methods) needed to achieve it. This makes it possible to examine the foundation of conventional design theory and to see how it was originated.

It is assumed that a control system operates in an *environment* that generates the input for the system. See Figure 1.1. The *input* is a vector of time functions, the components of which occur at corresponding *input ports* of the system. The input is transformed by the system into a response, which is a set of time functions. Some of the individual responses are classified as *outputs*, which occur at corresponding *output ports*. Some of the output ports are classified as error ports. The response at an error port is called an *error*. Unless an explicit distinction is made, the word *system* means either the concrete physical situation, which is a primary focus of interest, or its mathematical model. The word *environment* also has similar dual meaning. The term *port* means a location, either on the physical system or on a corresponding block diagram model, where a scalar input or a scalar response occurs. The notion of a port provides a convenient way of taking into account the fact that the input and the output are, in general, vectors. The system is comprised of two subsystems, the plant and the controller, connected together by mutual interaction; that is to say, in a feedback arrangement.

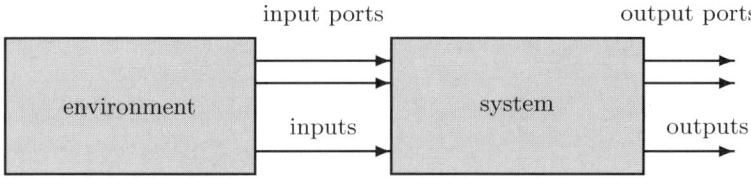

Fig. 1.1. Environment-system couple

The way that certain output ports of the system are classified as error ports, depends on the design situation. An error, which is the response at the corresponding error port, is required to be small. How small it is required to be is one of the crucial aspects of control that is considered in the new design framework proposed in this chapter.

The environment is modelled by a set of generators and, if necessary, corresponding filters. Each generator produces a scalar function of time, called an input, which feeds a corresponding filter, the output of which is a scalar function of time that feeds into an input port of the system. In some cases, a filter is not necessary and is replaced by the identity transformation. The term filter-system combination will denote such an arrangement but this will sometimes be abbreviated to the simpler term system, when there is no risk of confusion. Similarly, the term input port will refer either to the input of the system or to the input of the filter, depending on the context. This terminology could be simplified by defining the system so as to include the filters but that would obscure the fact that the filters are part of the environment. It is important to maintain a clear distinction between the environment and the control system, especially when the model of the environment is considered in detail (see Section 1.5).

Unless otherwise stated, it is assumed that the filter-system combination can be represented by ordinary linear differential equations with constant coefficients, expressed in the standard state-space form by the two equations: $\dot{x} = Ax + Bf$, $e = Cx + Df$. Here, as usual, x is the *state* vector, f is the input vector of dimension n, produced by the generators, and e is the system output vector of dimension m. The integers n and m are the numbers of input and output ports, respectively. A *response* is any linear combination of states and inputs. An output is a particular linear combination of states and inputs defined by the matrices C and D. The value of the state vector at time zero is assumed to be zero. The matrix D characterises the non-dynamic part of the input-output behaviour of the system and is called the *direct transmission matrix*.

The assumption that the standard state-space equations represent the filter-system combination is made here partly to simplify the presentation. Many of the ideas in this chapter are applicable to very general systems (see Sections 1.2 and 1.3) or to the more general linear time-invariant systems (see Sections 1.4 and 1.5), notably those having time delays, and can also be translated to make them applicable to sampled-data systems. Also, the ideas can be extended to any *vague system*, the input-output transformation of which is characterised by a known set of rational transfer functions, which can be used to characterise a time-varying or non-linear system or a linear time invariant system whose parameters are not precisely known.

As shown in Section 1.5, there are two ways of determining the environment filters. In one way, if all the environment filters are chosen appropriately (in some cases every filter can be chosen to be the identity transformation) then the direct transmission matrix D is equal to zero. In the other way, every filter is chosen to be the identity transformation and suitable restrictions are imposed on the derivative of the input.

The *characteristic polynomial* of the system is defined by $\det(sI - A)$, and its zeros are called the characteristic roots of the system. For each characteristic root α_i, there is a *mode*, of the form $t^{k_i} \exp(\alpha_i t)$, which characterises the behaviour of the system.

A mode is said to be *controllable* if it can be excited by some input. If every mode is controllable then the system is said to be controllable. If every mode can be excited by some nonzero input from the environment then the environment is said to be *probing*. Although, for design purposes, an adequate working model of the environment might not be probing, it is safe to assume that an accurate model would be probing because an accurate model might take into account small parasitic inputs that are ignored in the working model. Such parasitic inputs can be significant, as will be seen, however small they might be. Notice that, an environment is probing with respect to a given system, implies that the system is controllable. This emphasises that the properties of the system are dependent on the properties of the environment and *vice versa*. Thus, for design purposes, the environment and the system must also be considered as a single unit, called the *environment-system couple*, and not only as two separate entities. The notion of environment-system couple plays a major role in the framework presented in this chapter. Accordingly, the environment and the ways it can be modelled play a central role in the new framework.

For each input-output pair of ports there is an input-output transformation (operator) that, for the purpose of analysis, can be considered in isolation from the system. Under the assumption that the system can be represented by standard state space equations, this transformation can be represented by a rational transfer function, which is proper (numerator degree not greater than the denominator degree) and without common factors between the numerator and denominator. If the filter-system combination has zero direct transmission matrix D then, for every input-output pair of ports, the transfer function is strictly proper (has denominator degree greater than the numerator degree).

Informal definition of control: Suppose that the environment-system couple is such that the environment is probing. Then the environment-system couple is said to be under control if the following three conditions are satisfied. First, for every input generated by the environment, all the states are bounded. Second, for every error port, the response (the error) stays close to zero for all time. Third, for some specified output ports (other than the error ports), the response is not too large for all time. □

This definition involves only input-response concepts. However, the definition is purely qualitative because the second and third conditions for control are not quantified. That is to say, how small the errors are required to be is not stated in quantitative terms and, for the remaining output ports, what is considered to be too large a response, is again not stated in quantitative terms. Usually, the responses at those output ports that are not error ports, represent the behaviour of actuators or other physical devices, whose range of operation is limited, often because of saturation but sometimes also for other reasons, such as limits imposed on the consumption of power. Notice also that it is not just the system that is under control but the environment-system couple. This is because the environment produces the input that, together with the system, determines the size of the responses. However, the environment is not specified quantitatively and, consequently, the responses at the output ports cannot be quantified.

For the purpose of analysis, Maxwell considered the system in isolation from the environment and defined a practical algebraic concept of stability. According to this, a system is, by definition, *stable* if all its modes are stable and each mode of the system is, by definition, stable if its characteristic root has negative real part. Hence the absolute value of a stable mode is bounded by a constant multiplied by an exponential function that decays with time. Maxwell's condition for stability is relevant and necessary for control. To see this, suppose that the environment is restricted so that, for every input generated by the environment, the states are bounded if the filter-system combination is stable (in fact, stability of the filter-system combination is necessary and sufficient to ensure that, for every bounded input, all the states are bounded). This implies that, provided the filter-system is stable, any environment that generates only bounded inputs causes only bounded states and hence bounded outputs. These ideas are generalised somewhat in Section 1.4 for filter-system combinations with zero direct transmission matrix D.

It may be noted that, by convention, modes that cannot be excited by any input (that is, uncontrollable modes) are assumed to be quiescent and therefore to generate zero, and hence bounded, responses. However, if the system is not stable then the controllable but unstable modes become unbounded, for some non-zero bounded input, and the uncontrollable and unstable modes, although theoretically quiescent, become unbounded if some stray or parasitic non-zero bounded inputs, however small they might be, are introduced into those modes. Hence Maxwell's condition for stability is necessary to achieve control in the sense of the informal definition. The condition provides the necessary assurance that the states are bounded if the environment is probing even if the working model of the environment, with which practical designs are obtained, is not probing.

A transfer function, if it is rational and proper, is said to be stable if the real parts of all its poles are negative. If the transfer functions of all the input-output transformations of the system are stable, this does not imply that all the modes of the system are stable. The reason is that those modes, which do not contribute to the transfer functions, can be stable or unstable. If some of those modes are not stable the system is said to be internally not stable. This is to distinguish from the input-output or external (involving ports) stability, which is determined by the stability of the corresponding transfer functions. A system is said to be input-output stable if the transfer functions of every input-output transformation is stable. The concept of system stability is equivalent to the concept of input-output stability if and only if all the modes of the system contribute to the input-output transformations.

The foundation of conventional theory for the design of control systems contains two primary concepts. The first is the stability of the system. The second is a notion of *sensitivity* of a transfer function that, in its original form, is well known in terms of the concepts of phase margin and gain margin of the Nyquist diagram. This notion follows from Nyquist's practical condition for the stability of a transfer function. Roughly, sensitivity of a stable transfer function is one or more non-negative numbers that measure the extent to which a non-zero input is magnified (or attenuated, depending on whether the number is greater than or less than one) in the process of becoming the response. As sensitivity tends to become infinite, so magnification becomes infinite and hence, for some bounded input, the response tends to become unbounded, in which case the transfer function and hence the system to which it belongs, tends to become unstable. It is convenient to assign the value infinity to the sensitivity of an unstable transfer-function. Thus, any way of quantifying or measuring sensitivity provides a way of quantifying the stability of a transfer function. Terms such as 'degree of input-output stability' and 'margin of stability' have also been used elsewhere to mean an inverse measure of sensitivity. The term input-output sensitivity is used to mean the sensitivity of a transfer function or, more generally, the sensitivity of an input-output transformation, from an input port to an output port, or the combined sensitivities of all the input-output transformations from all the input ports to one output port or to all the output ports. The precise meaning will be obvious from the context.

The notions of stability and sensitivity merge together to form the following definition of control. This definition constitutes the core of the current paradigm of control. In effect, mainstream theories and methods of design are all those that aim to achieve control in the sense of this definition. No other theories are, by general consensus, part of the mainstream. The definition therefore characterises the foundation of conventional control theory.

Conventional definition of control: A system is said to be under control if the following three conditions are satisfied. First, the system is stable. Second, for every error port, every input-output transformation feeding the error port has low sensitivity or minimal sensitivity. Third, for every output port that is not an error port, every input-output transformation feeding the output port has sensitivity that is not too large. □

This definition is partly equivalent to the informal definition but is much more convenient in practice, because its requirements of stability and minimal input-output sensitivity are more easily achieved than the corresponding requirements of bounded states and small errors resulting from a probing environment. In fact, stability implies that all the states are bounded, whether or not the modes are excited by the input, provided that the environment produces bounded inputs. More generally, if the direct transmission matrix D of the filter-system combination is zero and provided that the environment produces inputs whose p-norms are finite then all the states are bounded if the system is stable (see Section 1.4).

The extent to which the conventional definition simplifies the design problem is worth emphasising. The point to note is that the definition does not involve a model of the environment. In fact, it is obvious that the system can be designed to satisfy this definition of control, without taking into account the environment. However, as will be seen, this neglect of the environment is sometimes an oversimplification of the real design problem.

Minimal input-output sensitivity implies that the output is made as small as possible, in some sense. But again, like the informal definition, how small and in what sense the output is required to be small, is not stated. This lack of quantification and precision implies that, provided a system can be made stable, control can always be achieved by minimising the appropriate sensitivities. Clearly, the only firm constraint imposed by the conventional definition of control is the stability of the system. As will be seen, this constraint is not sufficiently stringent and does not represent the notion of control needed in some important situations.

A further difficulty with the conventional definition is its third condition, which is intended to limit the size of the responses at the corresponding output ports. The difficulty arises because the condition is stated in qualitative terms that are not easy to quantify, even if the meaning of an output being too large is defined quantitatively. Evidently, it is not possible to specify quantitatively when the sensitivity of a transfer function is too large, even if a restriction on the corresponding output is specified quantitatively, without taking into account the magnitude of the input.

The concept of sensitivity is central to control theory but has been given various, somewhat arbitrary, mathematical interpretations, each leading to a separate branch of mainstream control theory and design. Although sensitivity is a way of quantifying the stability of a transfer function, there appears to be no universally agreed way of defining this concept and the various definitions that have been adopted are arbitrary. This lack of agreement will be seen to have significance in motivating the introduction of the new framework for control systems design.

One well known interpretation of sensitivity of a transfer function, derived from its definition for stability, is to measure sensitivity by the size of the real parts of all its poles, assuming that these poles are confined to a wedge-shaped region of the left-half plane, to ensure that any oscillations of the corresponding modes decay quickly. The methods of design called the root locus and pole placement are based on this interpretation of input-output sensitivity.

As already mentioned, the original meaning of sensitivity was defined by the phase margin or the gain margin of the Nyquist diagram. As these two quantities become smaller, so the sensitivity becomes larger. As the margins tend to zero, so some of the real parts of the poles of the transfer function tend to zero.

Classical methods of design, such as those of Nyquist and root-locus, are characterised by the use of measures of sensitivity that are derived naturally from their respective practical conditions for stability of a transfer function. However, many other well-known measures of sensitivity, which are not derived from a practical criterion of stability of a transfer function, have been defined.

These various well-known measures of sensitivity include the characteristics (settling time and undershoot) of the error due to a step input, as well as certain q-norms (usually $q = 1$ or 2, see Section 1.4) of the (possibly weighted) error resulting from a step input or delta function input. Well-known examples of this are, for the $q = 1$ norm, the integral of the absolute error (IAE or, for the $q = 2$ norm, the square root of the integral of the square of the error $\sqrt{ISE}$. Another measure of sensitivity is provided by the $\mathcal{H}^\infty$-norm of the frequency response. All these measures of sensitivity are defined when the transfer function is stable. However, in some cases, for example the step-response characteristics or the $\mathcal{H}^\infty$-norm, if the transfer function is unstable then the measure of sensitivity is not defined by the same process that defines its value for a stable transfer function but is defined by assigning to it the value infinity.

Yet other measures of sensitivity are obtained by considering the transfer function as an input-response operator, defined by a convolution integral, and deriving certain functionals, in some cases representing the operator norm (a q-norm of an impulse response), that depends on the p-norm ($p^{-1} + q^{-1} = 1$) used to characterise the input space, to act as measures of sensitivity (see Section 1.4).

Using positive weights, any weighted sum of different measures of sensitivity, related to one transfer function, defines another sensitivity of that transfer function. Also, a weighted sum of sensitivities, which correspond to different transfer functions of a system, defines a composite scalar sensitivity for those transfer functions considered all together. This scalar composite type is characteristic of certain optimal control methods, which minimise a scalar composite measure of the sensitivities of the system.

As has been noted, the concept of sensitivity provides a useful measure of the stability of a transfer function. However, if the transfer function is unstable then the sensitivity is infinite, whatever the extent of instability. Clearly therefore, sensitivity does not provide a measure of the extent of instability of a transfer function. It follows that, whereas a stable transfer function can, for the purpose of design, be represented by its sensitivity, an unstable transfer function cannot be so represented. This also points to the difference between design and tuning. If a system is stable, all its sensitivities can be measured or computed and, by some means, tuned (adjusted) to the required values, without knowing the transfer functions of the system. Otherwise, stability has to be achieved first.

Design, in the conventional sense, therefore involves achieving stability first and then tuning the sensitivities to the required values. This emphasises further the central role played in design by the two concepts of stability of a transfer function and stability of a system. However, stability of a system, which is what is required by the conventional definition of control (and also by the new definition given below), can be achieved in different ways. One particular way[1] is to employ numerical methods to satisfy the inequality that

---

[1] Another well-known way (usually found in the literature under the term *internal stability*; see, for example, Boyd and Barratt, 1991) is to consider certain transfer functions that, if they are all stable then the system is stable. Then, by some

states that the abscissa of stability (this is also called the spectral abscissa of the matrix A of the system and is defined as the largest of the real parts of all the characteristic roots) is negative (see Section 1.6).

Design, in the conventional framework, involves selecting one system, from a given set of systems, called the system design space Σ, so that control is achieved, in accordance with the conventional definition of control. Because the sensitivities can be tuned only when the system is input-output stable and because all the modes of the system are required to be stable, it is useful to have a convenient characterisation of the stable subset Σ_{Stable}, comprising every element of the set Σ such that the system is stable. An initial step in design involves determining one element of the stability set Σ_{Stable}. This can be done by defining stability either in terms of the abscissa of stability (see Section 1.6) or in terms of the concept of internal stability. This aspect of design is here called the *principle of uniform stability*, because every element of the set Σ_{Stable} is a stable system and the search for a satisfactory design is restricted to this uniformly stable set. This principle is an obvious extension of the concept of stability and it is named in this way to emphasise its importance in design.

1.1.2 Crisis in Control

Although some strong preferences have existed among practitioners, the many versions of the concept of sensitivity, and the corresponding distinct design methods that are used to achieve control in the sense of the conventional definition of control are, by this very definition, essentially equivalent. That is to say, the conventional framework for design includes all design methods that achieve control, in the sense of the conventional definition of control, where each method is characterised by a distinct way of defining the concept of sensitivity. All such design methods are therefore equivalent. Some design methods might have advantages, with respect to ease of modelling or computations, but these aspects are concerned with the means and not with the ends of design.

This overabundance of distinct, but essentially equivalent, versions of the same theory suggests that the practitioners of the subject are making futile attempts to transcend its limitations. After Nyquist's work, each new way of defining sensitivity has been introduced on grounds that somehow, unlike previous versions, it captures more accurately the real meaning of control. The historian of science, Kuhn (1970), has pointed out that this is a symptom

---

convenient means, proceed to ensure that these transfer functions are stable. The various means include the use of the Nyquist's condition for stability of a transfer function or, alternatively, a purely computational method for stabilising transfer functions (Zakian, 1987b). The term internal stability of a system is synonymous with the term stability of a system. The former is used to indicate that a system can be stabilised by means of techniques that stabilise transfer functions.

of crisis in a subject. The following quotation from Page 70 of his influential book illustrates the point: "By the time Lavoisier began his experiments on airs in the early 1770s there were as many versions of the phlogiston theory as there were pneumatic chemists. That proliferation of versions of a theory is a very usual symptom of crisis. In his preface, Copernicus complained of it as well."

The current mainstream approach to control is the product of a merger between control (servomechanism and regulator) theory and amplifier (circuit) theory that took place somewhat hastily during the wartime period of 1940-1945. This merger gave rise to the conventional definition of control, as stated above. However, too restricted a focus on the concept of feedback, which is shared by both subjects, has sometimes obscured the differences between them. The long-term validity of the consensus that followed the merger has been questioned by Bode (1960). Although his well-known book appeared in 1945, Bode contributed to feedback theory up to but not after, the year 1940, which is just before the merger. He expressed his "misgivings" about the "fusion" of the two fields by means of incisive metaphors, used delicately and with humour but nevertheless with serious intent, when he came to the conclusion that control theory and amplifier theory are "quite different in fundamental intellectual texture" and the "shotgun [that is to say, hasty and forced] marriage between [these] two incompatible personalities" (that took place during the Second World War), which resulted in the current mainstream approach to control, should perhaps be dissolved with an "amicable divorce". There has since been ample time to reconsider the long-term wisdom of that merger. However, although the analysis given below provides added reasons for Bode's conclusions, the reasons given by him were perhaps not sufficient for his conclusions to be acted upon, also because the nature of the crisis in control was yet to be clarified.

The conventional definition of control has been accepted, as characterising the foundation for mainstream theory and design of control systems, since the year 1945. Although this has been largely a fruitful move, it has also been insufficient because, like the informal definition of control, the conventional definition is not quantified. The consequences of this are now considered.

1.1.3 Factors that Deepen the Crisis

To quantify the informal definition of control, it is necessary to state more precisely what is meant by the errors remaining close to zero. It can, with some justification, be argued that such precision is not necessary in some practical problems of control systems design and therefore that design methods based on the conventional definition of control are likely to remain useful for the foreseeable future. It can also be argued, with even greater justification, that additional precision is not needed in the design of feedback amplifiers. There are, however, some important problems of control system design where precision and quantification, in the formulation of the design problem, is dictated by the nature of the problem. In one such kind of problem, the system contains what are called critical output ports, defined below, where it is important to ensure that the output at such a port remains bounded throughout time by a specified tolerance level. Similarly, there are systems that have responses that can saturate and preventing saturation is an important approach to their design because the linear model of the system remains valid and therefore linear theory can be employed. It will become clear that what is needed are design methods that can cope, not only with the usual problems based on the conventional definition of control, but also with critical systems and conditionally linear systems, defined below. Such methods must be based on a more general and more precise foundation for control theory.

Consider the n-dimensional vector of input ports, receiving input f. Consider the m scalar output ports and let e_i denote the transformation from the vector input port to the ith output port. This can be written as $e_i : f \mapsto e_i(f)$. Let $e_i(f)$, which denotes the output at the ith output port, be a real function that maps the time axis $\mathbb{R}$ into the real line $\mathbb{R}$. The value of this function at time t, which is denoted by $e_i(t, f)$, is the output at time t. Suppose that the system, and hence the output, depends on a design parameter σ. Accordingly, whenever necessary, the more explicit notation $e_i(f, \sigma)$ is used to denote the ith component of the output. Thus the output is the vector $e(f, \sigma)$ with components $e_i(f, \sigma)$. The output at time t is denoted by $e(t, f, \sigma)$. The set of all system design parameter values σ is denoted by Σ and, as already noted above, is called the system design space.

Definition of specifically bounded output: For a given input f, the corresponding output $e_i(f, \sigma)$ is said to be *specifically bounded* if, for a specified positive number ε_i, called the tolerance (elsewhere called bound or margin or limit), and for all time t, the absolute output at that port does not exceed its tolerance; that is to say, the following condition is satisfied

$$|e_i(t, f, \sigma)| \leq \varepsilon_i \tag{1.1}$$

□

The notion of specifically bounded output has long been well known, especially in the process control industries. It was explicitly introduced into control systems design, because it leads to a more accurate representation of certain design problems (Zakian, 1979a) and, more significantly, because the design facilities provided by *the method of inequalities* (Zakian and Al-Naib, 1973; Zakian 1979a; 1996; see Sections 1.2 and 1.6) made its introduction practical. Obviously, a specifically bounded output is more than just a bounded output, because it is bounded by a specified tolerance. In contrast, a bounded output is bounded by some unspecified constant. Clearly, therefore, requiring that the output be specifically bounded represents a more stringent constraint than the requirement of stability.

Definition of possible set: Let the input f, produced by the generators in the environment, be known only to the extent that it belongs to a set of inputs called the *possible set P*. □

Definition of peak output: For a given possible input set P, the *peak output* at the ith output port is defined by

$$\hat{e}_i(P,\sigma) = \sup\{|e_i(t,f,\sigma)| : t \in \mathbb{R}, f \in P\} \tag{1.2}$$

□

Here, $\mathbb{R}$ denotes the real line and P the set of all possible inputs. The peak output functional $\hat{e}_i : e_i(f,\sigma) \mapsto \hat{e}_i(P,\sigma)$ maps the set of all output functions into the extended half-line $[0,\infty)$, so that the peak output is infinite if the output is unbounded and is finite otherwise.

Definition of specifically bounded peak output: The peak output at the ith output port is said to be specifically bounded if

$$\hat{e}_i(P,\sigma) \leq \varepsilon_i \tag{1.3}$$

□

Clearly, the peak output at every output port is specifically bounded if and only if

$$\hat{e}_i(P,\sigma) \leq \varepsilon_i \text{ for all } i = 1,2,\ldots,m \tag{1.4}$$

This conjunction of inequalities expresses design criteria as required by the *principle of inequalities* (see Section 1.2).

An input f is said to be *tolerable* (Zakian, 1989) if, for every output port, the resulting output is specifically bounded. The set of all tolerable inputs is denoted by T and is called the *tolerable set*.

The conjunction of inequalities (1.4) is a necessary and sufficient condition for every possible input to be tolerable; that is to say, for the possible input set P to be a subset of the tolerable set T. In that case, the environment-system couple is said to be *matched*. However, the inequalities (1.4) provide only necessary conditions for the environment-system couple to be *well-matched*; that is to say, for the set of all the tolerable but not possible inputs to be small. Of particular interest are systems that, in some sense, maximise the extent to which the tolerable set T is inclusive. Such systems are, in that sense, optimally tolerant. These topics are considered in Section 1.3.

The *inverse problem of matching* is defined as follows. Given a system that is input-output stable, determine a useful expression for a subset of the tolerable set T. A subset of the tolerable set is considered to be useful if it can be used as the possible input set P that characterises an environment of the given system. It may be recalled here that conventional design methods are specifically intended to yield a design that is, to a large extent, input-output stable. The inverse problem of matching can be solved by means of a concept called a *linear couple* (see Section 1.3). Such a solution provides a bridge between conventional design methods and the design of a matched environment-system couple.

However, it is also shown (see Section 1.3) that a design that satisfies the conventional definition of control does not, except in the case of systems having one input and one output, give an environment-system couple having certain well-defined and desirable properties.

The above four paragraphs constitute a very sketchy outline of the *principle of matching* (Zakian, 1979a, 1989, 1991, 1996), which is considered in some detail in Section 1.3. The concept of matching was first made explicit in Zakian (1991) and is elaborated in Section 1.3 to include the new concepts of *augmented possible set*, *perfect matching*, *extreme tolerance to disturbances* and a *well-designed environment-system couple* . The following definition indicates the practical need for these concepts.

Critical output port: An output port is said to be critical if the peak output is required to be specifically bounded and if the consequences of the absolute output exceeding its tolerance are strictly unacceptable. A system is said to be critical if it contains one or more critical output ports. □

Despite their ubiquity, critical systems, although well known to some practitioners, have been largely unnoticed in mainstream practice, because they do not conform to the conventional definition of control and also because mainstream theory and methods are inadequate to deal with the resulting design problems. It is a known fact that perception of a situation requires appropriate cognitive equipment and motivation (a cat might not notice a mouse, unless it is hungry or playful and its brain contains, perhaps from birth, the idea of small prey). Without the explicit notion of critical systems and some adequate tools to design them, such systems have largely been ignored. Also, the official status of the conventional foundation of control, together with the extensive superstructure built upon it, has hindered the view of critical systems.

However, the cognitive means to perceive critical systems were progressively introduced (Zakian, 1979a, 1987a, 1989) with ideas that built, into the method of inequalities, new design criteria, requiring the peak outputs of the system to be specifically bounded, as shown in (1.4). As already mentioned, the notion of a specifically bounded output, as represented by inequalities of the form (1.1), is essentially the criterion employed, as a standard practice, by plant operators, to assess the performance of process control systems. Despite this, the criterion of a specifically bounded output does not form part of the conventional theoretical framework of control. This disparity between conventional control theory and the practice in some industries was perceived as significant in 1968 and influenced subsequent work.

The framework built around the method of inequalities, including criteria that require the peak outputs to be specifically bounded, has since been the subject of continuous developments that culminated in 1989 in the principle of matching. This principle is further developed in Sections 1.3 and 1.4 and is elaborated and applied in other parts of the book. Even with this cognitive equipment, it took some years to perceive clearly the existence of critical systems. Then, in order to draw attention to this important class of control problems, the term *critical system* was introduced and work was started in 1988 to demonstrate how the framework could be used to design critical systems. Several case studies, showing such designs, are included in Part IV of the book. In parallel with this, various computational techniques, needed to achieve designs, were developed more fully (see Parts II and III).

1.1.4 Towards a New Consensus

The importance of critical control systems and the fact that their design cannot be achieved by any means that are based on the conventional definition of control, but can be achieved otherwise, clarifies the nature of the crisis in control theory.

Following Kuhn's (1970) analysis of historical aspects of science, the official theories and methods and the set of all problems that can be solved by these methods, constitute what is called the official paradigm. It is official in the sense that it is accepted, by general consensus, within the community of practitioners. Thus, the current control paradigm is the generally accepted conventional framework, characterised by the conventional definition of control, together with the set of all design problems that can successfully be solved within that framework and all its possible extensions. Critical systems are excluded from the official paradigm. However, if other methods can be used to solve important problems, not solvable within the official paradigm, such as those involving critical systems, then a crisis exists. To resolve the crisis requires a new consensus and hence a new paradigm. Therefore, once an alternative and more powerful paradigm exists, having a significantly larger range of applications, the ensuing crisis is a social phenomenon that can be resolved, not by additional scientific work, but only by a new consensus within the community. Kuhn emphasises that achieving a consensus is an arduous process.

It is perhaps more than pure coincidence that critical systems have contributed to a crisis in control theory. The two words *crisis* and *critical* share the same Greek root *krinō*, that is translated as *decide* (The Concise Oxford Dictionary, fourth edition).

The same methods, that allow a critical system to be designed successfully, on the assumption of a linear model of the plant, can also be used to ensure that the linear model is a valid representation of the plant, during the design process and in the subsequent operation of the control system. Obviously, if a model departs significantly from the way a plant does operate then designs, based on the model, might not predict the actual operation of the control system, with the possible consequence that critical tolerances at some response ports are exceeded.

A linear model remains valid only within a limited range of operation of the plant. If the operation of the plant is restricted to that range then the assumption of linearity is justified. A system, or its model, that is linear within such restrictions, but is not linear otherwise, is said to be *conditionally linear* (Zakian, 1979a). Accordingly, some of the outputs (not necessarily errors) of a conditionally linear system are specifically bounded outputs, satisfying inequalities of the form (1.1) and, for each such inequality, the corresponding tolerance is the largest number that ensures that the plant operates in its linear region, for all the inputs that can be generated by the environment. A conditionally linear model is particularly valuable when the plant has saturation type non-linearity.

Whereas the concept of critical system is of primary significance, that of conditionally linear system is not. This is because the concept of critical system represents physical and engineering situations of importance. In contrast, the concept of conditionally linear system is useful only because it allows control systems to be designed within the limitations of linear theory. There are also other known reasons why it might be convenient to require that an output be specifically bounded and these include, for example, constraints such as limiting power consumption. Such a limitation is often not critical and corresponds to what is sometimes referred to as a soft constraint, because the tolerance is not dictated by strict necessity but is negotiable to some extent.

It is now obvious that the informal definition of control can be quantified by requiring all the outputs to be specifically bounded. Although such quantification is essential when the system is critical or is subject to other similar hard constraints, it remains useful even in cases that are less than critical and require softer constraints.

Before a new definition of control can be given, that embraces the above ideas, it is necessary to provide a more specific definition of the possible set. The following concept leads to such a definition.

Appropriately restricted environment: Suppose that every filter within the model of the environment is an identity transformation, so that the output of the generators feeds the control system directly. Then the environment is said to be *appropriately restricted* if its generators produce functions that are restricted in the following way. For every input port $j = 1, 2, \ldots, n$, $f_j = f_j^{per} + f_j^{tra}$, where f_j^{per}, f_j^{tra} are, respectively, the persistent and the transient components of the input f_j, and for some finite non-negative numbers $d_j^{per}, \dot{d}_j^{per}, d_j^{tra}, \dot{d}_j^{tra}$ independent of the input, let

$$\left\|f_j^{per}\right\|_\infty \le d_j^{per}, \quad \left\|\dot{f}_j^{per}\right\|_\infty \le \dot{d}_j^{per} \tag{1.5}$$

$$\left\|f_j^{tra}\right\|_2 \le d_j^{tra}, \quad \left\|\dot{f}_j^{tra}\right\|_2 \le \dot{d}_j^{tra} \tag{1.6}$$

□

As usual, the notation employed here is defined by

$$\|z\|_\infty = \sup\{|z(t)| : t \in \mathbb{R}\}, \quad \|z\|_2 = \left\{\int_{-\infty}^{+\infty} |z(t)|^2 dt\right\}^{1/2} \tag{1.7}$$

and $\dot{z}$ denotes the derivative of z.

Suppose that the possible set P is a subset of the set of all the inputs that can be generated by this appropriately restricted environment. Expressing the input as a sum of persistent-transient and transient components results in a more efficient way of modelling the environment in the sense that, for the same physical environment, the peak outputs given by the environment-system model are smaller. The restrictions imposed on the derivative of the input ensure that the peak outputs depend in a useful way on the system design parameter. This topic is discussed in Section 1.5, where the notion of a well-constructed environment-system model is developed. The restriction on a derivative can be removed by setting the derivative bound equal to infinity. If the direct transmission matrix D is zero then the restrictions on the derivatives are not necessary but might be retained in order to minimise the size of the possible set and hence reduce the peak outputs.

The two restrictions (1.5) and (1.6) ensure that, for every output port, the peak output is finite if the system is input-output stable (see Section 1.5). This means that the finiteness of the peak output can be used as an indication of the input-output stability of the system. For, if the peak output is not finite then the system is not input-output stable.

Definition of control: Suppose that the environment is appropriately restricted. Suppose also that the possible set P is a subset of the set of all the inputs that can be generated by this environment. Then an environment-system couple is said to be under control if the system is stable and, for every output port, the peak output is specifically bounded. □

On comparing this definition with the informal definition of control, it can be seen that they differ because, in this definition, the outputs, some of which are the errors, are required to be specifically bounded and not just vaguely small (the errors) or vaguely not too large (the outputs that are not errors). Also, this definition, which assumes an appropriately restricted environment, replaces the requirement of bounded states with the requirement of system stability. In this respect, it is analogous to the conventional definition of control. However, unlike the conventional definition, which requires the input-error sensitivity, and hence the errors, to be somewhat small, and other input-output sensitivities to be not too large, all the peak outputs in this new definition are required to be specifically bounded. A further difference is the extent to which the environment is specified. The conventional definition of control does not specify the environment, although it is generally accepted that the environment is implicitly restricted to a set of possible inputs such that the output is bounded, provided that the system is input-output stable. Thus, the shift from the conventional definition of control to this new definition brings the necessary precision required for the range of design problems that includes critical systems.

The need for a design theory based on this definition of control has led to the adoption of two complementary concepts of system design, *the principle of inequalities* (see Section 1.2) and *the principle of matching* (see Section 1.3). In combination, these two primary concepts replace the concept of input-output sensitivity, which becomes a derivative of the two primary concepts (see Section 1.4).

The new definition of control can be restated, with important consequences, in terms of the concept of matching by noting that the requirement that, for every output port, the peak output is specifically bounded is equivalent to the requirement that the environment-system couple is matched. The advantages of stating the definition of control in terms of the principle of matching are discussed in Section 1.3. In particular, it leads to a framework within which the two notions of over-design and extremely tolerant system can be defined and this in turn leads to designs that are more economical and more efficient.

The principle of inequalities and the principle of matching, together with the principle of uniform stability, which is a straightforward application of the idea of system stability, form a theory of design that embraces the new definition of control. The theory is the foundation that, together with a superstructure of design methods, constitutes the new *design framework*.

1.2 The Principle of Inequalities

The principle of inequalities underlies the approach to design called the method of inequalities (Zakian and Al-Naib, 1973; Zakian, 1979a, 1996; see Section 1.6). The principle and its origin are explained in this section.

Conventional characterisation of good design: Any theory of design must provide a mathematical way of characterising a good design. It has long been recognised that a good design is usually specified by means of several criteria that have to be satisfied simultaneously. Finding an appropriate way of representing such criteria mathematically has been a challenge.

The approach conventionally taken is to derive a vector of objective functions

$$\phi = (\phi_1, \phi_2, \ldots, \phi_M) \tag{1.8}$$

Each objective function ϕ_i maps a set C, called the design space, into the extended real line $(-\infty, \infty] = \mathbb{R} \cup \{\infty\}$. As usual, $\mathbb{R}$ denotes the real line. The objective functions are then divided into two classes. One class contains the performance objectives and the other class contains objectives that are to be constrained.

In the conventional framework of control systems design, a typical performance objective is the sensitivity of a transfer function, defined in some particular way, and $c = \sigma$ where c denotes the parameter that characterises the performance objective. In contrast, when the principle of matching is employed, a typical performance objective is a peak output $\hat{e}_i(c)$, $c = (P, \sigma)$. As is noted in Section 1.1, P denotes the possible input set and σ denotes the system design parameter. Thus, under the principle of matching, c is the design parameter of the environment-system couple.

In the case of a constraint, the objective $\phi_i(c)$ has to satisfy an inequality. In the case of a performance objective, the conventional approach to design makes the assumption that, for each design $c \in C$, the value $\phi_i(c)$ of the performance objective function ϕ_i represents a cost or an undesirable aspect of the design and that, as the value of the performance objective $\phi_i(c)$ gets larger, so the design gets worse. Accordingly, a good design c minimises, in some sense, all the performance objectives, while satisfying the constraints.

More significantly, the performance objectives are not to be bounded in any specified way and thus no minimal acceptable performance is specified. This is of no practical consequence in cases where only a single performance objective exists because minimising the scalar objective would show if the system meets any required minimal level of performance. However, in cases where there are more than one performance objective, this approach does not guarantee to satisfy *a priori* specifications placed on each performance objective separately.

As will be seen, the above dichotomy between performance objectives and constraints is somewhat artificial and introduces unnecessary complications. One complication is the need to define in what sense the minimisation of the performance objectives is to be done. This is because, in general, there exist many elements of the design space C, called Pareto minimisers that, in distinct but similar ways, minimise the vector of performance objectives.

Suppose that $L < M$, and the first L objective functions represent performance and the remaining objective functions satisfy inequality constraints. Let $\bar{C}$, called the constrained design space, denote the set of all designs in C that satisfy the inequality constraints. Let the symbol $\tilde{x}$ denote the set $\{1, 2, \ldots, x\}$. A Pareto minimiser $c \in \bar{C}$ has the property that, for every

$i \in \widetilde{L}$, the number $\phi_i(c)$ can be decreased (by varying the design within the constrained design space) only by increasing $\phi_i(c)$ for some other $i \in \widetilde{L}$.

The concept of a Pareto minimiser can be restated using the following notation and general definition. Let x, y denote two vectors of dimension q. Then $x \prec y$ means that no component of x is greater than the corresponding component of y and some components of x are less than the corresponding components of y; that is,

$$(\forall i \in \widetilde{q},\ x_i \leq y_i)\ \&\ (\exists i \in \widetilde{q},\ x_i < y_i) \tag{1.9}$$

Pareto minimiser: Let X denote a subset of the design space C and let ϕ denote a vector of objectives. A design $c \in X \subseteq C$ is said to be a Pareto minimiser in the set X if there is no other design $c^* \in X$ such that $\phi(c^*) \prec \phi(c)$. The set of all Pareto minimisers in X is called the Pareto set in X and is denoted by P_X. □

Comments: In the conventional formulation of the design problem, the set X is the constrained design space and the objective vector is the performance objective vector. The above definition highlights the fact that the Pareto set of interest is P_X and this is a subset of X, which is a subset of the design space C. It follows, obviously, that P_C, the Pareto set in the design space C, is in general not the same as $P_{\bar{C}}$, the Pareto set in the constrained design space $\bar{C}$.

Clearly, every Pareto minimiser defines a limit of design. Although some Pareto minimisers may be considered good designs, not every Pareto minimiser is a good design and most Pareto minimisers are usually not considered to be good designs. For example, a Pareto minimiser that minimises one component of the performance objective vector $\phi(c)$, while at the same time it maximises another component, may not represent a good design because it may be better to choose another Pareto minimiser such that both components are not too large.

It is therefore clear that the performance objective vector $\phi(c)$ (comprising the first L components of (1.8)) does not, without additional information, represent the design problem adequately.

A conventional way of providing additional information is to associate a positive weight λ_i, with each performance function ϕ_i, such that the weight represents the relative importance of the function. Any design c, within the constrained design space $\bar{C}$, that minimises the weighted sum

$$\psi(c) = \sum_{i=1}^{L} \lambda_i \phi_i(c) \tag{1.10}$$

is then taken to characterise a good design. Evidently, a design c that minimises the weighted sum $\psi(c)$ is a Pareto minimiser in the constrained design space because there is no other element of the constrained design space that reduces the value of any component of the performance objective vector without increasing some of the others.

One drawback of this approach is that a design situation seldom suggests the values of the weights λ_i. These weights are therefore either chosen to satisfy other criteria or are chosen somewhat arbitrarily. Typically, in the conventional framework for control systems design, the weighted sum shown above is a composite measure of sensitivity. Usually, the weights are chosen to satisfy criteria that are not stated explicitly. Alternatively, the weights are chosen to satisfy criteria that are in accordance with the principle of inequalities (Whidborne et al, 1994; see Chapter 11).

The inequalities approach: The principle of inequalities provides a way of formulating control problems appropriately. It treats all the objectives to be constrained and also all the performance objectives, mathematically in the same way. This principle asserts that a design problem is appropriately stated in the form of the conjunction of all the inequalities

$$\phi_i(c) \leq \varepsilon_i, \quad i \in \widetilde{M} \tag{1.11}$$

Here, each constant ε_i, called the tolerance (elsewhere also called the bound, the margin or the limit), is the largest acceptable or permissible value of the objective $\phi_i(c)$. Any design c, within the design space C, that satisfies all the inequalities is a solution of the design problem and is called an admissible design. Using vector notation, (1.11) can be restated as

$$\phi(c) \leq \varepsilon \tag{1.12}$$

The set of all admissible designs is called the admissible set and is denoted by C_a. In some cases, an admissible design does not exist and this means that the admissible set is empty.

Clearly, each inequality in (1.11) can represent either a distinct performance criterion or a design constraint. They all take the same mathematical form in the principle of inequalities. At first sight, this appears to involve a trivial change to the conventional way of formulating a design problem. This change is simply to treat all performance objectives mathematically in the same way the constraints are formulated. But, as is well known, a change in a single axiom of a mathematical system can result in a new system with very different properties. Nonetheless, whatever new mathematical properties are introduced, the change must be evaluated by the advantages it bestows on the practice of design. Part IV of this book demonstrates these advantages with case studies showing how control systems can be designed using the principle of inequalities.

Obviously, the formulation (1.11) obliges the designer to give physical or engineering meaning to all the tolerances; even those associated with the performance objectives. That is to say, it obliges the designer to quantify the design problem in a meaningful way.

Notice that, unlike the conventional approach, it is not assumed that, as those objectives that represent performance become smaller, so the design improves. But only that each objective is required to be not greater than its tolerance. As will be seen, this is a more satisfactory representation of most design problems. What is sought primarily is satisfaction of the design criteria expressed in the form of inequalities and not the minimisation of performance objectives. This means that, for any scalar objective $\phi_i(c)$, provided that the objective does not exceed its tolerance ε_i, smaller values of the objective are not preferable to larger values.

Every tolerance is chosen to be as large as the situation will allow. In this way, no objective function is favoured unduly and the design freedom is not used too narrowly on some performance objectives at the expense of other aspects of the design problem. Once the tolerances are fixed, every admissible design is equally good and equally acceptable.

The designer is required to determine the value of each and every tolerance ε_i, either as part of the way that the design problem is modelled or the way that the designer wishes to express the design specifications. Some tolerances, including those usually associated with constraints, are fixed relatively rigidly by the design situation. These are called strict tolerances. Others are less rigid and are called flexible tolerances. It is important to be aware that the tolerances, even those connected with performance objectives, have concrete (that is to say, physical, engineering or economic) interpretations. Any changes that might be made to the tolerances should therefore be made with clear understanding of any concrete implications. Changes made to a strict tolerance would involve important concrete changes (perhaps to hardware) while changes made to soft tolerances would involve relatively less important concrete changes. In particular, any increase in the value of a strict tolerance may correspond to a more expensive concrete change than a change made to a less strict tolerance.

To illustrate these points, consider a conditionally linear system, as defined in Section 1.1. The function of interest in this situation is a peak output $\hat{e}_i(c)$, $c = (P, \sigma)$. This objective is not to be minimised but only to be bounded by a tolerance. The tolerance for this objective is the saturation level, which is the largest value the variable can have without saturating. This example illustrates one kind of constraint involving a strict tolerance and therefore its representation by an inequality is in accordance with conventional practice.

Another, and more fundamental, example occurs when the system is critical. Here, the objective of interest is a peak error $\hat{e}_i(c)$, $c = (P, \sigma)$, which must not exceed a given tolerance in order to avoid unacceptable, and possibly catastrophic, consequences. Because this situation involves an error, conventional wisdom might require that the peak error, which is a performance objective, be minimised. But, actually, it is obvious that the peak error should primarily be bounded by its tolerance, which in this case is strict. In this respect, it is to be treated in exactly the same way as a constraint. Moreover, there might be no advantage in making the peak error smaller than its tolerance. Too precipitous a decision to minimise the peak error would mean that all the design freedom is consumed, leaving nothing for other requirements.

Thus, instead of minimising the performance objectives, it might be better to keep them within their respective tolerances and to utilise any remaining design freedom to improve the whole design.

To illustrate one advantage of the principle of inequalities, consider a control system where the actuator saturates and the error is critical. In this case, it is essential to include two objectives in the design. These two objectives are: $\hat{e}_1(c)$, which denotes the peak actuator response and $\hat{e}_2(c)$, which denotes the peak error. For a given hardware, the respective tolerances of the two objectives are strict and these tolerances are associated with corresponding costs in hardware. However, suppose that, with these tolerances, the admissible set is empty for all permissible controllers. Increasing either or both the tolerances to make the admissible set non-empty might involve more expensive hardware. Finding the least expensive change of hardware, would require that the costs associated with both tolerances be considered, using the same currency. It may, for example, be that replacing the actuator, to give a larger tolerance for the peak actuator response, is the most economical way of ensuring a non-empty admissible set.

Venn formulation: The principle of inequalities is a special case of a more general formulation of design problems. For every $i \in \widetilde{M}$, let S_i denote a subset of the design space C. Let S_i define a criterion of design such that the criterion is satisfied if and only if $c \in S_i$. It follows that a design c satisfies all the design criteria if, and only if, it is in the intersection of all the sets S_i. Figure 1.2 illustrates this by a Venn diagram, where the intersection of the first three sets is shaded. This general formulation of the design problem is called the Venn formulation.

Evidently, the principle of inequalities is a special case of the Venn formulation, where each set S_i is defined by an inequality as follows:

$$S_i = \{c \in C : \phi_i(c) \leq \varepsilon_i\} \tag{1.13}$$

The Venn formulation of the design problem was originated by the author in 1966 and became the basis of work in his laboratory on computer aided design. Its special case, the principle of inequalities, was found to be sufficient for most problems of control systems design and subsequently formed the basis of the design method called the method of inequalities (see Section 1.6).

An important aspect of the principle of inequalities can be seen with the help of the Venn diagram in Figure 1.2. This shows that, as the number of sets S_i taken into account is progressively increased from 1 up to 4, so the admissible set, which is the intersection of those sets that are taken into

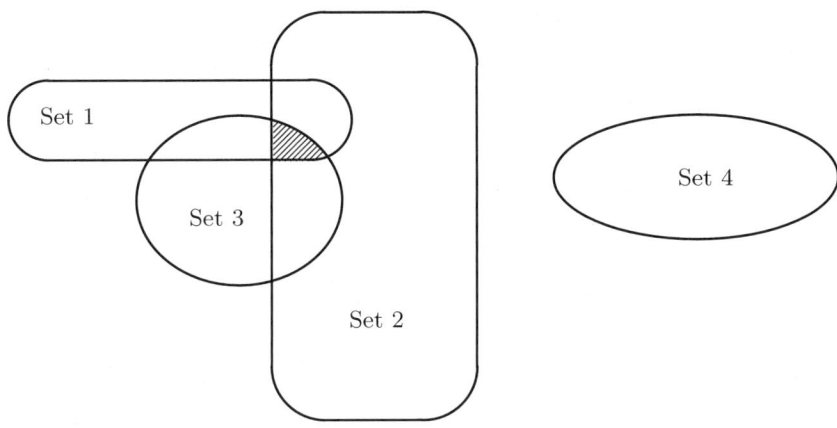

Fig. 1.2. Venn diagram

account, becomes progressively less inclusive until, in this example, it becomes the empty set when all four sets are considered. In this example, one way of avoiding an empty admissible set is to discard the fourth set.

More generally, the following process of negotiation can be employed to ensure that the design problem is formulated appropriately, while ensuring the admissible set is not empty. In this process, the objectives $\phi_i(c)$ may be partially or wholly discarded by increasing the tolerances ε_i and, moreover, the design space may be enlarged if necessary.

Process of negotiation: The purpose of the process of negotiation is to obtain a formulation of the design problem that includes the largest number of significant objectives, with their appropriate tolerances, while still ensuring that the admissible set is not empty. Suppose that the sets S_i are ordered in decreasing importance of the corresponding objectives $\phi_i(c)$. In particular, an objective involving a strict tolerance might be considered more important than an objective involving a flexible tolerance. Consider the non-empty admissible set which is the intersection of the largest number K of consecutive sets S_i, ordered as indicated above. In the example of Figure 1.2, $K = 3$. Now discard the objectives $\phi_i(c)$, for $i > K$, and redefine K so that $K = M$. Discarding an objective is equivalent to making the corresponding tolerance infinitely large. Evidently, any member of this admissible set can be considered a good design, provided that the discarded objectives can be so neglected. If such total neglect is not permissible then various strategies can be used to include some or all of the neglected sets. The most obvious strategy is to consider some of the tolerances ε_i, for $i \leq K$, and increase them, thereby making the corresponding sets S_i more inclusive, so as to avoid having to discard totally what are some important objectives. In effect, when a tolerance is increased then the corresponding objective becomes progressively more neglected. Thus adjusting the tolerance can be seen to be a way of altering the influence of an objective. Obviously, a flexible tolerance can be increased more readily than a strict tolerance.

Another strategy is to increase the inclusiveness of the design space C, for example, by increasing the efficacy (and usually also the complexity or order, see Section 1.6) of the controller. □

Negotiated inequalities: The M inequalities of the form (1.11) that result from the process of negotiation are called the negotiated inequalities. □

Comments: As mentioned above, increasing the value of any tolerance ε_i may have significant physical and engineering implications and should only be done with full awareness of these implications. For example, the tolerance may represent the limit of linear operation of the actuator of a plant that would otherwise saturate. Increasing the tolerance may correspond to the use of a more expensive actuator. Or, as in a critical system, the tolerance may represent a critical bound that, if exceeded, would result in unacceptable performance, unless either some other part of the system or some of the design specifications are altered. It follows that, if the tolerances have to be increased then, the smallest possible increase should be considered first. Evidently, some increase in the tolerance vector is necessary when the admissible set is empty and the design space cannot be enlarged. In which case, the smallest increase in the tolerance that results in a non-empty admissible set should be made. A new tolerance obtained in this way corresponds to a Pareto minimiser (see Section 1.7).

Controller efficiency and structure: In control system design, the higher the complexity of a controller the more efficient it can be. Increasing the complexity of the controller structure can result in a more inclusive design space. For example, when increasing the complexity of the controller from proportional control to proportional plus integral control, while still ensuring that the controller can be implemented. To see this, consider that, for a proportional controller, the design space is a section of the real line and, for a proportional plus integral controller, the design space is a rectangular region of the plane, which includes the real line section.

Conventional versus inequalities formulation: In contrast to the principle of inequalities, the conventional way of formulating the design problem, as the weighted sum of objectives shown in (1.10), allows all the objectives $\phi_i(c)$, however many there are, to be included in the design in a seemingly straightforward manner. Moreover, a design problem formulated in this way always has a solution, however large the number L of performance objectives.

At first sight, these might appear to be significant advantages of the conventional method but, on closer examination, the conventional method can be seen to have drawbacks.

The principle of inequalities, which includes the process of negotiation, requires the designer to prioritise the design objectives $\phi_i(c)$ by ordering them in decreasing importance. Decisions as to whether to increase the tolerances of some objectives can then be considered in the light of the emptiness or otherwise of admissible sets, as discussed above in the process of negotiation. An empty admissible set indicates that the design problem has been over-specified and that either some tolerances have to be increased, or, preferably and if possible, the design space C has to be made more inclusive. Such reformulation of the problem is aimed at obtaining an admissible set that is not empty.

On the other hand, the notion of an over-specified problem does not exist in the conventional formulation because the notion of an admissible set does not exist. There is no criterion, in the conventional formulation, to indicate a level, below which the quality of design is not acceptable. To compensate for this lack, many of the more modern conventional methods of control systems design (especially those called optimal control methods) aim to obtain the minimum, of the weighted sum of sensitivities, over the most inclusive class of controllers (typically, the class of all linear time invariant controllers). The minimum corresponds to the most complex (high order) controller, which often cannot be implemented. Numerous case studies have demonstrated that simple controllers can usually be employed to achieve what is required in practice (see Part IV). In conclusion, the conventional formulation is a simpler but less discriminating and less useful approach to design.

As mentioned before, there is also another significant and equally decisive reason for preferring the principle of inequalities. This is that it is not always clear how the weights are to be chosen in the conventional formulation, because their physical or engineering meaning is not always obvious. In contrast, the tolerances ε_i in the principle of inequalities have a more direct physical or engineering interpretation. In particular, for the class of control problems that involve critical systems, the tolerances that define criticality are dictated in an obvious manner by the problem and the conventional formulation is totally inappropriate, unless the performance criterion is treated in the same way as a constraint, which is what is done with the principle of inequalities.

The principle of inequalities can be summarised as follows.

Principle of inequalities: This principle requires that the design problem be formulated as the conjunction of M negotiated inequalities of the form (1.11). Any member of the admissible set belonging to the negotiated inequalities is taken to be a good design. □

Extreme design: It may happen, for example, in the process of negotiation described above, that the tolerances ε_i are chosen so that the admissible set is not empty and any reduction in the value of any of the tolerances results in an empty admissible set. Such an admissible set is said to be minimal. This is illustrated in Figure 1.3, which shows a Venn diagram of two sets S_i that intersect at two points only. These two points form the admissible set. If the tolerance ε_i associated with either of the two sets is reduced then the corresponding set becomes less inclusive and hence two sets cease to intersect and the admissible set becomes empty.

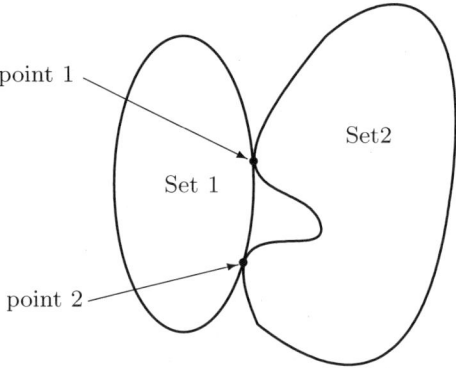

Fig. 1.3. Venn diagram with a minimal admissible set

For a more formal definition of minimal admissible set, let $C_a(\varepsilon)$ denote the admissible set corresponding to the vector tolerance ε.

Minimal admissible set: The admissible set $C_a(\varepsilon)$ is said to be minimal if it is not empty and it becomes empty if ε is replaced by any ε^* such that $\varepsilon^* \prec \varepsilon$. □

Comments: The reader is likely to have noticed an unusual absence of formal mathematical arguments, such as theorems, in the above discussion. This is because the purpose of the discussion is to establish mathematical definitions and axioms that represent, in a useful way, concrete design situations and problems. This is not unlike what the early geometers did when they used their experience and observation of the physical world to abstract the notions of point and line and lay down the axioms (assumptions about the nature of physical space) of Euclidian geometry. Care had to be taken in forming this geometrical framework of definitions and axioms to ensure that the resulting mathematical system was of use in such practical sciences as navigation, astronomy and land surveying. Once the definitions and axioms are decided upon, their more interesting implications can be obtained and stated in the form of theorems. It is emphasised here that the theory of design is not simply a body of implications that can be derived by careful and ingenious mathematical discovery but is also a body of carefully abstracted definitions and axioms that reflect the concrete reality of practical design situations. The more appropriately the reality is captured by the definitions and axioms, which represent aspects of the nature of design, the more useful some of the mathematical implications will be.

These remarks apply not only to the above discussion, which has led to the axiom called the principle of inequalities but also to discussions in other sections of this chapter and, in particular, to that in Section 1.3 and leading to the axiom called the principle of matching.

As discussed in Section 1.1, regarding the way critical systems have been perceived, it is necessary for some initial cognitive capacity to exist before the axioms of a useful design framework can be perceived. It is not unreasonable to suppose that man, the tool maker, has an innate capacity for design and can therefore cultivate the capacity to perceive the required axioms. This is not unlike the widely accepted notion that man has an innate capacity for language, which accounts for why a young infant can learn to speak in a relatively short time.

Proposition 1.1. *(Zakian, 1996) The following three conditions are equivalent:*

(a) The admissible set is minimal.
(b) The admissible set is not empty and, for <u>every</u> admissible design c, the equation $\phi(c) = \varepsilon$ holds.
(c) The admissible set is not empty and <u>every</u> admissible design is a Pareto minimiser in the set C.

Proof. Notice that if $\phi(c) = \varepsilon$ and $\varepsilon^* \prec \varepsilon$ then $\phi(c) \succ \varepsilon^*$.

(a)⇐(b). Suppose that (b) holds. Replace ε by ε^* such that $\varepsilon^* \prec \varepsilon$. Then, for every admissible design $c \in C_a(\varepsilon)$, the relation $\phi(c) \succ \varepsilon^*$ holds and hence the resulting admissible set $C_a(\varepsilon^*)$ is empty. This implies that the admissible set is minimal.

(a)⇒(b). Let the admissible set $C_a(\varepsilon)$ be minimal. Then the admissible set is not empty and hence any member of it c satisfies $\phi(c) \leq \varepsilon$. But, because the admissible set is minimal, replacing ε by ε^* makes the resulting admissible set $C_a(\varepsilon^*)$ empty. Hence, for every $c \in C_a(\varepsilon)$, it is true that $\phi(c) \succ \varepsilon^*$ which, together with $\phi(c) \leq \varepsilon$, gives $\phi(c) = \varepsilon$.

Notice that $\phi(c^*) \prec \phi(c)$ and $\phi(c) \leq \varepsilon$ imply that $\phi(c^*) \prec \varepsilon$. Hence to say that c is an admissible Pareto minimiser implies that there is no $c \in C$ such that $\phi(c) \prec \varepsilon$.

(a)⇐(c). Suppose the admissible set is not empty and every admissible design c is a Pareto minimiser. Then, to say that c is admissible means that $\phi(c) \leq \varepsilon$ and, because c is a Pareto minimiser, there is no $c \in C$ such that

$\phi(c) \prec \varepsilon$. Hence, for every admissible design, $\phi(c) = \varepsilon$ and, by Part (b) of the Proposition, the admissible set is minimal.

(a)$\Rightarrow$(c). Suppose that c is an admissible design. Then $\phi(c) \leq \varepsilon$. Suppose also that c is not a Pareto minimiser. Then there is a design $c \in C$ such that $\phi(c) \prec \varepsilon$; that is to say, there is a $c \in C$ and an ε^* such that $\phi(c) = \varepsilon^* \prec \varepsilon$. This means that there is an $\varepsilon^* \prec \varepsilon$ such that the admissible set $C_a(\varepsilon^*)$ is not empty. Hence the admissible set $C_a(\varepsilon)$ is not minimal. □

Comments on Pareto minimisers: Although every element of the minimal admissible set is a Pareto minimiser, the converse is not true. In general, most Pareto minimisers are not in the minimal admissible set. Most Pareto minimisers might not even be admissible designs. Clearly, if a design is a Pareto minimiser then it is not necessarily a good design. Also, there is no reason for preferring an admissible Pareto minimiser to any other admissible design. These points should be noted because there appears to be a misconception that a Pareto minimiser is somehow always superior to other designs. From the point of view of design, the more significant concept is not that of Pareto minimiser but the concept of minimal admissible set, since each member of this set, which also happens to be a Pareto minimiser, characterises a particularly good design. The concept of minimal admissible set emerges from the principle of inequalities.

Proposition 1.1 gives necessary and sufficient conditions for a set of designs to be a minimal admissible set. It is also of practical interest to derive necessary and sufficient conditions for a design c to be an element of a minimal admissible set.

Proposition 1.2. *A design c is an element of a minimal admissible set if and only if the following two conditions are satisfied:*

(a) $\phi(c) = \varepsilon$
(b) The design c is a Pareto minimiser in C

Proof. It might be easier to understand the proof by drawing simple Venn diagrams such as that in Figure 1.3. Suppose that a design c is in the minimal admissible set. Then, by Proposition 1.1, the two conditions (a) and (b) are satisfied. Conversely, let $\varepsilon^* \prec \varepsilon$. Hence, there is a design $c \in C$ such that $\phi(c) \leq \varepsilon^*$ implies that there is a design $c \in C$ such that $\phi(c) \prec \varepsilon$. If the design c satisfies the two conditions (a) and (b) then there is no $c' \in C$ such that $\phi(c') \prec \varepsilon$ and therefore (by the contra-positive of the implication in the previous sentence) there is no $c' \in C$ such that $\phi(c') \leq \varepsilon^*$. This means that the admissible set $C_a(\varepsilon^*)$ is empty. Therefore, since $C_a(\varepsilon)$ is not empty, it must be minimal. □

Comments on limits of design: This proposition shows again that a Pareto minimiser is only a necessary condition for a particularly good design. It has been shown above how the principle of inequalities provides a characterisation of good design. Specifically, a good design is an admissible design corresponding to a negotiated set of inequalities. If the process of negotiation yields a minimal admissible set then any one of its members can be considered an extremely good design and is characterised by the two conditions of Proposition 1.2.

Obviously, every member of a minimal admissible set is a limit of design. In fact it is simultaneously three limits of design, characterised respectively by the process of negotiation, which gives an upper limit on the number of inequalities, and the two conditions of Proposition 1.2. Such limits are sometimes called optimal designs. However, the use of the word optimal is avoided here. The word means best. But, as every experienced design practitioner knows, there is usually, in practice, no such thing as a best design in an absolute sense. A design is best only in some relative sense and with respect to a particular way of modelling or expressing the design problem. Conventionally, a design is considered to be best if it minimises a weighted sum of the performance objective functions, which is a scalar expression of the form (1.10). The danger with the use of the word optimal is that newcomers to the field might interpret the word in an absolute sense. A more experienced designer will consider in what way a particular definition of best is appropriate to the problem at hand.

A limit of design satisfies some extreme conditions. Attention has been focused on extreme designs that result from the process of negotiation and satisfaction of Proposition 1.2. Not all limits, for example most Pareto minimisers, characterise a good design. But when a limit does characterise a good design it may be appropriate to call it an extremely good design. This term does not make as strong a claim as the term optimal, since an extremely good design does not preclude the existence of a better design.

Which of the above alternative ways of defining a limit of design is preferable depends on other aspects of the engineering problem. One role of theory is to discover various limits of design and to derive their respective properties. It is then for the practitioner to choose which limit to aim for in practice.

Ordering relations: Finally, it is noted here that the principle of inequalities induces three ordering relations on the elements design space C (Zakian, 1996; see Section 1.7). These relations define when two elements of the design space are equivalent, one is non-inferior to the other or one is superior to the other. As explained in Section 1.7, these relations are useful for the construction of methods for computing an admissible design (the admissibility problem), if one exists. Or, in case an admissible design does not exist, for determining one or more tolerance vectors that characterise a useful nonempty admissible set.

1.3 The Principle of Matching

1.3.1 The Concept of Matching

The concept of matching considers the system in relation to the environment in which it is to operate and requires the designer to choose, from a set of environments and a set of systems, one environment and one system such that the environment-system couple is under control, in the sense of the new definition of control given in Section 1.1, and is well designed in the sense defined below.

It is recalled (see Section 1.1) that the environment generates the input f for the system (which, for convenience, is assumed to include any filters that belong to the environment) and that the input f is known only to the extent that it belongs to a set P, called the possible set.

To characterise the system, the concept of tolerable input is used. An input is said to be tolerable if the output it causes is tolerably small, according to some well-defined criterion of smallness. The set of all tolerable inputs, called the tolerable set T, characterises the input-output transformations (for example, transfer functions or convolution integrals) of the system.

Definition of matching: An environment-system couple is said to be *matched* if every possible input is tolerable; that is, if the possible set P is a subset of the tolerable set T. The couple is said to be well matched if it is matched and if the possible set P is, in some sense, close to the tolerable set T. □

The possible set P represents the stresses or challenges that the environment imposes on the system, while the tolerable set T represents the capabilities of the system to withstand such challenges successfully. There are various ways that the two sets P and T can be made to be close to each other and these ways are considered below. The significance of the concept of a good match is that the extent to which the tolerable set is more inclusive than a set, called the augmented set, which is related to the possible set, indicates the extent to which the environment-system couple is not well designed because the capabilities of the system exceed, to that extent, the challenges imposed by the environment. The following statement of the principle of matching can now be made.

Principle of matching: The principle of matching requires the designer to select one environment P from a given set Π of permissible environments and one system σ from a given set Σ of permissible systems, such that the environment-system couple is well matched. □

Notice, in this definition, that it is not only the system that is subject to design but also the environment, although in some cases the set of environments contains only one element. Conventionally, control theory restricts attention to the design of the system only. However, a more complete approach to design must include the environment. Such an approach is often implicit in practice. The concept of design is understood to be a process of selection of one element from a given set of elements, in accordance with given criteria of preference. The environment and the system can be selected from their respective sets of possible elements so that the environment-system couple is well matched.

In the development so far, the concept of tolerable input is open to interpretation, depending on the criterion used to define smallness of an output or an error. In selecting a mathematical criterion that conforms to the informal definition of control in Section 1.1, many obvious possibilities present themselves. Although some of these possibilities might be worth exploring, the one adopted here is based on the inequality (1.1) and is chosen for its mathematical simplicity and because it results in a framework within which critical systems can be designed methodically.

Definition of tolerable input: An input f is said to be *tolerable* if, for every output port $i \in \widetilde{m}$ (whether critical or not), the output $e_i(f, \sigma)$ satisfies the inequality $\|e_i(f, \sigma)\|_\infty \leq \varepsilon_i$. □

Accordingly, the tolerable set T is the set of all inputs such that, for every output port, the input is tolerable. This is expressed by

$$T = \{f : \|e_i(f, \sigma)\|_\infty \leq \varepsilon_i, \; \forall\, i \in \widetilde{m}\} \tag{1.14}$$

Let the design space C be equal to the Cartesian product $\Pi \times \Sigma$. This is the set of all ordered pairs (P, σ) such that the first element is in the set of all possible environments Π and the second element is in the set of all possible systems Σ. Thus, every element of C is an environment-system couple, denoted by c, and $c = (P, \sigma) \in \Pi \times \Sigma$.

On recalling (see Section 1.1) that the peak output at the ith output port is defined by $\hat{e}_i(c) = \sup \{\|e_i(f, \sigma)\|_\infty : f \in P\}$, it is now obvious, using the definitions of matching and tolerable input, that the conjunction of the following m inequalities

$$\hat{e}_i(c) \leq \varepsilon_i, \quad i \in \widetilde{m}, \; m \leq M \tag{1.15}$$

is a necessary and sufficient condition for the environment-system couple to be matched. That is to say, the environment-system couple is matched if and only if, for every output port, the output is specifically bounded. This shows how the principle of inequalities and the principle of matching are merged together to provide necessary and sufficient conditions for a match.

As discussed in Section 1.2, M denotes the number of inequalities involved in formulating the design problem in accordance with the principle of inequalities. If the number of output ports m is such that $m < M$, the remaining $M - m$ inequalities may be used to represent other constraints in the design situation. However, to simplify the presentation, it is henceforth assumed in this section that $m = M$.

It is also recalled that the admissible set of the conjunction of inequalities (1.15) is the set of all couples $c \in C$ that satisfy the inequalities. It follows that the admissible set of (1.15) is not empty if and only if it is possible to find a couple $c \in C$ that is matched.

The concept of matching is implicit in earlier work (Zakian 1978, 1979a, 1986a, 1986b, 1987a, 1989) where the condition (1.15) is the central theme. To make the concept of matching explicit requires the notion of tolerable input (Zakian, 1989). However, the principle of matching (Zakian, 1991, 1996) includes the additional concept of a well-matched environment-system couple. Yet other concepts that emerge in this section from the principle of matching are that of a perfect match and that of an extremely tolerant system. These two ideas together give rise to the concept of a well-designed environment-system couple. These additional concepts are explained below.

1.3.2 Conditions for a Good Match

A matched environment-system couple may still leave room for improvements in design. The reason for this and how this room can be exploited are now considered. A match is only a necessary condition for a good match. The extent to which the tolerable set T is more inclusive than the possible set P might indicate the extent to which the capabilities of the system exceed the challenges imposed on it by the environment. Actually, as will be seen, it is not, in general, the possible set that gives this indication but a related set, called the augmented set or the augmented possible set. If the capabilities of the system greatly exceed the challenges generated by the environment then the environment-system couple is poorly matched and therefore poorly designed.

As already noted above, in general, both the environment and the system are subject to design. If the system is not fully challenged by the environment then it can either be made to operate in a more challenging environment, thus making the augmented possible set more inclusive, or the performance required of the system can be increased, thus making the tolerable set less inclusive. In either case, it is necessary that a match be maintained, so as to ensure that every possible input is tolerable. The aim of design is to make the augmented possible set and the tolerable set as close to each other as possible, subject to every possible input being tolerable. If the tolerable set includes all the possible inputs and also many other inputs that are not possible, the match is conservative. This concept becomes clearer below.

Increasing the number of output ports or errors might be one way of increasing the performance required of the system. As the number m of output ports is increased, so the tolerable set becomes less inclusive. The process of negotiation that forms part of the principle of inequalities (see Section 1.2) can now be reinterpreted to show how a less conservative match can be obtained.

Process of negotiation: Suppose that the output ports are ordered in decreasing importance. Let K denote the largest number of consecutively numbered output ports such that, for some design $c \in C$, the first K inequalities in (1.15) are satisfied. If $m > K$ then consider only the first K ports and disregard the remaining $m - K$ ports so that the set of ports is redefined and number of ports is redefined to be $m = K$. The possible and tolerable sets are now as equal as they can be made, by adjusting only the number m of output ports. If some or all of the remaining $m - K$ ports, in the original set of output ports, cannot be disregarded then either some of the tolerances ε_i are increased, so as to avoid disregarding some or any output ports, or the inclusiveness of the design space C is increased as discussed in Section 1.2. □

Notice that the value of m, obtained by the above process, can be considered to be a limit of design. Such a limit gives the least conservative environment-system couple with respect to the number of output ports m. This means that the difference between the possible input set P and the tolerable set T has been minimised by adjusting the number of output ports, using the process of negotiation.

The number of output ports m is only one of three variables that can be adjusted to make the possible set as equal as possible to the tolerable set. The other two are contained in the design variable c, which has the two components P, σ. It is therefore important to consider how, for any given value of m, the two components P, σ of the design variable c can be adjusted to minimise the difference between the possible set and the tolerable set, while maintaining a match. To this end, it is useful to start with the concept of an ideal match.

Definition of ideal match: For a given design $c \in C$, an environment-system couple is said to be *ideally matched* if the possible set P is equal to the tolerable set T. □

Evidently, an ideal match (previously called a perfect match; see Zakian, 1991) represents a desirable, if not always attainable, limit of design. However, as will be seen, there is another and more generally attainable limit of design that is not an ideal match but is called a *perfect match*. With the concept of a perfect match, defined below, it is possible to derive necessary and sufficient conditions on the two components P, σ of the couple c to ensure that the possible set P is either equal to or is as close as possible to the tolerable set T, while still ensuring that the couple is matched.

The deeper aspects of the principle of matching follow from the fact that both the possible set P and the tolerable set T are sets of inputs and therefore can be compared. It is certainly true to say that the extent to which the control system can successfully withstand challenges imposed by the environment is represented by the extent to which the tolerable set T is inclusive. But, as will be shown, the extent of the challenge imposed by the environment is represented not by the inclusiveness of the possible set P but by the inclusiveness of a related set $\bar{P}$, called the *augmented set*.

Definition of augmented set: Suppose that, for a given system σ and for every output port $i \in \tilde{m}$, the peak output $\hat{e}_i(P,\sigma)$ is finite. Then the *augmented set* is defined by

$$\bar{P} = \{f : \|e_i(f,\sigma)\|_\infty \leq \hat{e}_i(P,\sigma),\ \forall\, i \in \tilde{m}\} \tag{1.16}$$

□

This shows that the definition of the augmented set can be obtained from that of the tolerable set by replacing, in the latter, each tolerance by its corresponding peak output. It follows from this that the possible set is a subset of the augmented set; that is

$$P \subseteq \bar{P} \tag{1.17}$$

It also follows, from (1.14) and (1.16), that the peak output is not altered if the possible set is replaced by the augmented set; that is

$$\hat{e}_i(P,\sigma) = \hat{e}_i(\bar{P},\sigma) \tag{1.18}$$

Hence, from the definition of tolerable and (1.18), it follows that an environment-system couple is matched if and only if the augmented set is a subset of the tolerable set. This establishes the following lemma.

Lemma 1.1. *The environment system-couple is matched if and only if the augmented set is a subset of the tolerable set; that is*

$$P \subseteq T \Leftrightarrow \bar{P} \subseteq T \tag{1.19}$$

The significance of equation (1.18) and this lemma in design is that there is no additional design cost in replacing the possible set P by the augmented set $\bar{P}$. The augmented set might contain elements that cannot occur or are unlikely to occur in reality but replacing the possible set by its augmented version entails no additional challenge to the system but can simplify the design of the couple. It follows that a convenient rule for modelling the environment or designing the couple is to replace the possible set by its augmented version if this simplifies the design concepts and procedures (see Section 1.4). The notion of augmented set also leads to other useful concepts, such as the following.

Definition of perfect match: For a design $c \in C$, an environment-system couple is said to be *perfectly matched* if the augmented set $\bar{P}$ is equal to the tolerable set $T(\sigma)$. □

Notice here the difference between an ideal match and a perfect match. The first requires the possible set to be equal to the tolerable set while the second requires the augmented set to be equal to the tolerable set. A perfect match is a more general and more useful notion of a limit of design than the concept of an ideal match. This is because the concept of a perfect match allows greater flexibility in modelling the environment by means of a suitable set P. For example, the possible set P might, in extreme cases, contain only a single element, in which case an ideal match might be impossible because the tolerable set cannot be made to contain that one element only. However, the augmented set can always be defined, even for such a case, and it is not unreasonable to seek a design that results in a perfect match.

As with an ideal match, a perfect match means that the system is not over-designed with respect to the environment. The concepts of ideal match and perfect match coincide if, and only if, the possible set is the same as the augmented set.

Theorem 1.1. *An environment-system couple is perfectly matched if and only if the couple $c = (P, \sigma)$ satisfies the equation $\hat{e}(P, \sigma) = \varepsilon$.*

This theorem restates the condition for a perfect match in terms of peak output and tolerance, which are the quantities involved in numerical computations during the design process. The proof of the theorem is based on results that have to be established first. To this end, the following definition is needed.

Definition of saturated set: Let A denote a set of inputs. Then A is said to be *saturated* if

$$A = \{f : \|e_i(f, \sigma)\|_\infty \leq \hat{e}_i(A, \sigma),\ \forall i \in \tilde{m}\} \tag{1.20}$$

$\square$

Evidently, the augmented set $\bar{P}$ is saturated. So is the tolerable set $T(\sigma)$, because of the following identity:

$$\hat{e}(T(\sigma), \sigma) = \varepsilon \tag{1.21}$$

Lemma 1.2. *Let A and B be two saturated input sets. Then the following two conditions are equivalent:*

(a) $\hat{e}(A, \sigma) = \hat{e}(B, \sigma)$ (1.22)

(b) $A = B$ (1.23)

Proof. Equations (1.20) and (1.22) give

$$A = \{f : \|e_i(f, \sigma)\|_\infty \leq \hat{e}_i(B, \sigma),\ \forall i \in \tilde{m}\} \tag{1.24}$$

Hence, since B is a saturated set, the right hand member of (1.24) is equal to B and therefore condition (1.23) holds. Conversely, it is obvious that (1.23) implies (1.22). $\square$

Proof of Theorem 1.1: If the environment-system couple is perfectly matched then $\bar{P} = T(\sigma)$ and hence (1.18) and (1.21) give $\hat{e}(c) = \varepsilon$. Conversely, if c satisfies the equation $\hat{e}(c) = \varepsilon$ then, by the equations (1.18) and (1.21), it follows that

$$\hat{e}(\bar{P}, \sigma) = \hat{e}(T(\sigma), \sigma) \tag{1.25}$$

Because the two sets $\bar{P}$, $T(\sigma)$ are saturated, the application of Lemma 1.2 to equation (1.25) gives $\bar{P} = T(\sigma)$. This means that the environment-system couple (P, σ) is perfectly matched. □

Although, according to Theorem 1.1, the equation $\hat{e}(c) = \varepsilon$ is a necessary and sufficient condition for the environment-system couple to be perfectly matched, it is, according to Proposition 1.2, a necessary but not sufficient condition for the couple c, that satisfies the equation, to be a member of a minimal admissible set (see Section 1.2).

1.3.3 Well Designed Couple

When a couple is perfectly matched, there is still some design freedom left, since the design parameter c contains two components (P, σ) and a perfect match imposes only one condition. The remaining design freedom can be utilised by requiring the system alone (not the couple) to have the property, called extreme tolerance, defined as follows.

Definition of extremely tolerant system: A system $\sigma \in \Sigma$ is said to be *extremely tolerant* if there is no other system $\sigma^* \in \Sigma$ such that $T(\sigma^*) \supset T(\sigma)$. □

Thus an extremely tolerant system is a limit of design, in the sense that there is no other system that can make the tolerable set more inclusive. At this limit, the capabilities of the system are made fully available.

The above definition of extreme tolerance is given for a system, without regard for the environment to which it is coupled. What is of interest is to obtain conditions for extreme tolerance, in terms of the vector peak output function $\hat{e} : (P, \sigma) \mapsto \hat{e}(P, \sigma)$, which characterises the environment-system couple. To this end, the following mild condition is required.

Definition of alpha condition: The vector peak output function $\hat{e} : (P, \sigma) \mapsto \hat{e}(P, \sigma)$, is said to satisfy the *alpha condition* if, for any two saturated sets A and B, and for every system $\sigma \in \Sigma$,

$$A \supset B \Leftrightarrow \hat{e}(A, \sigma) \succ \hat{e}(B, \sigma) \tag{1.26}$$

□

Lemma 1.3. *Let the alpha condition be satisfied. Then for any two distinct systems* $\sigma, \sigma^* \in \Sigma$,

$$\hat{e}(T(\sigma), \sigma^*) \prec \varepsilon \Leftrightarrow T(\sigma^*) \supset T(\sigma) \tag{1.27}$$

Proof. The left-hand member of (1.27), together with the identity (1.21), imply that

$$\hat{e}(T(\sigma), \sigma^*) \prec \hat{e}(T(\sigma^*), \sigma^*) \tag{1.28}$$

Since, on replacing σ by σ^*, (1.28) is changed into an equation, it follows, by virtue of the alpha condition, that (1.28) implies the right-hand member of (1.27).

Conversely, by virtue of the alpha condition and the identity (1.21), the left-hand member of (1.27) is implied by the right-hand member. □

Lemma 1.4. *Let the alpha condition be satisfied and let the couple be perfectly matched. Then, for any two distinct systems* $\sigma, \sigma^* \in \Sigma$,

$$\hat{e}(P, \sigma^*) \prec \varepsilon \Leftrightarrow T(\sigma^*) \supset T(\sigma) \tag{1.29}$$

Proof. The result follows from Lemma 1.3 and because, for a perfectly matched couple, $\hat{e}(P, \sigma^*) = \hat{e}(T(\sigma), \sigma^*)$. □

Definition of well-designed couple: An environment-system couple is said to be *well designed* if it is perfectly matched and the system is extremely tolerant. □

This definition brings together the two limits of design: perfect matching of a couple and extreme tolerance of the system to disturbances. Thus a well-designed couple represents an extremely good design for the couple as a whole. In contrast, a perfectly matched couple, if it is not well designed, is an inferior couple because the system is not as tolerant as it could be. The environment of a well-designed couple is the most challenging environment that still satisfies the condition of matching, because the augmented possible set is equal to the tolerable set, which is maximally inclusive.

Theorem 1.2. *Let the vector peak output function* $\hat{e} : (P, \sigma) \mapsto \hat{e}(P, \sigma)$ *satisfy the alpha condition. Then an environment-system couple* (P, σ) *is well designed if and only if it satisfies the following two conditions:*

(a) $\hat{e}(P, \sigma) = \varepsilon$
(b) There is no system $\sigma^* \in \Sigma$ *such that* $\hat{e}(P, \sigma^*) \prec \varepsilon$.

Proof. Condition (a) follows immediately from Theorem 1.1 and is equivalent to the couple being perfectly matched. Condition (b) follows from Lemma 1.4 and the definition of an extremely tolerant system. □

1.3.4 Linear Couples

In the development so far, the only assumption that has been made about the environment-system couple is that it satisfies the alpha condition. Here a stronger assumption is made, called a *linear couple*, and it is shown that a linear couple implies the alpha condition. In Section 1.4, it is shown that a specific linear couple provides a practical tool for design.

Definition of linear couple: An environment-system couple is said to be a *linear couple* if the following three conditions are satisfied:

(a) The peak output vector is given by a linear algebraic equation of the form
$$\hat{e}(d, \sigma) = N(\sigma)d \qquad (1.30)$$

Here, $N(\sigma)$ is an $m \times n$ matrix whose elements are non-negative numbers that represent sensitivities of the input-output transformations and d is an n vector of non-negative numbers, called the environment parameter, that characterises the possible set P so that, for every d there is well defined possible set P.

(b) Let d, d^* be two distinct values of the environment parameter and P, P^* the corresponding possible sets. Then $d \succ d^* \Leftrightarrow P \supset P^*$.

(c) For every value of environment parameter d, the corresponding possible set P is saturated. □

It is shown in Section 1.4 how the concept of a linear couple can be substantiated by appropriate definition of the sensitivity matrix $N(\sigma)$ and the environment parameter d. In some cases, every element of the sensitivity matrix is the sensitivity of the corresponding element of the transfer function matrix.

Note the minor change of notation in (1.30), where the peak output is shown to depend on the two parameters d, σ instead of P, σ.

Lemma 1.5. *A linear couple satisfies the alpha condition.*

Proof. Note that $d \succ d^* \Rightarrow N(\sigma)d \succ N(\sigma)d^*$. Hence, from Conditions (a) and (b) of the definition of a linear couple, it is deduced that $P \supset P^* \Rightarrow \hat{e}(d, \sigma) \succ \hat{e}(d^*, \sigma)$, which is one part of the alpha condition. To establish the converse of this implication, which is the other part of the alpha's condition, suppose that $\hat{e}(P, \sigma) \succ \hat{e}(P^*, \sigma)$. Then, by Condition (c) of a linear couple, P and P^* are saturated and hence $P \supset P^*$. □

Theorem 1.3. *A linear couple is well designed if and only if it satisfies the following two conditions:*

(a) $\quad N(\sigma)d = \varepsilon \qquad (1.31)$

(b) $\quad$ *There is no system $\sigma^* \in \Sigma$ such that $N(\sigma^*)d \prec \varepsilon$.*

Proof. The result follows from Theorem 1.2 and Lemma 1.5. □

Comments: It is recalled here that, according to the conventional definition of control (see Section 1.1), some measure of the sensitivity matrix $N(\sigma)$ is minimised. It is therefore interesting to consider the extent to which conventional methods of design can be used to achieve a well-designed environment-system couple, assuming that the sensitivity matrix satisfies equation (1.31) and thereby defines a linear couple. To this end, consider the simplest case of a linear couple having a system with one input port and one output port; that is, $m = n = 1$. Condition (b) of Theorem 1.3 is satisfied by choosing a value of the system design parameter σ that minimises the scalar sensitivity $N(\sigma)$. After which, the parameter d of the environment is determined uniquely by equation (1.31), thus satisfying Condition (a) of the theorem. The environment-system couple is then well designed. This shows that, in the case of systems with one input port and one output port, conventional methods of design, which involve the minimisation of a scalar sensitivity that satisfies equation (1.31), do yield a well-designed couple, provided that the environment parameter d satisfies equation (1.31). In essence, this result was established in previous work for specific linear couples (Zakian, 1989, 1991, 1996).

However, the same conclusion does not hold for multivariable systems in general where the sensitivity matrix is not a scalar. The reason is that, unlike the one-input one-output case, the two conditions of Theorem 1.3 are seldom satisfied by minimising an arbitrary norm of the sensitivity matrix $N(\sigma)$. It is therefore concluded that, although it might be possible to devise methods of design that utilise some of the techniques of conventional methods, so as to yield a well designed couple, such methods would have to conform to ideas that are derived from the principle of matching. In particular, the sensitivity matrix has to be minimised in accordance with the two conditions of Theorem 1.3.

Nonetheless, although it is, in general, not possible to obtain a well-designed couple by designing the system in accordance with the conventional definition of control, it is in many cases possible to obtain a perfectly matched couple by using some conventional techniques. The following problem shows how this can be done.

The inverse problem of matching: The inverse problem of matching can be stated as follows. Given a value of the system parameter σ (perhaps this parameter satisfies the conventional definition of control), such that every element of the sensitivity matrix $N(\sigma)$ is a finite number, determine a value of the environment parameter d such that the linear couple is matched or is perfectly matched.

Evidently, the inverse problem of matching provides one practical justification for the concept of a linear couple. This is because if the system parameter σ is determined by some conventional design method to give a sensitivity matrix with minimal norm and if there is an environment parameter d that satisfies the linear equation (1.31) then the couple (d, σ) is perfectly matched. Notice however that a perfectly matched couple is only a necessary condition for a well-designed couple.

Evidently, given the tolerance ε, every solution (d, σ) of equation (1.31) represents a perfectly matched couple. For a given system σ, every value of the environment parameter d, that satisfies the linear equation (1.31), represents a solution to the inverse problem of matching.

If, for a given system σ, equation (1.31) does not have a solution then a value of the environment parameter d may be sought that minimises a norm of the difference $\varepsilon - N(\sigma)d$. The corresponding pair (d, σ) represents a couple that is as close as possible, in some sense, to a perfectly matched linear couple.

The concepts of linear couple and the inverse problem of matching are generalisations of ideas contained in (Zakian, 1989, 1991, 1996).

An illustration: Some of the main ideas about matching can be illustrated by means of the following example. Consider the design of a transportation system with a roadway to enable a vehicle to travel at constant speed despite disturbances (inputs) caused by changing road gradients and wind conditions. The possible set P is defined by considering all road and wind conditions in which the vehicle is intended to operate. For an initial design, the tolerable set T is assumed to comprise all inputs, possible or not possible, that do not cause the speed of the vehicle to exceed its prescribed tolerance. The contents of the tolerable set depend on the power of the engine (the power plant) driving the vehicle and the type of speed control. Suppose that an initial design results in a system that is matched to the environment, so that all possible disturbances are tolerable, but that the tolerable set is much more inclusive than the possible set. This means that the system is over designed with respect to the environment because the capabilities of the system far exceed the challenges imposed on it by the environment. There are two ways to improve the design of the environment-system couple. One way is to make the tolerable set less inclusive either by tightening the specified tolerances on the speed or, perhaps preferably, by adding further output ports which give rise to further conditions, in the form of inequalities, to ensure that the vehicle moves so as to keep the passengers in sufficient comfort, thereby making the definition of tolerable more stringent. With the tolerable set thus redefined and made less inclusive, a matched environment-system couple ensures, not only that the vehicle speed remains within the specified tolerances but also that the passenger ride is comfortable to the specified extent. The other way of reducing over-design is to make the possible set more inclusive, preferably by making the augmented set equal to the tolerable set, by allowing the vehicle to operate under more difficult road and wind conditions. Moreover, all the capabilities of the system can be made fully available by satisfying the conditions of Theorem 1.3, thereby ensuring that the system is extremely tolerant. Some aspects of this example are treated in detail by Rutland (1992, 1994a). Other illustrations of the principle of matching can be found in Parts II and IV.

Some of the concepts of the principle of matching, in particular those reported in this section for the first time, including the notions of: augmented possible set, perfect matching and extreme tolerance to disturbances, have yet to be applied to practical problems of control.

1.4 A Class of Linear Couples

1.4.1 Preliminaries

The purpose of this section is to substantiate the concept of a linear environment-system couple, also referred to as a linear couple, as defined in Section 1.3. It is shown here that certain linear couples provide a way of characterising, and hence constructing models of, the environment-system couple in a practical manner, bearing in mind that such modelling must be done for the couple as a whole and not for the environment and the system as separate entities. The results obtained here also give an explicit formula for computing the peak output vector $\hat{e}(d, \sigma)$.

To these ends, it is assumed that the block diagram model of the environment is, for each input port of the system, comprised of one generator that produces the scalar input f_j, which is a component of the vector input f. Each scalar input feeds a filter, the output of which feeds one corresponding input port of the system. The filters and the control system form a composite system that is linear and time-invariant. It is also assumed that, for a given constant that characterises a generator, every scalar input f_j has a p-norm that is uniformly bounded by that constant. With these assumptions, a practical expression for the peak output vector is derived. This expression also provides a convenient way of defining sensitivity and input-output stability. This section is based on previous work (Zakian, 1987a).

Let the input f be the vector $(f_1, f_2, \ldots, f_n)$ where each component f_j is a piecewise continuous function that maps the real line into itself and $f_j(t) = 0$, $t \leq 0$.

It is assumed that the filter transfer function is either equal to unity or is a rational function that is stable and strictly proper (the numerator degree is less than the denominator degree). The filter transfer function, if not equal to unity, acts to smooth the input when the impulse response of the system is equal to a delta function plus a piecewise continuous function. There are two related reasons for introducing the filter. The first is to make the impulse response, of the combined filter and system, piecewise continuous (see comment following Condition (b) below) in order to make use of certain p-norms (defined below) to characterise the output of the generators of the environment. The second purpose is considered in detail in Section 1.5, which deals with modelling the environment-system couple.

1.4.2 Derivation of Linear Couple

Let F denote the set of all real piecewise continuous functions x such that $x(t) = 0$, $t \leq 0$. Consider the function $\|\cdot\|_p$, which maps the set F into the extended half-line $[0, \infty]$, defined by

$$\|x\|_p = \left\{ \int_{-\infty}^{\infty} |x(t)|^p \, dt \right\}^{1/p}, \, p \in [1, \infty) \tag{1.32}$$

$$\|x\|_\infty = \sup \{|x(t)| : t \in R\} \tag{1.33}$$

The function $\|\cdot\|_p$ is called a p-norm. The purpose of making the range of the p-norm equal to the extended half-line is so that if the function x is unbounded then $\|x\|_\infty = \infty$. That is to say, if x is unbounded then the least upper bound of the absolute value of x is defined and is equal to infinity. In this respect, the definition of norm used here differs from the usual.

Let d denote the column vector $d = (d_1, d_2, \ldots, d_n)^T$ and let $P_j(d_j)$ denote the possible set, that characterises the jth generator in the environment, be defined by

$$P_j(d_j) = \left\{ f_j \in F : \|f_j\|_{p_j} \leq d_j \right\} \tag{1.34}$$

Condition (a): Let the possible set that characterises the generators within environment be

$$P(d) = \left\{ f : f_j \in F, \, \|f_j\|_{p_j} \leq d_j, \, j \in \tilde{n} \right\} \tag{1.35}$$

□

Notice that the dependence of the possible set on the vector d, which is called the bound of the possible set, is shown explicitly in the above notation. Notice also that this possible set is a set of vector inputs.

Condition (b): For every $(i, j) \in \tilde{m} \times \tilde{n}$, the response at the ith output port, caused by the input f_j, is defined by the sum of convolution integrals

$$e_i(t, f, \sigma) = \sum_{j=1}^{n} \int_0^t f_j(t - \lambda) e_{i,j}(\lambda, \delta, \sigma) d\lambda \tag{1.36}$$

It is assumed that the impulse response function $e_{i,j}(\delta, \sigma) : \lambda \mapsto e_{i,j}(\lambda, \delta, \sigma)$ is piecewise continuous. □

Here δ denotes the delta function occurring at $\lambda = 0$ and the impulse response function $e_{i,j}(\delta, \sigma)$ is the response at the ith output port caused when the input at the jth input port is the delta function while the input at all the other ports is zero.

Condition (b) implies that the system, together with the filters, is linear and time invariant. If necessary, the piecewise continuity of the impulse response at every output port is achieved by the use of an environment filter with smoothing characteristic. Such continuity can be achieved for a wide class of systems; for example, all those filter-system combinations characterised by the state space equations $\dot{x} = Ax + Bf$, $e = Cx$.

Let $N_{i,j}(\sigma)$ denote the sensitivity of the transformation from the jth input to the ith output. The following condition defines the sensitivity to be the operator norm of the transformation.

Condition (c):

$$N_{i,j}(\sigma) = \|e_{i,j}(\delta,\sigma)\|_{q_j}, \quad p_j^{-1} + q_j^{-1} = 1 \tag{1.37}$$

$\square$

Let the $m \times n$ matrix $N(\sigma)$ have its elements defined by (1.37).

Theorem 1.4. *Suppose that the Conditions (a), (b) and (c) are satisfied. Then the environment-system couple is linear and the peak output is given by the equation*

$$\hat{e}(d,\sigma) = N(\sigma)d \tag{1.38}$$

Proof. By an application of Hölder's inequality to (1.37) and making use of (1.35) and (1.38), it follows that

$$\hat{e}_i(d,\sigma) \leq \sum_{j=1}^{n} N_{i,j}(\sigma)d_j \tag{1.39}$$

For every $t \geq 0$ and every $(i,j) \in \tilde{m} \times \tilde{n}$, consider the input function $f_{i,j}^t$ that is zero outside the interval $0 \leq \lambda \leq t$ and is defined as follows inside the interval

$$f_{i,j}^t(t-\lambda) = d_j \left(\|e_{i,j}(\delta,\sigma)\|_{q_j}\right)^{1-q_j} |e_{i,j}(\lambda,\delta,\sigma)|^{q_j-1} \operatorname{sgn} e_{i,j}(\lambda,\delta,\sigma) \tag{1.40}$$

Bearing in mind that $p(q-1) = q$, it follows that $\|f_{i,j}^t\|_{p_j} = d_j$, which implies that the input function $f_{i,j}^t$ belongs to the possible set $P_j(d_j)$. Let $f_i^t = (f_{i,1}^t, f_{i,2}^t, \ldots, f_{i,m}^t)$ then, from (1.36) and (1.40), it follows that

$$e_i(f_i^t, \sigma) = \sum_{j=1}^{n} \|e_{i,j}(\delta,\sigma)\|_{q_j} d_j \tag{1.41}$$

This means that the ith peak output must be at least as large as this. Hence expressions (1.39) and (1.41) imply that, for every $i \in \tilde{m}$,

$$\hat{e}_i(d,\sigma) = \sum_{j=1}^{n} N_{i,j}(\sigma)d_j \tag{1.42}$$

$\square$

For the single input case $n = 1$, the theorem reduces to a well-known result that can be found in standard books on mathematical analysis (see, for example, Royden, 1968, page 119). For the multi input case $n > 1$, the theorem is a straightforward generalisation of the well-known result and hence the sensitivity matrix can be regarded as a generalisation of the concept of operator norm for the specific p-norms employed here to measure the size of the input and for the input-output transformation represented by (1.36).

For an extension of the theorem, involving sensitivities of the form $\|e_{i,j}(\delta,\sigma)\exp\alpha(\cdot)\|_{q_j}$, see Zakian (1987a). This more complex sensitivity, which is the norm of the impulse response weighted by an increasing exponential function, allows the possible input set to contain both transient and persistent elements (see comments below and Section 1.5).

The theorem can be extended in a straightforward fashion to situations where each input port of the system receives input from more than one generator, with each generator followed by its own filter, and the outputs of the filters are summed to make the input for one input port of the system. Such an input is said to be complex and an important example of a complex input is discussed in Section 1.5.

The computation of sensitivity $N_{i,j}(\sigma) = \|e_{i,j}(\delta,\sigma)\|_{q_j}$, $1 \leq q_j < \infty$, is usually carried out, in practice, by standard numerical methods. These involve computing the impulse response and then computing the integral (1.33) by standard methods of numerical integration. When designing a system by the method of inequalities, a sequence of impulse responses corresponding to a sequence of values of the design parameter σ is computed. The interval of integration is chosen to be sufficiently large for an initial value of σ and this interval is usually sufficiently large for all subsequent values of σ encountered in the design process. In the cases where the sensitivity is defined by the 1-norm or the 2-norm and the Laplace transform of the impulse response is a rational function then an alternative method is provided (Rutland and Lane, 1995).

1.4.3 Input-output Stability

Evidently, $N_{i,j}(\sigma)$ defines a sensitivity of its corresponding linear input-output transformation $e_{i,j}(\sigma) : f_j \mapsto e_{i,j}(f_j,\sigma)$ and hence it measures the extent of input-output stability of the transformation. By the definition of input-output stability given in Section 1.3, it follows that the linear transformation $e_{i,j}(\sigma) : f_j \mapsto e_{i,j}(f_j,\sigma)$ is input-output stable if and only if $N_{i,j}(\sigma) < \infty$. Also, the system is input-output stable if and only if every element of the sensitivity matrix $N(\sigma)$ is a finite number. Suppose that the linear transformation is characterised by a strictly proper rational transfer function, which is defined to be stable if all its poles have negative real parts. It can easily be shown that the transfer function is stable if and only if its sensitivity is finite. This means that, for such a transfer function, the definition of its stability in terms of its poles is equivalent to the definition of its stability in terms of the notion of sensitivity. However, the notion of sensitivity is applicable also to transfer functions that are not rational and therefore provides a more general way to define input-output stability. This idea leads to a computational method for stabilising transfer functions (Zakian, 1987b; Section 1.6).

A well-known concept in control is that of input-output stability in the sense of every bounded input resulting in a bounded output. Accordingly, an input-output transformation is said to be BIBO stable if, for every bounded input, the output is bounded. Evidently, by Theorem 1.4, BIBO stability of a scalar input-output transformation, having impulse response $e(\delta, \sigma)$, is equivalent to the condition that the sensitivity $N(\sigma) = \|e(\delta, \sigma)\|_1$ is finite. This concept of stability does not take into account the possibility that the input might be unbounded. It is therefore not suitable for transient inputs (see Section 1.5) that might have arbitrary amplitude and are therefore unbounded, but have a finite 2-norm. The BIBO concept of stability is generalised by Theorem 1.4. This more general concept may be called p-NBO (finite input p-norm, bounded output) stability. Accordingly, an input-output transformation is p-NBO stable if, for every input such that its p-norm is finite, the output is bounded. In particular, BIBO stability is equivalent to ∞-NBO stability. Also, for example, for every transient input such that its 2-norm is finite then the output is bounded, if and only if the sensitivity of the input-output transformation that, as given by Theorem 1.4, is the 2-norm of the impulse response, is finite. If the Laplace transform of the impulse response of an input-output transformation is a strictly proper rational function and all the poles have negative real parts then the transformation is p-NBO stable for every $p \geq 1$. Notice that, for a system with several input generators, the norm can vary with the generator. For example, the functions produced by the first generator might be measured by the 2-norm, while that of the second might be measured by the ∞-norm. If all the input-output relations of the system are p-NBO stable then the system is said to be p-NBO stable. Hence a system represented by the state space equations $\dot{x} = Ax + Bf$, $e = Cx$ is input-output stable in the p-NBO sense if all the characteristic roots of the system have negative real parts. The condition is also necessary if all the roots are involved in the input-output transformations.

1.4.4 Example

Suppose that a control system has $m = n = 2$. The response at the first output port is the error in the controlled variable, which is required to be bounded by the tolerance ε_1, while the response at the second output port is the actuator variable, which is required to be bounded by ε_2 so as to avoid saturation. The first source input port generates transient source input while the second source input port generates persistent source input so that the set of all possible source inputs that can be generated satisfies

$$\|f_1\|_2 \leq d_1, \quad \|f_2\|_\infty \leq d_2 \tag{1.43}$$

It follows from Theorem 1.4 that the peak output of the linear couple is expressed by

$$\begin{pmatrix} \hat{e}_1(P(d),\sigma) \\ \hat{e}_2(P(d),\sigma) \end{pmatrix} = \begin{pmatrix} \|e_{1,1}(\delta,\sigma)\|_2 & \|e_{1,2}(\delta,\sigma)\|_\infty \\ \|e_{2,1}(\delta,\sigma)\|_2 & \|e_{2,2}(\delta,\sigma)\|_\infty \end{pmatrix} \begin{pmatrix} d_1 \\ d_2 \end{pmatrix} \quad (1.44)$$

Now suppose that, as discussed in Section 1.5, one of the inputs is complex so that $f_1 = f_1^{per} + f_1^{tra}$ and $\|f_1^{per}\|_\infty \leq d_1^{per}$, $\|f_1^{tra}\|_2 \leq d_1^{tra}$.

1.5 Well-constructed Environment-system Models

This section is concerned with three main themes. The first is the modelling of the environment, including some alternative ways of achieving effective models. The second is that modelling the environment has to be considered in the context of the model of the system. The third is the consequences of the various ways of modelling the environment on the computation of the vector peak output.

Modelling the environment is a relatively undeveloped aspect of control theory, mainly because much of conventional theory treats the control system as if it were somewhat separate from the environment. Since conventional theory aims to provide a stable system, which minimises an arbitrary measure of sensitivity of the input-output transformation, a precise model of the environment is not required. However, an accurate model of the environment is essential to the new framework for design, especially because the design of critical control systems requires an accurate determination of the peak outputs.

Besides accuracy, an environment model has to satisfy certain compatibility conditions that depend on the model of the system, so that the resulting environment-system model is useful in design. Thus, accuracy and compatibility are the two conditions that ensure a well-constructed environment-system model.

1.5.1 Persistent and Transient Input

In order to obtain a useful characterisation of a generator within the environment, it is necessary to consider two distinct types of input functions: persistent functions and transient functions. All input functions can then be categorised as either persistent or transient or the sum of persistent and transient (Zakian, 1989, 1996).

A scalar input f is said to be persistent if it is piecewise continuous and is bounded so that $\|f\|_\infty \leq d^{per}$, for some non-negative number d^{per}. A generator within the environment is said to produce persistent functions bounded by the number d^{per} if all the functions it produces belong to the possible set

$$P^{per} = \{f \in F : \|f\|_\infty \leq d^{per}\} \quad (1.45)$$

Here F denotes the set of all real piecewise continuous functions such that $f(t) = 0, t \leq 0$. Evidently, this possible set contains repetitive and quasi-repetitive functions, random processes that continue indefinitely and also transients that are bounded by the number d^{per}. A transient function is a change that lasts for a finite time (such as a pulse) and then either becomes zero thereafter or decays to zero (for example, exponentially) with time. Thus, for convenience, the term persistent applies to some functions that are transient.

A scalar input $f \in F$ is said to be transient if $\|f\|_2 \leq d^{tra}$, for some non-negative number d^{tra}. A generator within the environment is said to produce transient functions bounded in the 2-norm by d^{tra} if all the functions it produces belong to the possible set

$$P^{tra} = \{f \in F : \|f\|_2 \leq d^{tra}\} \tag{1.46}$$

Obviously, some transient functions are members of the set P^{per}. Notice, however, that a transient function can be unbounded, in which case it is not a member of the set P^{per}. Similarly, there are persistent functions, for example a random process, that do not belong to the set P^{tra}. Although the two sets P^{per} and P^{tra} overlap, they have distinctive properties that are essential for modelling the environment.

An input $f \in F$ is said to be complex if $f = f^{per} + f^{tra}$ and $f^{per} \in P^{per}$, $f^{tra} \in P^{tra}$ for some pair d^{per}, d^{tra}. A generator is said to produce complex functions bounded by the pair d^{per}, d^{tra} if all the functions it produces belong to the possible set

$$P^{complex} = \{f : f = f^{per} + f^{tra}, f^{per} \in P^{per}, f^{tra} \in P^{tra}\} \tag{1.47}$$

This possible set of complex functions provides a flexible and accurate way of modelling the generator within the environment. The set can include functions that are persistent throughout positive time and superimposed on such a function are a few transients of short duration but with unbounded amplitude.

Suppose that a system with impulse response $e(\delta, \sigma)$, which is piecewise continuous and is zero for negative time, is coupled to the generator of an environment characterised by the possible set $P^{complex}$. Then the peak output is given by

$$\hat{e}(P^{complex}, \sigma) = \|e(\delta, \sigma)\|_1 d^{per} + \|e(\delta, \sigma)\|_2 d^{tra} \tag{1.48}$$

This equation is easily obtained by using the methods of Section 1.4. Note, however, that the usefulness of this equation is based on the assumption that the impulse response is piecewise continuous. If this assumption is not satisfied then, as will be shown, the environment-system model is not well constructed. This is because, if the assumption is not satisfied, the peak output does not change in a useful manner with changes in the system design parameter.

The stability of the input-output transformation is defined in terms of the finiteness of the two sensitivities $\|e(\delta,\sigma)\|_1$, $\|e(\delta,\sigma)\|_2$ that characterise the transformation. Accordingly, the transformation is said to be stable if these two sensitivities are finite. In the most common case, where the transformation is defined by a rational transfer function, the sensitivities are finite if and only if all the poles of the transfer function have negative real parts.

For any physical environment, modelling a generator within the environment requires that its two parameters d^{per}, d^{tra} be estimated from observations of a sample of functions f generated by the environment.

Such observations might contain a sample function f with a transient pulse of peak magnitude d^{tra} added onto a persistent change bounded by d^{per}. It would therefore be possible to model the possible set as comprising all persistent functions bounded by $d^{per} + d^{tra}$. However, such a possible set would not be an accurate representation of the environment because it would contain many persistent functions that cannot occur in reality. Such a model of the environment might, at best, lead to conservative designs and, at worst, to failed attempts at design. This is the principal reason for modelling every generator within the environment by means of a complex possible set $P^{complex}$.

It is noted, in passing, that another method, that makes use of exponential weights, for enabling both persistent and transient inputs to be modelled simultaneously, has been considered (Zakian, 1987a; see also Zakian, 1979a).

For the purpose of analysis, it is usually convenient to replace the single generator, characterised by the complex possible set $P^{complex}$, by two generators characterised, respectively, by the possible sets P^{per}, P^{tra}; with each generator feeding the same input-output transformation, the outputs of which are summed as shown in (1.48). In this way, possible sets, of the form P^{per}, P^{tra}, can be used to model the environment and the environment-system couple can be cast in the form of a linear couple, as discussed in Section 1.4.

Nonetheless, depending on the nature of the input-output relation of the system, the possible set $P^{complex}$ is not, by itself, always a suitable model of the environment. The main reason for this is that some of the functions it can generate have stepwise discontinuities at some points in time, which means that the rate of change at those points is infinite. Also, some of the transients it can generate have unbounded magnitude. Both of these features might not reflect accurately the behaviour of a physical environment and might lead to poor designs if they are not compensated in some way. Such compensation is achieved either with the use of filters to obtain a satisfactory model of the environment or, alternatively, by restricting the rate of change of the function produced by the generator. Both these approaches are considered here and both result in an environment model that restricts, in some way, the rate of change of the input to the system. However, some emphasis is placed on the use of filters because such an approach facilitates mathematical analysis, which brings with it a number of benefits.

1.5.2 Necessity of Imposing Some Restriction on Rate of Change

It is useful to make the distinction between an input function f produced by a generator within the environment and the corresponding function v at an input port of the system. These two functions are identical unless the generator is coupled to the input port of the system by means of a filter. Any such filter is assumed to be linear time-invariant and non-anticipative and to have smoothing properties. This implies that $v \in F$. Evidently, it is implicit that the system input function v depends on the input f produced by the generator and hence this can be written more explicitly as $v(f)$ but either notation is used. The analysis presented below is, with a few added refinements, essentially that given previously (Zakian, 1986a, 1996).

If the system is linear, time-invariant and non anticipative then every scalar input-output relation of the system is, in general, characterised by equations of the form

$$e(t, v, \sigma) = \alpha(\sigma)v(t) + y(t, v, \sigma) \tag{1.49}$$

$$y(t, v, \sigma) = \int_0^t y(\lambda, \delta, \sigma) v(t - \lambda) d\lambda \tag{1.50}$$

Here, $\alpha(\sigma)$ is a real parameter and the function $y(\delta, \sigma)$ is piecewise continuous and represents the impulse response from the input port to the response port that produces $y(t, v, \sigma)$. It follows that if the input function v is piecewise continuous then the function $y(v, \sigma)$ is continuous. The two components, $\alpha(\sigma)$ and $y(\delta, \sigma)$, involved in the input-output transformation (1.49)–(1.50) from the input v to the output $e(v, \sigma)$ characterise, respectively, the instantaneous and the dynamical components of the transformation. This kind of input-output relation is said to involve some instantaneous transmission because the parameter $\alpha(\sigma)$ modifies the input without time lag. If the Laplace transform of the impulse response $y(\delta, \sigma)$ is a rational function then it is strictly proper.

If $\alpha(\sigma) = 0$ and $f = v$ then the input-output relation of the system is in the form required by Theorem 1.4 and hence the conclusions of that theorem are applicable. In particular, (1.48) holds because $y(\delta, \sigma) = e(\delta, \sigma)$.

However, it is often the case that $\alpha(\sigma)$ is not zero and cannot be made negligible by appropriate choice of the system design parameter σ. For example, if $\alpha(\sigma) = 1$ then the response function $y(v, \sigma)$ can be interpreted as a feedback intended to cancel part of the input function v. Clearly, this situation occurs very commonly and it represents, for example, the standard feedback control system or servomechanism, with one input port, where the input v is the reference function and the response $y(v, \sigma)$ is the controlled function that is fed back.

If $|\alpha(\sigma)|$ is not negligibly small then it is necessary to restrict the derivative of the input v so as to ensure that the design problem is well posed, in the sense that it is possible to adjust the design parameter σ of the system, with the aim of reducing the size of the peak output and thereby achieve a specifically bounded peak output.

In order to see this, let $\alpha(\sigma) = 1$. If the input v is piecewise continuous and, in consequence, the response function $y(v,\sigma)$ is continuous then, at every point in time where a stepwise discontinuity occurs in the input v, the same stepwise discontinuity occurs also in the output function $e(v,\sigma)$. The size of the discontinuity in the input function v is independent of the design parameter σ of the system and hence it is not possible to reduce the size of the output at the points of discontinuity, to less than the size of the discontinuity of the input, by changing the value of the design parameter of the system.

To take the above analysis further, let $v = f$, let $\alpha(\sigma) = 1$ and let P^{per} be the possible set that characterises the generator within the environment. Evidently, the input f can, for all positive time, be any arbitrary piecewise constant function taking either of the extreme values d^{per} or $-d^{per}$. This means that, at any positive time t and whatever the value of $y(t, f, \sigma)$, the input can switch instantaneously from one extreme value to the other. Consider the two possibilities: $y(t, f, \sigma)$ is positive and the input switches from d^{per} to $-d^{per}$ or $y(t, f, \sigma)$ is negative and the input switches from $-d^{per}$ to d^{per}. In either case, by virtue of (1.49), the absolute output $|e(t, f, \sigma)|$ is not less than d^{per} immediately after the switch. It follows that $\hat{e}(P^{per}, \sigma) \geq d^{per}$. In fact, the peak output $\hat{e}(P^{per}, \sigma)$ is minimised to the value d^{per} when the function $y(f, \sigma)$ is zero. Thus, if this function represents a feedback intended to reduce the peak error then its action is counterproductive, unless it serves to stabilise the system. Hence, under these conditions, the output cannot be made to be less than or equal to a specified tolerance, if the specified tolerance is less than d^{per}.

This difficulty can be overcome if the function v at the input port of the system is restricted in its rate of change. As will be seen, appropriate restrictions on the rate of change of the function v ensure that the response $y(v,\sigma)$ can be made, by appropriate choice of the system design parameter σ, to cancel some of the input function. Another way of interpreting this cancellation is to view the response function $-y(v,\sigma)$ as an approximant of the input function v. Imposing a bound on the rate of change of the input function v ensures that this function does not have stepwise discontinuities and therefore can be approximated by the response function $-y(v,\sigma)$ by appropriate choice of the design parameter σ.

It is therefore concluded that it is necessary to impose some restriction on the rate of change at the input port to ensure that the environment-system model is not badly constructed. The precise nature of the restriction depends on the possible set that characterises the generator within the environment.

To demonstrate that an appropriate restriction on the rate of change of v is sufficient to make the peak output dependent in a useful way on the system design parameter σ, the input-output relation that characterises (1.49) and (1.50) is rewritten in the form

$$e(t, v, \sigma) = \int_0^t [\alpha(\sigma) + y(\lambda, h, \sigma)]\, \dot{v}(t - \lambda) d\lambda \qquad (1.51)$$

Here $\dot{v}$ denotes the rate of change of the function v. Also, h denotes the unit step function, stepping at time $t = 0$, and $\alpha(\sigma)h + y(h, \sigma)$ is the output resulting from the unit step input at the input port. That is to say, $e(h, \sigma) = \alpha(\sigma)h + y(h, \sigma)$.

Now, suppose that $f = \dot{v}$ and $f \in P^{complex}$. This means that the ∞-norm and the 2-norm of the rate of change $\dot{v}$ are bounded by the constants d^{per}, d^{tra}, respectively. Then, (1.48) holds with

$$e(\delta, \sigma) = \alpha(\sigma)h + y(h, \sigma) \tag{1.52}$$

Notice that the right hand member of (1.52) is the output of (1.49)–(1.50) when a step is applied at the input port. It is therefore the step response of the system alone but it is also the impulse response from generator to output. Evidently, as shown by (1.51), the rate of change $\dot{v} = f$ does not have to be uniformly bounded to make (1.48) meaningful. In particular, if the bound $d^{tra} \neq 0$ then the transient component of the input can be unbounded but the peak error remains finite, provided that the 1-norm and 2-norm of (1.52) are finite.

This shows that the restriction $\dot{v} \in P^{complex}$ on the rate of change $\dot{v}$ is sufficient to ensure that the design problem is well posed, in the sense that the peak output is dependent in a useful way on the system design parameter σ. Empirical confirmation of this conclusion is obtained by noticing that the 1-norm and 2-norm of the function (1.52), appearing in (1.48), have been used extensively in design. In fact, bearing in mind that (1.52) is the step response of the input-output transformation (1.49)–(1.50), the 1-norm is the Integral Absolute Error (IAE), which is a well-established objective function employed in design. Similarly, the 2-norm is the square root of the Integral Square Error (ISE), which is another well-established objective function. Thus the expression (1.48) is a weighted sum of two useful objective functions, with the weights determined, not arbitrarily, but by the nature of the possible set $P^{complex}$. For certain simple systems, (Zakian, 1986a, 1996), the IAE and the ISE can be made arbitrarily small by appropriate choice of system design parameter and this means that the peak error can be made arbitrarily small. Accordingly, such environment-system couples are said to be fully designable (elsewhere called fully controllable (Zakian, 1986a, 1996)). Further relations between the peak output and modified versions of the IAE and ISE objective functions are shown in Section 1.5.4 below.

Obviously, the assumption that $\dot{v} = f$ means that $v(t) = \int_0^t f(\lambda)d\lambda$. This means that a filter, in the form of a pure integrator, is interposed between the input port of the system and the generator that produces the input f.

For the peak output (1.48) to be finite it is necessary and sufficient that the 1-norm and the 2-norm of the system step response $\alpha(\sigma)h + y(h, \sigma)$ are finite. If the Laplace transform of the step response $\alpha(\sigma)h + y(h, \sigma)$ is a rational function then a necessary and sufficient condition for the peak output to be finite is that the poles of this transform have negative real parts. For this to be so, a zero at the origin of the complex plane must cancel the pole at the origin, introduced by the integrator filter. For a control system of standard form, such a zero must be provided by a pole at the origin in the transfer function of the controller (integral controller) or of the plant.

It is therefore seen that interposing a filter that is a pure integrator, so that $f = \dot{v}$, between the system and the generator within the environment that is characterised by the possible set $P^{complex}$, means that the rate of change at the input port of the system is bounded in exactly the same way as the input produced by the generator. It is therefore concluded that such bounds on the rate of change at the input port makes the peak output dependent in a useful way on the design parameter of the system.

1.5.3 The Use of More Complex Smoothing Filters

A pure integrator provides the simplest possible filter, with the significant advantage that it has no undetermined parameters. However, the use of a filter in the form of a pure integrator is not without its shortcomings. These shortcomings are now examined and it is found that they can be eliminated by the use of more complex filters.

It is useful to note at this point that a filter linking the generator and the system might be situated either within the environment or within the system. If within the system, then it has to be realised physically, perhaps as part of the controller (and this might not always be possible), but the modelling of the environment is thereby simplified to the modelling of the generator within the environment, by estimating its two parameters d^{per}, d^{tra}. Any free parameters of the filter (a pure integrator has no undetermined parameters) can be included in the design parameter σ of the system and their values determined so as to make the output specifically bounded. Alternatively, if the filter is considered to be part of the environment then the modelling of the environment becomes more complicated because it involves estimating, from observed samples of the output of the environment, not only the two parameters d^{per}, d^{tra} of the generator but also the parameters of the filter. If a filter is a part of the environment model then, for the purpose of designing the system, it exists only as a mathematical entity and does not have to be realised in physical form, although it does represent some physical aspects of the environment.

The use of a filter to restrict the rate of change at an input port is an indirect way of achieving such restriction, while using the possible set $P^{complex}$, which does not involve restrictions on the rate of change, to model the generators within the environment. The principal advantage of using a filter is that it makes it possible to derive analytical expressions for the peak output, of the form shown in (1.48).

A filter in the form of a pure integrator is usually considered only as a mathematical entity that models an aspect of the environment and, for the purpose of design, is included in the overall model of the combined environment-system. Such a filter has two potential drawbacks. The first is that the integrator has a pole at the origin of the complex plane, which has to be cancelled by a zero in the transfer function from input port to output port of the system. Such a zero exists, in particular, when, either the controller, with integral action, or the plant, has a pole at the origin, provided that a zero in the plant transfer function does not cancel this pole. In an extreme situation, the generator produces a step function so that, after integration by the filter, the environment produces a ramp input to the system, which becomes unbounded if the step persists. An environment model of this kind might be unrealistic because, in most physical situations, such a severe input to the system does not occur. Moreover, with this input, if the system transfer function does not have a zero exactly at the origin then the output is unbounded. The second drawback is that it is not always possible to arrange for the pole of the pure integrator filter to be cancelled. This can be seen in the case of the standard feedback system with a reference input. If the output is the error then a controller with integral action provides the necessary cancellation of the filter integrator pole. However, if the output is the input of the plant then cancellation cannot be achieved unless, fortuitously, the plant transfer function has a pole at the origin. It follows that it might be more appropriate, in general, to choose a *stable* smoothing filter and a convenient form of such a filter has a transfer function that is rational, strictly proper and stable.

If a stable smoothing filter is used and the filter is considered as part of the system (and is realised within the controller or otherwise) then, as mentioned above, its parameters are determined, during the design process, together with the other parameters of the system, to ensure that the peak outputs are specifically bounded.

Alternatively, if the filter is considered a part of the environment model then its parameters are determined from samples of functions produced by the physical environment.

One way of doing this has been considered in relation to a stable filter in the form of a simple lag, which has the transfer function $(s+a)^{-1}$, where the parameter a is a positive constant (Rutland, 1994a, 1994b).

Another way is to consider two stable filters of the form $(1+s\tau)^{-1}(1+sk\tau)^{-1}$, as part of the environment model, where $0 < k < 1$. The pole $(1+sk\tau)^{-1}$ is introduced to ensure that the output of the filter has continuous first derivative and consequently the constant k can be made arbitrarily small. One filter $(1+s\tau^{per})^{-1}(1+sk\tau^{per})^{-1}$ is fed from the generator that produces persistent functions and characterised by the possible set P^{per}. The other filter $(1+s\tau^{tra})^{-1}(1+sk\tau^{tra})^{-1}$ is fed from the generator that produces transient functions and characterised by the possible set P^{tra}. The outputs of these filters are the functions denoted, respectively, by $v^{per}(f^{per},\tau^{per})$, $v^{tra}(f^{tra},\tau^{tra})$. The corresponding first derivatives are continuous and are denoted by $\dot{v}^{per}(f^{per},\tau^{per})$, $\dot{v}^{tra}(f^{tra},\tau^{tra})$. The two filter outputs are summed to produce the input function v for the system. Equivalently, it might be more convenient to allow the output of each filter to be transformed by the input-output relation of the system and then effecting the summation. This allows the environment-system couple to be expressed as a linear couple. Unlike a filter in the form of a pure integrator, these filters are stable and hence do not have a pole at the origin.

Let M^{per}, D^{per} respectively, denote the estimated values of the largest magnitude and the largest absolute rate of change of the output of the filter, with their size measured with the ∞-norm, when fed with persistent input from the generator characterised by the possible set P^{per}. Similarly, let M^{tra}, D^{tra} denote, respectively, the estimated values of the largest magnitude and the largest rate of change of the output of the filter, with their size measured with 2-norm, when fed with persistent input from the generator characterised by the possible set P^{tra}. The estimates are obtained from samples of output from the physical environment.

Given the estimates M^{per}, D^{per}, the generator bound d^{per} and the filter time constant τ^{per} are determined so that they satisfy the two equations

$$\hat{v}^{per}(P^{per}, \tau^{per}) = M^{per} \tag{1.53}$$

$$\hat{\dot{v}}^{per}(P^{per}, \tau^{per}) = D^{per} \tag{1.54}$$

To this end, the computation of the peak output and peak derivative of output can be facilitated with the use of

$$\hat{v}^{per}(P^{per}, \tau^{per}) = \|v^{per}(\delta, \tau^{per})\|_1 \, d^{per} \tag{1.55}$$

$$\hat{\dot{v}}^{per}(P^{per}, \tau^{per}) = \|\dot{v}^{per}(\delta, \tau^{per})\|_1 \, d^{per} \tag{1.56}$$

Similarly, given the estimates M^{tra}, D^{tra}, the generator bound d^{tra} and the filter time-constant τ^{tra} are determined so that they satisfy the two equations

$$\sup\left\{\left\|\hat{v}^{tra}(f^{tra}, \tau^{tra})\right\|_2 : f^{tra} \in P^{tra}\right\} = M^{tra} \tag{1.57}$$

$$\sup\left\{\left\|\hat{\dot{v}}^{tra}(f^{tra}, \tau^{tra})\right\|_2 : f^{tra} \in P^{tra}\right\} = D^{tra} \tag{1.58}$$

It is easy to show that equations (1.57)–(1.58) hold if the superscript *per* is replaced by *tra* and the 1-norm is replaced by the 2-norm. This implies that the transients at the output of this filter are uniformly bounded in magnitude and rate of change. Thus, although the generator characterised by P^{tra} can produce unbounded functions, the model of the environment produces transients that are uniformly bounded in magnitude and rate of change. In many cases, this might represent a more accurate model of the environment.

Notice that, with the parameters of the environment determined in this way, when the environment and system are coupled, an input that drives the system to its peak output does not necessarily drive the filter to attain the estimated values M^{per}, D^{per}, M^{tra}, D^{tra}.

1.5.4 Restricting the Rate of Change at Generator

Instead of using a filter, a more direct way of restricting the rate of change at the input port of the system is to model the generator so that it is characterised, not simply by bounds on norms of the magnitude of the input generated, as is done with the possible set $P^{complex}$, but also a bound on norms of the rate of change of the input generated. In the absence of a filter, the generator is coupled directly to the input port of the system and hence $f = v$. The advantage of this approach is that there are no filter parameters to be determined. The disadvantage is that it seems unlikely that the corresponding peak output can be expressed by an analytical formula and therefore the peak output has to be computed numerically. In this section it is assumed that $f = v$.

For a possible set that characterises a generator to be a part of a well-constructed environment-system model, it has to satisfy two conditions. First, the peak output, resulting from the possible set, has to be finite for some values of the system design parameter. Second, the peak output must depend in a useful way on the system design parameter.

Define the set

$$\dot{P}^{per} = \left\{ f \in \dot{F} : \left\| \dot{f} \right\|_\infty \leq \dot{d}^{per} \right\} \tag{1.59}$$

Here, $\dot{F}$ denotes the set of all functions in F such that the rate of change $\dot{f}$ is piecewise continuous. It follows that

$$P^{per} \cap \dot{P}^{per} = \left\{ f \in \dot{F} : \|f\|_\infty \leq d^{per}, \left\| \dot{f} \right\|_\infty \leq \dot{d}^{per} \right\} \tag{1.60}$$

To determine the suitability of $P^{per} \cap \dot{P}^{per}$ as a possible set, consider the relation

$$e(t, f, \sigma) = \rho(\sigma) f(t) + \int_0^t [e(\lambda, h, \sigma) - \rho(\sigma)] \dot{f}(t - \lambda) d\lambda \tag{1.61}$$

that follows from (1.51), if $v = f$.

An application of Hölder's inequality to (1.61) gives (Zakian, 1979a)

$$\hat{e}\left(P^{per} \cap \dot{P}^{per}, \sigma\right) \leq \rho(\sigma) d^{per} + \|e(h, \sigma) - \rho(\sigma)\|_1 \dot{d}^{per} \tag{1.62}$$

Let

$$\rho(\sigma) = \lim_{t \to \infty} e(t, h, \sigma) \tag{1.63}$$

This means that $\rho(\sigma)$, provided that it exists and is finite, is the well established objective function, which is the steady state output resulting from a step input, while $\|e(h, \sigma) - \rho(\sigma)\|_1$ is a modified version of the well established *IAE* objective function employed in conventional design, where the steady state output $\rho(\sigma)$ has been subtracted from the step response because this is necessary, in order to make the improper integral, that defines the 1-norm, converge to a finite value. Both these objective functions depend in a useful way on the system design parameter. The modification to the IAE ensures that the objective function $\|e(h,\sigma) - \rho(\sigma)\|_1$ is applicable even in cases where the output is not an error and it might therefore not be necessary or achievable to require that $\rho(\sigma) = 0$.

Notice, that if $\rho(\sigma) = 0$ and $\|e(h,\sigma)\|_1$ is finite then the peak output is finite even if the input is unbounded, provided that the rate of change $\dot{f}$ of the input is bounded.

However, suppose that $\rho(\sigma) \neq 0$ and $\|e(h,\sigma) - \rho(\sigma)\|_1$ is finite and the rate of change of the input is bounded. Then, by virtue of (1.61)–(1.62), the integral in (1.61) is bounded and the output is bounded if and only if the input f is bounded (Zakian, 1996). This shows that, under these conditions, it is not sufficient to require that the rate of change of input be bounded and it is necessary to impose a restriction in the form of a bound on the magnitude of input, to ensure that peak output is finite. Nonetheless, as will be seen, depending on the restriction imposed on the rate of change, an appropriate restriction on the magnitude of the input can take a different form.

It is therefore concluded that, for environments that generate persistent functions, the possible set $P^{per} \cap \dot{P}^{per}$ gives rise to a well-constructed environment-system model, even in case the input-output relation (1.49)–(1.50) involves some instantaneous transmission, so that $\alpha(\sigma)$ is not zero and is not negligible.

The problem of computing the peak output $\hat{e}(P^{per} \cap \dot{P}^{per}, \sigma)$, which is the peak output corresponding to the possible set of persistent functions that are bounded in magnitude and rate of change, has been a focus of attention for a long time, although for much of that time it was considered in an isolated setting and not in the context of the new framework of this book. Also, the reason for bounding the rate of change of the input was not known. However, it has become apparent that an analytical expression for this peak output is unlikely to be found in the near future. Accordingly, a numerical method for computing this peak output was proposed (Birch and Jackson, 1959) and other methods, which are computationally more satisfactory, have since become available (Satoh, Chapter 3).

Now define the possible set of transients by

$$P^{tra} \cap \dot{P}^{tra} = \left\{ f \in \dot{F} : \|f\|_2 \leq d^{tra}, \left\|\dot{f}\right\|_2 \leq \dot{d}^{tra} \right\} \tag{1.64}$$

This possible set is known in the context of circuit theory and a method for computing its corresponding peak output has been suggested (Papoulis, 1970). A more efficient method, based on the techniques of convex optimisation, has also been developed (Lane, 1995; see Satoh, Chapter 3).

1 Foundation of Control Systems Design

To see that the possible set (1.64) gives rise to a well-constructed environment-system model, consider again (1.61) and note that

$$f = \int_0^t e^{-(t-\lambda)}[f(\lambda) + \dot{f}(\lambda)]d\lambda \tag{1.65}$$

This identity can be verified by Laplace transformation. Hence, by Hölder's inequality,

$$\|f\|_\infty \leq \left\|e^{-(\cdot)}\right\|_2 (d^{tra} + \dot{d}^{tra}) \tag{1.66}$$

It follows from (1.61) that

$$\hat{e}\left(P^{per} \cap \dot{P}^{per}, \sigma\right) \leq \rho(\sigma)(2)^{-1/2}(d^{tra} + \dot{d}^{tra}) + \|e(h,\sigma) - \rho(\sigma)\|_2 \, \dot{d}^{tra} \tag{1.67}$$

Here, it can be seen that the first term on the right hand side of the inequality is proportional to $\rho(\sigma)$, which is the steady-state output caused by a unit step input to the system while the second term is proportional to $\|e(h,\sigma) - \rho(\sigma)\|_2$ which is a modified version of the well established *ISE* objective function, where the steady-state value has been subtracted from the step response to ensure that the improper integral, that defines the 2-norm, converges. Both of these objective functions are dependent in a useful manner on the system design parameter.

Evidently, (1.62) and (1.67) provide convenient upper bounds on the peak output corresponding, respectively, to the persistent and transient possible sets. These upper bounds, which can be called majorants because they hold for all values of the system design parameter, are easier to compute than the exact values of the peak output and can therefore be used in preliminary design. Obviously, if an upper bound does not exceed the specified tolerance for the peak output then there is no need to compute the exact peak output. Minimising an upper bound with respect to the system design parameter is an economical way of progressing towards, and sometimes achieving, a specifically bounded peak output.

The above considerations lead to the possible set of complex inputs defined as follows

$$\tilde{P}^{complex} = \left\{f^{per} + f^{tra} : f^{per} \in P^{per} \cap \dot{P}^{per}, \, f^{tra} \in P^{tra} \cap \dot{P}^{tra}\right\} \tag{1.68}$$

Thus the environment contains two generators, the outputs of which are summed, and the generators are, respectively, characterised by the possible sets (1.60) and (1.64).

It is useful to have an expression for the peak output that corresponds to this complex possible set (1.68), in terms of the peaks outputs that correspond to the possible sets (1.60) and (1.64). To this end, note that since the input-output transformation from the two generators to the output is linear then the output is given by

$$e(f^{per} + f^{tra}, \sigma) = e(f^{per}, \sigma) + e(f^{tra}, \sigma) \tag{1.69}$$

The definition of peak output implies that there are two persistent inputs $\pm f \in P^{per} \cap \dot{P}^{per}$ such that, for some time $t \in [0, \infty]$, $|e(t, \pm f, \sigma)| = \hat{e}(P^{per} \cap \dot{P}^{per}, \sigma)$. Clearly, if one of these inputs causes the output to be negative then the other causes the output to be positive. It follows that there is a persistent input that drives the output to its peak value. Similarly, there is a transient input that drives the output to its peak value. Hence, by virtue of the linearity of the input output transformation,

$$\hat{e}(\tilde{P}^{complex}, \sigma) = \hat{e}(P^{per} \cap \dot{P}^{per}, \sigma) + \hat{e}(P^{tra} \cap \dot{P}^{tra}, \sigma) \tag{1.70}$$

This shows that the peak output that corresponds to the complex possible set is the sum of the peak outputs that correspond, respectively, to the persistent possible set and the transient possible set. The summation of peak outputs to give a resultant peak output is suggested by previous work (Zakian, 1987a, 1989; Lane, 1995) and its proof is given here. The summation is generalised as follows.

1.5.5 Summation of Peak Outputs

Consider a vector input $f = (f_1, f_2, \ldots, f_n)$ such that $f_j \in P_j$. Let the scalar output $e(f, \sigma)$ be the sum of the outputs $e_j(f_j, \sigma)$ of n, possibly distinct, linear input-output transformations, each one fed by the corresponding scalar input f_j.

Proposition 1.3. *Assume that $f_j \in P_j$ implies that $-f_j \in P_j$. Then*

$$\hat{e}(P, \sigma) = \sum_{j=1}^{n} \hat{e}_j(P_j, \sigma) \tag{1.71}$$

Here, P is defined by the Cartesian product set $P = \{f = (f_1, f_2, \ldots, f_n) : f_j \in P_j\}$ and therefore represents the possible set of vector inputs. Alternatively, if all the input-output transformations are identical then the output remains unchanged if the inputs are summed into one scalar input before being fed into the transformation. In this case, the summed input is $f = \sum_{j=1}^{n} f_j$ and the possible set is $P = \{f = \sum_{j=1}^{n} f_j : f_j \in P_j\}$. Notice the two distinct meanings of the symbol P, which are sometimes inadvertently confused (Zakian, 1992). The second meaning is that used in (1.47) and (1.68).

Proof. It follows from the definition of peak output that, for some $f_j \in P_j$ and some time t, $|e_j(t, f_j)| = \hat{e}(P_j)$. Thus, an input that drives the absolute value of the output to its peak value either makes the output positive or negative when it reaches the peak value. If positive, then the input drives the output to its peak value. If negative then, by the assumption of the proposition, there is an input that drives the output to its peak value. □

1 Foundation of Control Systems Design   61

1.6 The Method of Inequalities

1.6.1 Preliminaries

The method of inequalities is a generic term for any useful method, of formulating and solving the problem of designing control systems, which is based on the principle of inequalities. Attention here is focused mainly on the original method of inequalities (Zakian and Al-Naib, 1973; Zakian 1979a, 1996), which has been used in practice for three decades (see Part IV). The method has the following characteristics:

- It is based on the principle of inequalities (see Section 1.2).
- It separates the two problems of formulating and solving the design problem.
- The design space is the set of all controllers that can be implemented in practice.
- It is based on the principle of uniform stability.
- It is applicable to two distinct ways of formulating the design problem: that based on the conventional definition of control and that based on the new definition of control (for definitions, see Section 1.1).
- It makes use of iterative numerical search methods for solving the inequalities (see Section 1.7 and Part III).

Henceforth, it is assumed that the design problem is formulated in accordance with the principle of inequalities and that the ensuing computational problem of solving the resulting conjunction of inequalities can be solved by the methods of Section 1.7 and Part III. Accordingly, the discussion is focussed on the remaining aspects of the method of inequalities and, in particular, on the ways that the design problem is formulated.

The design problem is expressed as the conjunction of the following inequalities

$$\phi_i(c) \leq \varepsilon_i, \quad i = 1, 2, 3, \ldots, M \tag{1.72}$$

Here, for each $i = 1, 2, 3, \ldots, M$, the objective function ϕ_i maps the N dimensional real space $\mathbb{R}^N$ into the extended real line $(-\infty, \infty] = \mathbb{R} \cup \{\infty\}$ and the constant ε_i is a bound or a tolerance. Some of the inequalities in (1.72) represent design criteria and the remaining inequalities represent constraints on the controller or the plant. Every component of the N dimensional vector c is an undetermined parameter of the controller. The vector c represents a controller and hence a design.

The design space C is defined as the subset of $\mathbb{R}^N$ such that all those inequalities that define constraints on the controller are satisfied. Since the constraints vary according to the design problem, it follows that the design space is tailored to the problem.

As will be seen, some of the constraints on the controller are not represented by inequalities but by setting equal to zero some of the off diagonal elements of the transfer function matrix of the controller.

By imposing all the relevant constraints on the controller, every element of the design space C is a physically realisable controller that is implementable in practice. It follows that membership of the design space C is a necessary and sufficient condition for a design to be implementable. Obviously, the admissible set, which is the set of all solutions of (1.72), is a subset of the design space. Accordingly, the search for an admissible design can be restricted to the design space.

1.6.2 Separation of Processes of Formulation and Solution

Any method of design involves two processes: the process of formulating the design problem and the process of solving the design problem. A significant feature of the method of inequalities is that it regards these two processes as distinct and separate.

Prior to the advent of the computer, it was necessary for a design method to provide for these two main aspects of design within a single integrated scheme. For example, Nyquist's method provides a way of formulating the design problem in accordance with the criterion of a good phase margin and gain margin, thereby ensuring that the sensitivity (as measured by these margins) of the input-output relation of the system is adequate. Moreover, Nyquist's method also makes it possible for the designer to use hand driven graphical computations to determine a controller that ensures an adequate margin of stability. In the absence of a computer, this in-built computational facility constitutes a very decisive practical feature and was the main reason for the rapid adoption and success of what was the first practical design method[2]. The same thinking prevailed with the early development of analytical design methods (Newton *et al*, 1957) and still dominates these methods. A limitation of all these methods is that the computational facilities built into them are restricted to achieving the narrow design objectives of the method.

---

[2] It might be recalled that Maxwell's criterion for the stability of the whole system (not merely input-output stability) had not led to a practical design method because of the absence of a suitable computational way of ensuring that the criterion is satisfied in practice. Maxwell was at least partly aware of this difficulty and he encouraged the work that led to Routh's test for stability. Much later, this led to a simple iterative method, making repeated use of the Routh test (Zakian, 1979b), which provides an effective way of computing the abscissa of stability. This, together with any method for solving the inequalities that require the abscissa to be sufficiently negative and the imaginary parts of the characteristic roots to be correspondingly small, is essentially all that is needed to design control systems in accordance with the conventional definition of control. The works of Maxwell and Routh provided almost all the elements of a design method, which had to wait for a whole century, until the arrival of the computer, to be implemented.

For example, Nyquist's method can give good phase and gain margins but is not easily adapted for other criteria of design; for example, those required for critical systems. Similar statements can be made for virtually all design methods that preceded the method of inequalities.

However, it is preferable to let the designer focus attention on obtaining an accurate and realistic formulation of the design problem, knowing that the resulting computational problem can be solved with the computer. Thus, it is possible to bring to bear the designer's expertise and creativity to formulate the design problem in accordance with the principle of inequalities and then to seek a solution by using a general method for solving inequalities (see Part III). This also allows the designer to select the design criteria from a wide repertoire of criteria or to devise his own criteria.

The separation of the two aspects of design allows great flexibility in the formulation of the problem and in the choice of computational methods for obtaining a solution. This is one of the essential aspects of the method of inequalities.

1.6.3 Design Space is the Set of All Implementable Controllers

The control system is considered, in the conventional manner, as two interacting subsystems: the plant and the controller. The plant model is given and fixed and the controller is to be determined.

An important practical characteristic of the method of inequalities is that the design space is the set of all implementable controllers. This implies that any design, within the design space, which satisfies the inequalities (1.72), is an admissible design that can be physically realised and implemented in practice.

For example, suppose that the plant has two inputs and two outputs and each output is fed back to its corresponding input through a proportional plus integral controller. In many industries, hardware realisations of such controllers are as much part of the given physical aspects of the design problem as is the plant to be controlled. All that is left to be determined are the numerical values of the parameters of the controllers. Thus the transfer function matrix of the controller has the dimensions 2×2, its off diagonal elements are equal to zero and each of its diagonal elements is the transfer function of a proportional-plus-integral controller. Here, the design space is a subset of the four dimensional real space $\mathbb{R}^4$ formed by the two parameters belonging to each of the pair of proportional plus integral controllers. This design space has the shape of a four-dimensional rectangle defined by the limits imposed on the parameters of the two proportional plus integral controllers. The search for an admissible design can be restricted to this design space or, alternatively, the limits on the parameters can be expressed as inequalities in (1.72).

Thus, the method of inequalities allows the designer to prescribe the structure and complexity (order) of the controller in accordance with the physical and engineering requirements of the problem. This facility is of considerable significance in practice. In many situations, scalar controllers having a fixed structure or order are specified or are provided as a part of the hardware of the plant. The proportional plus integral controller is the most common example. Its order is equal to one and its structure involves two parameters. All that is left to be determined are the values of the parameters, within their respective limits. In multivariable control, the freedom to choose controller structure and complexity in the design process is of even greater significance. Maciejowski (1989, page 325), has expressed it as follows. "In the process industries, for example, there are considerable costs associated with each non-zero element of the transfer-function matrix. Every such element represents a considerable amount of hardware, possibly extensive additional wiring, additional testing and maintenance procedures, and additional training for the plant's operating personnel. There are therefore strong pressures for keeping the number of 'cross-couplings' [off diagonal elements in the transfer function matrix of the controller] as low as possible. In aerospace applications, considerations of reliability may dictate an analogue realisation of the controller, with the smallest possible number of states."

Although, in the case of single variable control systems, it might be possible to simplify a high order controller so as to make it implementable, such a procedure might not be practical for multivariable systems, where some off diagonal elements (possibly a majority of such elements) of the transfer-function matrix of the controller are required to be zero. Such controllers are sometimes called non-centralised (elsewhere called decentralised) to reflect the fact that the elements of the controller can be in different geographical locations. Design methods have to take account of the need for non-centralised control in some situations. This necessitates that some elements of the transfer function matrix of the controller are set equal to zero, in order to represent the absence of cross coupling.

Another constraint on a controller arises when it is to be realised physically by means of passive analogue components (masses, springs, dashpots, resistors, capacitors etc). Such a controller has a stable transfer function and this stability property is expressed by means of an inequality in (1.72) that requires the largest real part of its poles to be negative.

This need to constrain the structure, complexity and other properties of the controller in the design process has long been widely recognised, and was taken for granted before 1960. Nyquist's and other methods allowed the designer to prescribe a controller satisfying the required constraints. After 1960, mathematically attractive methods (the so called analytical or 'optimal' control methods) for computing a controller, that ignore the need for constraints on the controller, gained prominence[3]. Consequently, the need to constrain

---

[3] This is an example of what might be called a technogenic problem, which is a kind of unintended side effect of the application of techniques intended to solve a primary problem. In the context of the design of control systems, the

the controller has been re-emphasised by experienced control practitioners. Rosenbrock, among others, pointed to the importance of designing 'simple controllers' and allowed for this in his Inverse Nyquist Array method (see, for example, Maciejowski, 1989).

Restricting the structure, complexity and other properties of the controller implies a narrowing of the design space so that every element of the design space can be implemented in practice. This means that, as indicated above, a useful design space is the set of all implementable controllers.

Another problem, which arises in practice, corresponds to a situation where the controller structure is subject to constraints, so that some elements of the transfer function matrix of the controller are zero, but the complexity of each non-zero element is not fixed *a priori* but is required to be minimal in some sense. These non-zero elements can be determined so that, for example, the resulting order of the minimal realisation of the controller is least or, alternatively, the order of every element is uniformly bounded by a positive integer that has least value. The computations start with the unconstrained controller transfer function elements being given zero order and then proceed to higher order, until the design inequalities are satisfied.

To sum up, for the concept of admissible design (see Section 1.2) to be of practical value, an admissible design must also be a design that can be implemented, because the controller is made to satisfy the required constraints on its structure, complexity and other properties, as well as all the inequalities that specify the performance criteria and other constraints on the control system.

Notice that the designer does not arbitrarily choose the constraints on the controller. The constraints reflect the given physical and engineering aspects of the design problem.

1.6.4 Controller Parametrisations

An appropriate parametrisation of the controller can be specified in a straightforward manner, with the aim of simplifying the computation of an admissible controller. The elements of the controller transfer matrix are usually assumed to be proper rational functions. The polynomials can be expressed either in the usual way or in terms of their factors. The latter way implies

---

technogenic problem is the difficulty some methods have to satisfy constraints on the controller and this difficulty is introduced by the methods, in particular those that are called analytical, intended to solve the computational problem of obtaining a controller. In essence, therefore, these methods are capable of solving a simplified problem that can be solved to its fuller extent by numerical methods. The term technogenic is suggested by analogy with the term iatrogenic, which refers to an illness produced by a medicine intended to cure a primary illness. The terms apply especially when the problem introduced by the technique or medicine is as severe as the primary problem.

that the poles and zeros of the transfer functions are the parameters of the design space.

An alternative way of parametrising the controller is to express it in state-space form. Here, with little loss of generality, the A-matrix can be chosen to be diagonal. This parametrisation makes it more difficult to constrain elements of the transfer function matrix of the controller to be zero. However, it does allow the entire control system to be expressed as an algebraic differential system (Zakian, 1979a) that has advantages in the computation of the time responses of the system.

1.6.5 The Roles of Iterative Numerical and Analytical Methods

Remarkable feats of design can be achieved by harnessing the capabilities of the computer, with the use of iterative numerical methods. This statement is not intended to devalue the role of analytical methods but simply to suggest appropriate roles for analytical and numerical methods, according their potential contributions to design. To this end, it is useful to recognise that some of the roles of analytical methods, which were established before the advent of the computer, are no longer appropriate today. This is not to say that analytical methods should be eschewed from the development of design methods but only that the part they play has to be reconsidered.

In fact, analytical methods have significant roles to play in the development of design methods, including that of rendering the design problem solvable by computer, while numerical methods have a role in the routine solution of the problems. The contents of this chapter illustrate these points.

The above assertions might seem unremarkable to many designers and to some developers of design methods. However, it appears that considerable attention is still given to the development of methods that have an in-built computational facility, thus in effect usurping or reducing the role of the computer and numerical methods, and thereby not making full use of their capabilities. As already mentioned, such methods involve a sacrifice of flexibility in the formulation of the design problem.

1.6.6 Design Criteria

The two main types of design criteria, of the form (1.72), that are used with the method of inequalities correspond, respectively, to the conventional definition of control and to the new definition of control, as discussed in Section 1.1.

For the conventional definition of control, some of the objective functions ϕ_i in (1.72) collectively represent the sensitivities of the input-output transformations of the control system. For example, the sensitivity might be measured by certain characteristics of the output of a transfer function, caused by a step input. Typically, these characteristics are the time taken for the output to recover from such a step input and any overshoot that occurs during that recovery.

For the new definition of control, the objective functions are the peak outputs.

Notice, however, that the design objectives are defined only if the input-output transformations are stable.

1.6.7 The Principle of Uniform Stability

The principle of uniform stability requires that a search for an admissible controller be carried out within the subset of the design space comprising all the controllers that correspond to stable systems. This is because if a system is stable then all its input-output transformations are stable and hence all the design criteria (1.72) are defined and, moreover, all the modes of the system are stable, whether or not they are involved in the input-output transformations.

Maxwell's condition for the stability of a system can be expressed by an inequality that requires the abscissa of stability of the system to be negative. That is to say,

$$\phi_0(c) \leq -\varepsilon_0, \quad \varepsilon_0 > 0 \tag{1.73}$$

$$\phi_0(c) = \max\{\operatorname{Re} s : \det(sI - A) = 0\} \tag{1.74}$$

Here, $\phi_0(c)$ denotes the abscissa of stability, $\det(sI - A)$ denotes the characteristic polynomial of the system and ε_0 is called the margin of stability.

Thus one interpretation of the principle of uniform stability is to require the search for an admissible controller to proceed within the set of all systems such that the abscissa of stability is negative.

As stated above, the concept of abscissa of stability holds for all systems characterised by standard state-space equations (see Section 1.1). Moreover, the concept of abscissa of stability is well known to hold also for a large class of linear time invariant systems involving delays (see Arunsawatwong, 1996).

For more general classes of linear time invariant systems, to which the concept of abscissa of stability has not been extended, the principle of uniform stability has to be interpreted differently. For such systems it appears that the concept of internal stability, developed by Desoer (Desoer and Chan, 1975), might provide a useful interpretation of the principle of uniform stability.

The idea of internal stability is to express the stability of the system in terms of the stability of certain input-output transformations of the system. Once these input-output relations are made stable then, equivalently, the system is stable. A useful computational method for stabilising an input-output transformation has been devised (Zakian, 1997b).

The method of inequalities proceeds initially by obtaining a controller, within the design space, that corresponds to a stable system. After that, a search is made for a controller that satisfies all the design inequalities (1.72).

1.6.8 Other Methods of Inequalities

More recent methods of inequalities include a method that uses convex optimisation (Boyd and Barratt, 1991), a method based on an analytical $\mathcal{H}^\infty$ approach (Whidborne *et al* 1994; Chapter 11) and one based on linear matrix inequalities (LMI) (Ono, Chapter 4). These methods are not intended for problems where the controller structure is subject to constraints and indeed the methods cannot take such constraints into account. Instead, the design space is defined somewhat broadly. For the method of convex optimisation, the design space is the set of all linear time invariant systems.

On the assumption that there are no constraints to be imposed on the controller, these methods offer some significant special features. The convex optimisation method solves the admissibility problem, because it always determines an admissible design if one exists, or it gives an indication if an admissible design does not exist. The $\mathcal{H}^\infty$ method always ensures a degree of robustness in the design. The LMI method does not solve the admissibility problem but solves a closely related problem such that an admissible solution of the related problem is also a solution of the actual problem.

The convex optimisation method can be used for a certain purpose of analysis, even in case the controller is subject to constraints. Specifically, if it is found that the admissible set is empty for the design space comprising all linear time invariant systems (that is, with a controller not subject to constraints) it can be concluded that an admissible controller satisfying constraints, and hence within a proper subset of this design space, does not exist. It then follows that the design problem has to be reformulated by the process of negotiation discussed in Section 1.2. A similar comment can be made with regard to the LMI method.

1.7 The Node Array Method for Solving Inequalities

1.7.1 Introduction

Because of the complexity of the problem of solving inequalities and the many alternative ways that can lead to a solution, the principal aim here is not to specify a definitive method in complete detail but to present ideas that can be used to construct and evaluate practical methods. The focus of attention is a class of globally convergent search methods, collectively called the *node array method*. Two other globally convergent methods are described in Part III, where, together with one specific node array method, their respective efficiencies are evaluated by numerical tests. The results of the tests suggest that the node array method is relatively the most efficient of the three.

The principle of inequalities (see Section 1.2) gives rise to the problem of computing a point $c \in \mathbb{R}^N$ that satisfies the conjunction of the following inequalities

$$\phi_i(c) \leq \varepsilon_i, \text{ for every } i \in \widetilde{M}, \ \widetilde{M} = \{1, 2, \ldots, M\} \tag{1.75}$$

1 Foundation of Control Systems Design 69

Here, each *objective function* ϕ_i maps the *design space*, which is the N-dimensional real space $\mathbb{R}^N$, into the extended real line $(-\infty, \infty] = \mathbb{R} \cup \{\infty\}$, ε_i is a constant bound, also called a tolerance, and the point c, also called a design or a design parameter, represents a control system or, more usually, the controller of a system. Every objective function ϕ_i belongs to the *universal set* comprising all piecewise-continuous functions. The ordered M-tuple $(\phi_1(c), \phi_2(c), \ldots, \phi_M(c))$ is called the *objective vector*, denoted by $\phi(c)$, and its components are called the *objectives*.

The universal set adopted here is sufficiently broad to include all objectives that are likely to arise in practice.

Any point $c \in \mathbb{R}^N$ that satisfies the inequalities (1.75) is said to be *admissible*. The set of all admissible points is called the *admissible set*.

The main problem is to determine an admissible point if one exists or, otherwise, to determine that an admissible point does not exist. This is called the *first admissibility problem*. If an admissible point does not exist, a further problem is to determine what minimal changes have to be made to the bounds ε_i so as to ensure that an admissible point does exist. This is called the *second admissibility problem*. Methods for solving these problems are the subject of this section.

A point c is said to be a *finiteness point* if, for every $i \in \widetilde{M}$, the objective $\phi_i(c)$ is a finite number in some neighbourhood of the point c. A sufficient condition for c to be a finiteness point is usually given by an inequality of the form

$$\phi_0(c) < 0 \tag{1.76}$$

Here, the objective function ϕ_0, called the *finiteness function*, maps the space $\mathbb{R}^N$ into the real line $\mathbb{R}$, which implies that, unlike some of the objectives in (1.75), the finiteness objective $\phi_0(c)$ is always a finite number, for every point c in the design space $\mathbb{R}^N$. The set of all finiteness points is called the *finiteness space*.

The finiteness function ϕ_0 is usually chosen to be the abscissa of stability of the system, which is the largest of the real parts of all the characteristic roots of the system (see Sections 1.1 and 1.6). As is discussed in Sections 1.1 and 1.6, a negative abscissa of stability is a necessary and sufficient condition for the entire system to be stable (that is to say, for all the modes of the system to be stable, including any modes that are not controllable or not observable). The condition is therefore sufficient for the input-output transformations of the system to be stable and hence for the objectives $\phi_i(c)$ to be finite (see also Sections 1.4 and 1.5). Moreover, if the abscissa of stability is continuous and negative then the point c is a finiteness point because the system is stable in some neighbourhood of the point c. The size of the neighbourhood increases, as the abscissa of stability is made more negative.

Before the inequalities (1.75) can be solved by a method that senses the size of the objectives in the vicinity of a current or base point c, to find another point that is closer to an admissible point, it is necessary to ensure that the current point is a finiteness point, which can be obtained either by solving the finiteness inequality (1.76) or otherwise[4]. This is because, in general, a non-finiteness point is part of a, possibly large, neighbourhood of such points; in which case, a direction of progress towards an admissible point cannot be found by exploring that neighbourhood. Hence, in general, a non-finiteness point cannot be used as the initial point in a search for an admissible point. Given a finiteness initial point, all subsequent trial points in the search, that do not result in finite objectives or that do not satisfy the finiteness condition (1.76), are treated as failures and rejected.

Each component of the vector c represents a parameter of the controller or of the system and is bounded from below and above by the known constants c_i^L, c_i^U, respectively. Hence, each component c_i of the design parameter c lies in the line segment $[c_i^L, c_i^U]$. These constraints can be expressed as inequalities in (1.75) but are more conveniently taken into account by defining the search space as follows. The *search space*, denoted by S, is the subset of $\mathbb{R}^N$ such that the constraints on the components of the vector c are satisfied. That is, the search space is the hyper-rectangle in the design space defined by

$$S = \left\{ c \in \mathbb{R}^N : c_i \in [c_i^L, c_i^U], \; i = 1, 2, \ldots, N \right\} \tag{1.77}$$

Accordingly, the first admissibility problem can be restated as follows.

First admissibility problem: Determine a point in the search space S such that the inequalities (1.75) are satisfied or determine that such a point does not exist. □

---

[4] An alternative (Zakian, 1987b) to the use of the finiteness inequality (1.76) for determining a finiteness point is to the start with an arbitrary plant model, having stable input-output transformations, that is progressively made to become the actual plant model in a sequence of computations. For each element of this sequence, a corresponding input-output stable system is determined, by ensuring that the objectives $\phi_i(c)$ remain finite (and perhaps even satisfy the inequalities (1.75)). The advantage of this approach is that it does not depend on having a finiteness function ϕ_0 and this feature might be exploited when such a function is not known for the system to be designed. However, unlike the method that interprets the finiteness function to be the abscissa of stability of the characteristic polynomial of the system, this approach does not ensure that the system as a whole (and not merely the input-output transformations) is stable, unless the concepts of internal stability (see, for example, Boyd and Barratt, 1991) are employed so as to make input-output stability equivalent to the stability of the system. From the perspective of the framework of this book, the practical usefulness of the concepts of internal stability appears to be mainly in situations where a finiteness function, analogous to the concept of the abscissa of stability of the characteristic polynomial of the system, is not known.

In the case of a control system having the usual feedback structure and a stable plant, a finiteness point is given by a controller having zero transfer function.

1 Foundation of Control Systems Design   71

The set of all finiteness points within the search space S is called the *finiteness search space*.

In practice, the search space S can be discretised to the level of the resolution with which the physical parameters, represented by the vector c, can be measured. This implies that, in practice, the *discretised search space* contains a finite number of meaningful points. Accordingly, for each of the N axes of the search space, the line segment $[c_i^L, c_i^U]$ can be replaced by a finite set of equidistant points. The total array of points within the search space thus obtained contains all the distinct meaningful points in the search space. A precise definition of the discretised space is given in Section 1.7.3, in terms of a node array of sufficiently high density.

A search that sets out to examine all the points in the discretised search space and stops when an admissible point is found, is called an *exhaustive search*. If, in the worst case, the search involves examining all the points in the discretised search space then it is said to be a *totally exhaustive search*. Even with the fastest modern computers, the labour involved in a totally exhaustive search becomes prohibitive, if the dimension N of the search space rises above a modest number, such as 4. For, assuming that each axis is discretised into 1000 points, the total number of points in the discretised search space is this number raised to the power N.

However, although the discretised search space gives a sufficiently accurate model of all the meaningful options available in design, it is convenient here to treat the search space in its discretised form only for some purposes and in its continuous (non-discretised) form for other purposes. Specifically, the node array (see Section 1.7.2 and Section 1.7.3) involves a discrete number of points in the search space whilst every local search (see next paragraph) is conducted in the continuous search space.

A sequence of points in the search space is said to be a *local search* if it proceeds, according to well-defined rules that define the *local search method* that generates the sequence, from a given finiteness initial point and stops after a finite number of evaluations of the objective vector. It is assumed that a local search method includes a rule to make the search stop when it finds an admissible point and a rule that makes the search stop when it makes no further progress and also a rule that makes the search stop when progress becomes too laborious.

The *labour* of a search is the number of evaluations of the objective vector involved in that search.

It is important to distinguish between a local search and a local search method. The characteristics of a local search depend not only on the local search method employed but also on the objective vector and on the initial point of the search.

For a given objective vector and a given local search method, the corresponding *convergence space* is defined as the set of all initial finiteness points in the search space, from which an admissible point can be reached (found) by a local search. The shaded region in Figure 1.4 shows a convergence space, located within the square search space. In this figure, which also shows an array with 9 nodes, the node labelled **4** is in the convergence space.

Obviously, the convergence space contains the admissible set as a subset. If the convergence space is equal to the admissible set then the local search has, in an obvious sense, the weakest possible search capacity. In this case, the local search is convergent in a trivial sense.

A local search is said to be *globally convergent* if the convergence space coincides with the entire finiteness search space. A local search method is said to be globally convergent for a specified class of objective functions if, for every member of that class, the convergence space coincides with the finiteness search space. In this extreme case, the term 'local search method' appears incongruous but is retained for convenience.

The convergence space is a closed set. This means that any local search, starting within the convergence space, remains within that space throughout the search. For, suppose that a search, starting within the convergence space, moves to a point outside that space, then the search cannot, by the definition of the convergence space, continue to find an admissible point, which means that it did not start within the convergence space. This is a contradiction and it is therefore concluded that the convergence space is a closed set.

A local search is said to be *non-divergent* if all points, that form part of the search sequence, are not worse than the initial point. The notion of a worse point is defined in Section 1.7.4 and is based on a suitable way of ordering the points of a search space. A local search method is said to be non-divergent if it is non-divergent from all initial finiteness points in the search space and for all objective vectors.

Any *moving boundaries process* (see Section 1.7.4) is a non-divergent local search method.

For any non-divergent search, it is possible to detect when the search makes no further progress. This happens when, after a sequence of non-deteriorating search points, no further new such point can be found. A useful criterion for terminating the search can be based on this concept.

A detailed analysis of the convergence properties of a local search method might consist of identifying special classes of the objective functions, having certain special properties like differentiability or convexity or mapping the search space into the not-extended (finite) real line, for which the convergence space is equal to the search space. This analysis is usually achieved by mathematical methods. For less restricted classes of objective functions, such as the universal set, it is, in general, more practical to determine the local convergence properties by numerical computations, by identifying the convergence space for some, specially selected and possibly challenging, objective functions.

The method considered here is the node array method. This method is based on the assumption that, for any objective vector that is likely to occur in practice, the search space contains localities that can be searched efficiently, for an admissible point, by some local search method, whereas, in many cases, the entire search space is too complex to be searched in this way. If an admissible point exists then one of these localities is the convergence space of the local search method. The method therefore employs local searches that start from different points, called *nodes*, on an array of points that are distributed uniformly throughout the search space. The local searches are allowed to range over the continuous (non-discretised) search space . The node array is a subset of the discretised search space and its density can be made as high as required. If sufficiently dense, the node array is the discretised search space. As will be seen, the node array method is a dual global-local search method, where the global part is provided by the node array and the local part by a local search method.

As is shown in Section 1.7.5, it can be deduced, in an obvious manner, that the node array method is *universally globally convergent* in the sense that, for every objective vector in the universal set of objective vectors, an admissible point is always found (if such a point exists) with a finite number of evaluations of the objective vector. This form of global convergence contrasts with that of a putative local search method, for which the global convergence can be demonstrated only for a restricted subset of the universal set of objectives, such as those having special properties, like convexity. However, the property of convergence alone does not ensure that a method is useful in practice. For, obviously, a totally exhaustive search of the discretised search space is convergent but is not practical, unless the dimension of the search space is small.

Next to convergence, the other important property of the node array method is its efficiency. This depends on the number of local searches needed, to find an admissible point, and also on the efficiency of the local searches. For a given objective vector, the number of local searches needed is inversely related to the size of the convergence space. A practical evaluation of efficiency of the node array method is obtained by using the method to solve a set of specially devised and challenging test problems.

1.7.2 The Node Array Method

Since the time, about the year 1970, when fast man-machine interactive computing facilities became widely available, because of the speed with which a search can be made, searches aimed at finding an admissible point anywhere in the finiteness search space have been carried out as a standard practice and with considerable success, by selection of an arbitrary initial finiteness point in each of various places in the search space and then searching for an admissible point, by starting from that initial point and using a local search method called the moving boundaries process (Zakian and Al-Naib, 1973). Many of the designs shown in this book were done in this way. Actually, this moving boundaries process is only one of a class of such local search methods, which differ only in the way the search points are generated. It is known, from extensive application of the method, that, for many objective vectors in the universal set, the convergence space of this particular moving boundaries process is significantly greater than the corresponding admissible set, for any choice of the tolerance vector that gives a non-empty admissible set.

A problem with this approach is that of selecting a finite set of initial points such that it includes at least one point in the convergence space. That is to say, the set of initial points is required to have a non-empty intersection with the convergence space. Given such a set of initial points, a finite sequence of local searches, conducted from these initial points, eventually contains one convergent local search, which finds an admissible point.

This problem is solved here by the simple device of covering the entire search space with a node array, which defines points that are distributed uniformly throughout the search space. These nodes are used as the starting points for local searches. The node array is a subset of the discretised search space and, when its density is sufficiently high, it becomes the discretised search space. At this stage, the search becomes exhaustive. For some density of the array and if an admissible point exists, the node array has at least one node in the convergence space; see, for example, Figure 1.4. This means that a sufficiently dense node array intersects the convergence space. Each node on the array is used, in sequence, as the initial point (provided that it is a finiteness point) for a search with a locally convergent method, until an admissible point is found. Each search from a node is carried out either until an admissible point is found or until, after a finite number of evaluations of the objective vector, the search makes no further progress, or until progress becomes too laborious (a precise meaning of this is given in Section 1.7.5), whichever occurs first. If an admissible point is not found after all the nodes are searched, the node array is made denser and the process is repeated, but only the new points defined by the denser node array are used as the starting points for local searches.

When a local search has been carried out from a particular node, that node is referred to as a *searched node*.

Note that, since the convergence space includes the admissible set, the global convergence of the node array method is ensured if the admissible set is non-empty. However, for the node array method to be efficient, it is necessary for the convergence space to be significantly larger than the admissible set. This is because, for a given objective vector, the larger the convergence space the less dense the node array is required to be and hence the more efficient the node array method becomes, provided that local searches can be carried out efficiently. In the extreme case, when the convergence space coincides with the search space, only one node is needed and the node array method becomes a local search method with global convergence. However, the efficiency of such a 'local' search depends on the particular objective vector and on the particular starting point for the search. In the other extreme, when the convergence space is equal to the admissible set, the global search carried out by the node array method is exhaustive and therefore is not efficient. In general, optimal efficiency of the node array method occurs somewhere between these two extremes.

If the inequalities (1.75) do not have a solution, this fact can also be determined without necessitating an exhaustive search of the discretised search space. Suppose that, for a given node array, a solution of the inequalities cannot be found. Then, instead of proceeding to a search with a denser node array, the node array is kept constant and the tolerances ε_i are replaced by larger numbers, so that the resulting inequalities (1.75) have solutions that can be determined by local searches from the nodes. After obtaining one solution, while maintaining the same node array, the bounds are reduced in small steps (see, for example, Section 1.7.6) towards their original proper values and at each step a solution of the inequalities is sought, starting first from the previous solution and then, if that fails, from the nodes. Suppose that, after a series of steps, at each of which a solution is found, a further step results in inequalities, for which a solution cannot be found. If some of the numbers, that replaced the bounds, are still larger than the original bounds then it is concluded that the inequalities have no solution. The last solution found in this way is a Pareto minimiser (see Section 1.2).

1.7.3 Specification of Node Array

This section contains a definition of a node array and a way of generating progressively denser node arrays.

A *node* is a point in the search space S defined above in (1.77). A *node array* is a set of points that are distributed uniformly throughout the search space. A node array can be made arbitrarily dense so that it covers the search space as thoroughly as is needed. The most uniform distribution of nodes is obtained by requiring the distance between any two adjacent nodes to be the same.

Let $l = 2^{-m} \min\{(c_i^U - c_i^L) : i = 1, 2, 3, \ldots, N\}$. Here m is a non-negative integer that denotes the density of the array.

For each of the N axes of the search space, consider the set of equidistant points in the segment $[c_i^L, c_i^U]$, defined by $\{p_{i,0}, p_{i,1}, p_{i,2}, \ldots, p_{i,k}\}$, where $p_{i,j} = c_i^L + jl$, and k is the largest integer such that $p_{i,k}$ is not greater than c_i^U.

Let A_m denote the node array of density m, defined by

$$A_m = \{c \in R^N : c_i \in \{p_{i,0}, p_{i,1}, p_{i,2}, \ldots, p_{i,k}\}, i = 1, 2, \ldots, N\} \qquad (1.78)$$

Thus the node array A_m is the set of all ordered N-tuples, of numbers taken from the set $\{p_{i,0}, p_{i,1}, p_{i,2}, \ldots, p_{i,k}\}$. Each N-tuple of numbers provides the Cartesian address of a node in the array.

To obtain a sequence of progressively denser arrays let $m = 0, 1, 2, 3, \ldots$ consecutively. Evidently, a given node array A_m contains, as proper subsets, all the node arrays with smaller density. For some, sufficiently high, density m, the node array becomes the discretised search space.

The same constant distance separates every adjacent pair of nodes in the node array A_m. In this sense, this particular node array has the most uniform distribution in the search space. Other node arrays with somewhat less uniformity can be defined and an example is given below.

For a given node array A_m, define the corresponding *incremental array* by

$$A_m - A_{m-1} \tag{1.79}$$

The second term is the union of all the arrays with density less than m. Thus the incremental array does not contain any nodes contained in all the arrays that are less dense than the array A_m.

Figure 1.4 shows a search space with $N = 2$, in the shape of a square with the bottom left hand corner of the square coinciding with the origin of the design space. Also shown are the nine nodes of the array for the case $m = 1$. The incremental array includes only the nodes labelled **1, 3, 4, 5, 7**. The nodes labelled **0, 2, 6, 8** belong to the less dense array, which corresponds to $m = 1$.

Each node in the node array is potentially the starting point for a local search. When a node has been searched, its address is recorded so that it is not searched again. In this way, when a denser array is considered, all nodes that have already been searched are not searched again. This is equivalent to searching only the nodes of the incremental array.

A *node sequence* is an ordered set of nodes, all belonging to one incremental array and subject to the restriction that every node of the incremental array occurs once and only once in the sequence. The order in which the nodes occur in a node sequence can be arbitrary or it can be subject to some rules. For example, a node sequence might be subject to the rule that every node in the sequence is not less likely to result in a successful local search than the next node in the sequence.

For the purpose of writing a simplified computer program for testing the node array method (see Satoh, Chapter 6) it is useful to define a node array, denoted by B_m, as follows. Consider the unit hypercube $[0, 1]^N$ in the design space $\mathbb{R}^N$, defined in the same way as the search space (1.77), but with $c_i^U = 1$, $c_i^L = 0$. For the case $N = 2$, the unit hypercube becomes the square shown in Figure 1.4. In general, the unit hypercube $[0, 1]^N$ has dimension N, is located in the positive sector of $\mathbb{R}^N$, has one corner that coincides with the origin of $\mathbb{R}^N$ and all its edges have unit length. Any point p in the unit hypercube corresponds to a point c in the search space S and the two are related by the simple transformation $c_i = p_i[c_i^U - c_i^L] + c_i^L$. The node array B_m is obtained by mapping, by means of this transformation, the node array A_m obtained on the assumption that the search space S is the unit hypercube, into the design space $\mathbb{R}^N$.

Notice, however, that for a given axis of the search space and for every pair of adjacent points in the array by B_m, the pair has the same separation only if it lies on a line parallel to the given axis of the search space. However, the separation between pairs does not, in general, remain constant as the axis is changed. This means that the array B_m is less uniform than the array A_m and it is therefore expected that a node array method based on the array B_m is less efficient, in general, than a node array method based on the array A_m.

A convenient way of labelling the nodes of the array A_m defined on the unit hypercube is given by the ordered-pair of non-negative integers m, χ. The number χ is expressed in arithmetic base $k+1$ form and is represented by an ordered set of N numerals, where each numeral η_j is an element of the set $\{0, 1, \ldots, k\}$. Here k is defined by $k = 2^m$. Hence the number χ can be written in the form

$$\chi = \eta_N \eta_{N-1} \ldots, \eta_1 = \sum_{j=1}^{j=N} \eta_j (k+1)^{j-1} \text{ and } \eta_j \in \{0, 1, \ldots, k\} \quad (1.80)$$

For a given density m, each node on the array is labelled by letting (for every $j = 1, 2, \ldots, N$) the number $l\eta_j$ be equal to the component of the Cartesian address on jth axis of the search space. Notice that the number χ is a non-negative integer not greater than $(k+1)^N$. Notice also that the label provided by the number χ depends on the density m. Thus, as the density is changed, the same node, if not on the origin of the search space, acquires a new label, which is another positive integer. It follows that the address of a node is specified uniquely by the ordered-pair of non-negative integers m, χ.

A property of the number χ is that, as it increases in increments of 1, from 0 to $(k+1)^N$, it moves in an obviously simple way, from the initial node at the origin, to scan the entire node array. This is one way of producing an arbitrary sequence of all the nodes in an array of a given density.

As an example, consider the case $N = 2, m = 1$, for which the node array is illustrated in Figure 1.4. Then $k = 2$, $l = 1/2$ and χ takes the values shown as trinary (arithmetic base 3) numbers, together with their decimal equivalent expressed in bold characters on the right

00=**0**, 01=**1**, 02=**2**
10=**3**, 11=**4**, 12=**5**
20=**6**, 21=**7**, 22=**8**

1.7.4 The Moving Boundaries Process

Any local search method proceeds from a current point to a point that is better, or a point that is not worse. This process is repeated so as to generate a local search sequence. An *ordering relation*, imposes an order (which depends on the objective vector) on the search space, so that all the elements of the

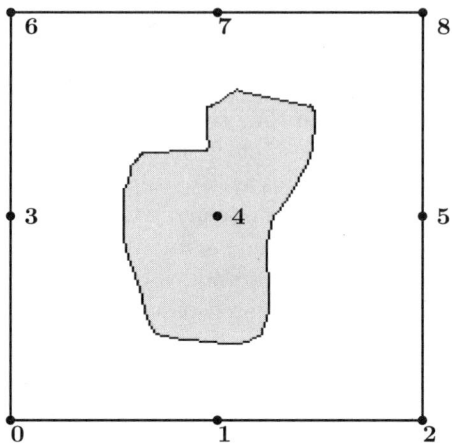

Fig. 1.4. Example node array

search space are ranked in some order of preference. This is what determines if a point in the search space is better or not worse than the current point. The ordering relation constitutes the conceptual core of the local search method. Thus, for any two points in the search space, one point is said to be preferable to the other if it is, according to the ordering relation, either closer to an admissible point, if such a point exists, or to a Pareto minimiser (that is to say, a point that is best in the sense of the Pareto ordering relation) if an admissible point does not exist. A moving boundaries process is any search method that is characterised by the specific ordering relation, defined below, which is induced by the principle of inequalities and which has the properties just stated above.

Another important component of a moving boundaries process is a rule for generating a trial point. This rule, called a *trial generator*, gives rise to a set of samples, called trial points, in a neighbourhood of the current point c, which are assessed by means of the ordering relation. By accepting only improvements, as defined by the ordering relation (alternatively, either improvements or no deteriorations), a sequence of progressively better points is obtained (provided that some improvements have occurred in the sequence of trials) that might converge to an admissible point.

There are many well-known trial generators that are used to find the local minimum of a scalar valued function and some of these can be used to be part of a moving boundaries process, which works as a local search method for solving inequalities. In particular, the Rosenbrock (1960) trial generator has been used as part of a moving boundaries process. This method has been found to work well as a local search method (Zakian and Al-Naib, 1973; see also Satoh, Appendix 6.A).

Thus, a trial point is accepted if and only if, according to the ordering relation, it represents a better point, or at least, a point that is not worse than the currently held point. This is called the *non-divergence rule*. The rule, which works like a ratchet mechanism, ensures that none of the points in a search sequence are worse than the initial point. Although it might appear somewhat rigid, the non-divergence rule can be made to contain great flexibility because a trial can be the end result of a subsidiary search that does not obey the non-divergence rule of the main search.

A trial generator is said to perform a *subsidiary local search* if it generates a sequence of current points, as is done by the main local search, by following rules that are different from the rules that guide the main local search. In particular, the non-divergence rule need not apply to a subsidiary search. An example of a subsidiary search is given in Section 1.7.6 below. It may be noted that a subsidiary local search involves a sequence of more than one distinct current point, whereas all the sampling performed by an ordinary trial generator is carried out from one current point. Accordingly, commonly known trial generators, such as the Rosenbrock generator, do not perform a subsidiary local search.

The moving boundaries process is characterised by the distinct multi-objective ordering relation (Zakian, 1996) that is induced by the principle of inequalities (see Section 1.2). This ordering relation is expressed as follows. Let c denote a current point and let c^* denote a trial point. Then the trial point c^* is, by definition, said to be superior to (preferable to, better than) the current point if

$$\phi_i(c^*) \leq \phi_i(c), \quad \text{for all } i \in \widetilde{M} \quad \text{such that } \phi_i(c) > \varepsilon_i \quad (1.81)$$

$$\phi_i(c^*) < \phi_i(c), \quad \text{for some } i \in \widetilde{M} \quad \text{such that } \phi_i(c) > \varepsilon_i \quad (1.82)$$

$$\phi_i(c^*) \leq \varepsilon_i, \quad \text{for all } i \in \widetilde{M} \quad \text{such that } \phi_i(c) \leq \varepsilon_i \quad (1.83)$$

The statements (1.81) and (1.82) above concern those objectives that, for the point c, do not satisfy their corresponding inequalities. In fact, those parts of these two statements bounded by the braces define a trial point c^* that is superior, in the Pareto sense, to the current point c. Thus, the Pareto relation of superiority is embedded in the more general ordering relation of superiority induced by the principle of inequalities. The two ordering relations (inequalities and Pareto) are identical only when all the inequalities are not satisfied at the point c. The two ordering relations are identical on the entire search space only if they are identical for every point in the search space and this happens only when the tolerances are so small that none of the inequalities can be satisfied anywhere in the search space. Otherwise, the inequalities ordering relation of superiority can be seen to be distinct and more general than the Pareto ordering relation of superiority.

If the statement (1.82) is deleted and only the two statements (1.81) and (1.83) are retained, these two statements define the relation of non-inferiority induced by the principle of inequalities, which means that the trial point c^* is, by definition, not worse than the current point c. And if, in addition, the equality sign holds in (1.81) then (1.81) and (1.83) define the equivalence relation induced by the principle of inequalities, which means that the trial point c^* is, by definition, equivalent to current point c; it is neither better nor worse. In contrast, the current and trial points are equivalent, in the Pareto sense, if the value of every objective at the trial point is equal to its value at the current point. Also, the trial point is non-inferior to the current point, in the Pareto sense, if the value of every objective at the trial point is not greater than its value at the current point.

Notice that the simpler conventional ordering relations (superiority, non-inferiority and equivalence), that are at the basis of conventional scalar (as distinct from multi-objective) methods of minimisation, are special cases of those induced by the principle of inequalities and can be obtained from the latter by setting $M = 1$, so that there is only one inequality, and by setting the bound of this inequality so that the admissible set is empty.

There is an important distinction between the Pareto ordering relations and those induced by the principle of inequalities. Confusing the two can result in misconceptions. In particular, the notion of admissible set is specific to the principle of inequalities and does not have a meaning in Pareto theory (if Pareto theory is derived as a special case of the inequalities theory then the admissible set can be defined but it is always empty). All elements of the admissible set are equivalent when judged by the ordering relations of the principle of inequalities. These admissible elements are also all equally the best when compared to other elements in the search space. However, in general, these same admissible elements are not all equivalent when judged by the Pareto ordering relation, because the Pareto minimisers inside the admissible set are then judged to be superior to the rest. Thus, by applying both types of ordering relations simultaneously, a contradiction arises whereby all admissible designs are equivalent but some admissible designs are superior to other admissible designs. This is called using a double standard. The application of this double standard has resulted in obscure statements like: it is 'wise' to choose the Pareto minimisers within the admissible set. This kind of wisdom should perhaps be avoided. Instead, it is better to choose the tolerances ε_i, by reference to the physical and engineering aspects of the design problem, so that the admissible set is minimal (Zakian, 1996; see Section 1.2) (or almost minimal). In which case, all the admissible designs are also Pareto minimisers (or almost so). In this situation, judgements, made by either standard, agree.

Using Venn diagrams, the concept of moving boundaries has, as its name implies, a useful pictorial interpretation as a sequence of applications of the ordering relations induced by the principle of inequalities. If the trial point is superior to or is equivalent to the current point, replace the tolerances, in those inequalities (1.75) that are not satisfied, by $\phi_i(c^*)$. The set of all solutions of the resulting inequalities is a subset X of the search space and X includes the admissible set as a subset. Thus every trial that is superior results in a new set X that includes less than the previous set X and therefore the boundary of the set X has contracted by moving towards the boundary of the admissible set. The process terminates when the set X is identical to the admissible set; that is to say, it terminates when the boundary of X has contracted to that of the admissible set.

Notice here that if two points are equivalent, it does not follow that they are identical. Thus a move to an equivalent but non-identical point can be made and this is intended to reduce the risk of the search becoming stuck.

The moving boundaries process was, apparently, the first search method to be based on the multi-objective ordering relation induced by the principle of inequalities. Since the ordering relation of a method is its most fundamental component, any useful method based on this ordering relation may be called a moving boundaries process. The trial generator embodied by such a method can be used to identify the particular method. The term multi-objective optimisation, although a valid way of describing such methods, does not distinguish between a moving boundaries process and a Pareto process.

The ordering relations defined by (1.81)–(1.83) have the property that they do not obscure the multi-objective nature of the problem. This is thought to have some advantages, and perhaps the efficacy of the moving boundaries process is partly due to this property. As has been pointed out (see, for example, Maciejowski, 1989), the more conventional scalar ordering relation could be used instead to solve inequalities, by restating the problem in a form suitable for scalar minimisation. There are many possible alternative forms of this kind (see, for example expression (1.86)), one of which is

$$\psi(c) = \max\left\{\phi_i(c) - \varepsilon_i : i \in \widetilde{M}\right\} \tag{1.84}$$

Clearly, the extent to which the objective $\psi(c)$ is positive is a scalar measure of the extent to which the inequalities (1.75) are not satisfied. In fact, this objective measures the largest extent to which the original objectives exceed their tolerance.

Any process for finding the least value of the objective $\psi(c)$ is called a minimax process. It follows from this that a point c is an admissible point if and only if it satisfies the inequality

$$\psi(c) \leq 0 \tag{1.85}$$

However, the inequality (1.85) obscures the multi-objective nature of the original problem represented by the inequalities (1.75). This might be a disadvantage because, for every value of the scalar objective $\psi(c)$, there is a large set of distinct values of the vector objective $\phi(c)$. This means that there is a large set of points in the search space that are not equivalent for the ordering relation induced by the principle of inequalities but are equivalent for the scalar ordering relation associated with the minimax process. This means that the former ordering relation is more discriminating. Therefore, any trial point that corresponds to improvements in all of the objectives, except the one most exceeding its tolerance, would not be considered an improvement in a minimax process. Nonetheless, there might be situations, in which the moving boundaries process becomes trapped, where this property of the minimax process has some advantages. In such situations, it might be useful for the moving boundaries process to be switched to a subsidiary search, in which inequality (1.85) is the problem to be solved. This makes the moving boundaries process a minimax process during the subsidiary search. Such a switch might enable the moving boundaries process to escape the trap by opening avenues of search that are closed to the standard moving boundaries process. This idea can be generalised so that a switch is made not necessarily to replace the inequalities (1.75) by the single inequality (1.85) but to other inequalities that are equivalent to the inequalities (1.75). If $M > 2$, many inequalities, equivalent to (1.75), are easily constructed by separating the members of the objective vector into groups containing more that one objective and representing each group by a single inequality as is done in (1.85).

The above discussion presents an opportunity to address a confusion that has arisen with regard to the word multi-objective. The confusion arises because there are two possible meanings to the word in the context of design. The first meaning concerns the fact that most design problems involve more than one objective function. Thus, in general, a design problem involves a multi-objective formulation of the problem, so that $M \geq 2$. Conventionally, this aspect of the formulation has been somewhat obscured, in order to simplify the ensuing computational problems, by restating the design problem in terms of one equivalent objective function. For example, either as discussed above in connection with the minimax process or, more commonly, by defining a linear combination (weighted sum) of the objectives. The minimum of the latter is a Pareto minimiser; that is to say, a point that cannot be improved upon, in the sense of the Pareto ordering relation of superiority. The second meaning concerns the ordering relation used in a search method to obtain an admissible design. If the ordering relation is the superiority or the non-inferiority relation induced by the principle of inequalities or is a corresponding Pareto ordering relation and if the search method is based on any of these ordering relations then the search method (as distinct to the formulation of the problem) is said to be multi-objective. Interest in the development and use of multi-objective search methods has intensified recently. There are some connections between a Pareto minimiser and an admissible point (Zakian, 1996; see Section 1.2).

Since the publication of the method of inequalities (see Section 1.6), the development of methods for solving inequalities has been an active area of research (see, for example, Mayne and Sahba, 1985; see also Part III).

1.7.5 Convergence and Efficiency of Node Array Method

It is recalled here that, for a given objective vector and a given local search method, a node array method is said to be globally convergent if it finds an admissible point (assuming that at least one exists) with a finite number of evaluations of the objective vector. It is also recalled that the convergence set is the set of all points in the finiteness search space, from which a local search finds an admissible point with a finite number of evaluations of the objective vector.

Global convergence: *For a given objective vector and a given local search method, let the admissible set be nonempty and let the convergence space contain a finite neighbourhood (however small) of some admissible point. Then the node array method is globally convergent.*

Proof. For some node array having finite density, there is a node in the convergence space. It follows that the number of nodes searched is finite and the total number of evaluations of the objective vector needed to find an admissible point is finite. □

It is noted that global convergence of the node array method requires that the local search method possesses some convergence capacity, even if this is minimal, since the local search is required to converge from an initial point that can be arbitrarily close to an admissible point. It is therefore seen that, unlike more conventional search methods, which depend on having a convergence space that is as large as possible and is preferably equal to the finiteness search space, the burden of global convergence is shifted away from the local search method. The advantage of this is that it enables the node array method to be globally convergent for all the objectives of the universal set. This contrasts with conventional methods, which are globally convergent for restricted classes of objectives, such as those that are convex.

It is also noted that, for any member of the universal set, it is readily possible to devise a local search method that finds an admissible point from an arbitrary initial point, for every member of the universal set, provided that the initial point is made arbitrarily close to an admissible point. With the use of such a local search method, the node array method becomes universally globally convergent, which means that it is globally convergent for every member of the universal set of objectives.

It is further noted that although, as indicated above, the use of an appropriate local search method makes the node array method globally convergent, this does not make the node array method efficient if convergence of the local search takes place only if the initial point (defined by a node of the array) is very close to an admissible point. For, in such a case the node array would have to be correspondingly very dense to ensure such an initial point.

As will be seen, an efficient node array method requires the use of a local search method that has, for any given objective vector, a convergence space that is significantly larger than the admissible set but not necessarily so large as to be equal to the finiteness search space. The moving boundaries process provides such a local search method.

The efficiency of a global search depends on having useful criteria for detecting when a local search is stuck or when its progress is too laborious. If either one of these two situations occurs, the local search is stopped and the global search is resumed by initiating a local search from a new point in the current incremental node array. It is recalled that the labour involved in a search is defined as the number of evaluations of the objective vector, that have occurred during that search. Such a number must take into account all trials made by the trial generator.

The following three stopping rules specify when a local search is stopped. Accordingly, a local search stops when any of the three conditions underlying the corresponding three rules, is satisfied. The global search is terminated only when the condition of Stopping Rule 1 is satisfied.

Stopping Rule 1: A local search is stopped when it finds an admissible point.

Stopping Rule 2: A local search is stopped when it is stuck (makes no further progress). A criterion for determining when a search is stuck can easily be defined by examining the detailed operation of the trial generator (for example, if the Rosenbrock trial generator is used, see Satoh, Chapter 6), provided that the local search method is non-divergent, as is any moving boundaries process. It is recalled here that, for a nondivergent method, no point in a local search sequence is worse than the initial point. Hence, if the local search method is non-divergent, it is possible to determine that the search is stuck when the trial generator cannot find a new point that is not worse than the current point.

Stopping Rule 3: A local search is stopped when progress becomes too laborious. Progress is said to be too laborious if the *putative labour* needed to complete the remaining part of the current local search is greater than a threshold value. The threshold value suggested here is $(k-1)\lambda$, where k is the number of nodes in the current search sequence (which is equal to the number of nodes in the current incremental array) and λ is the average labour per previous local search; that is, the total labour involved in all the previous local searches divided by the total number of previous local searches, if this number is greater than zero. If the number of previous local searches is zero then λ is equal to k or to the labour involved in the completed part of the current local search, whichever is largest.

1 Foundation of Control Systems Design

The putative labour needed for a search to progress from a given point in the search space to a nearest admissible point, depends on a suitable definition of the distance between them. One way of defining this distance is given by the following non-negative number δ

$$\delta(c) = \sum_{i \in X} \phi_i(c) - \varepsilon_i, \quad X = \{i : \phi_i(c) \geq \varepsilon_i\} \tag{1.86}$$

Consider the total labour involved (denoted by L), divided by the total change in δ, (denoted by $\Delta\delta$), over a fixed number (denoted by γ; usually, $\gamma = 3$ but this number is somewhat arbitrary and no attempt has yet been made to optimise it) of consecutive search points ending in the current search point. This is the inverse of the current rate of progress. Assuming that the local search is convergent and that this rate of progress remains constant for the remaining portion of the local search, the putative labour, required to progress from the current search point to a nearest admissible point, is obtained by linear extrapolation, and is therefore equal to $L\delta(c)/\Delta\delta$. Obviously, this extrapolation can be carried out only if the initial portion of the local search involves a number of steps not less than γ.

Thus, the idea is to compare the putative labour from the current search point, which is the work that might be needed to reach a nearest admissible point, with the work that might be needed to do local searches from all the other nodes in the current node sequence (which is the same as all the other points in the current incremental array), assuming that each of the other local searches would need an average amount of work (as deduced from all previous local searches). If to complete the current local search is expected to be more laborious than doing all the other local searches in the current node sequence (assuming that each requires an average amount of labour) then it is better to stop the current local search and start a new local search.

The threshold level suggested for stopping a local search depends on the number of nodes in the current incremental node array. This seems reasonable because proceeding to a node array of higher density involves a commitment to search what might turn out to be a relatively large number of nodes and it is therefore prudent to allow some scope for the search from the nodes of the current incremental array.

Test results (see Satoh, Chapter 6), with $\gamma = 3$, show that the stopping rule is effective in preventing excessively laborious local searches, thereby leading to more efficient global searches. The tests also show that the rule does sometimes prematurely terminate searches that would not be excessively laborious but such occurrences do not significantly affect the overall efficiency of the rule.

Recording end putative labour: If a local search, starting for a given node, is stopped, according to Stopping Rule 3, then the putative labour, from the end point of the search to a nearest admissible point, is recorded.

This is called the *end putative labour* of that node. Also, if the search is stopped according to Stopping Rule 2 then the end putative labour of that node is recorded as having the value infinity.

Size of convergence space: For a given objective vector, the efficiency of the search depends on the size or inclusiveness of the convergence space. Evidently, if the size of the convergence space is made larger by adopting a different local search method, so the highest density required of the node array becomes correspondingly smaller, while still ensuring global convergence of the node array method.

The smallest possible convergence space that ensures global convergence of the node array method is equal to the union of the admissible set with an arbitrarily small neighbourhood of one admissible point. The largest possible convergence space is equal to search space. Neither of these two extremes gives a node array method with optimal efficiency. In the first extreme, the global search is exhaustive and is therefore not efficient. In the second extreme, only one 'local' search is needed to make a global search but the labour needed for that is obviously greatly dependent on the starting point. An unfortunate starting point would maximise labour (with respect to all possible starting points in the current incremental node array) and that might involve more labour than the sum of labours from the local searches made from all the points in the current incremental node array. However, a 'local' search method with global convergence capacity could, with a suitable stopping rule, similar to Stopping Rule 3, be made to search all the points of the current incremental node array, without expending too much labour at any of them. The stopping rule would effectively restrict the size of the convergence space to be less than the search space.

Shape of objective functions and size of admissible set: There are also two other factors, both of which depend on the objective vector, that determine the efficiency of the node array method. The first is the size of the admissible set and the second is the intricacy or awkwardness of the shapes of the component objective functions.

The node array method is a dual global-local or coarse-fine or macro-micro search procedure, in which the global part is provided by the node array and the fine or micro part is provided by a local search. The global search serves to locate a point within the convergence space. This is a relatively coarse search task if the convergence space is a relatively large set compared to a single admissible point. A local search serves to find an admissible point, which is a fine detail within the convergence space, without searching the convergence space exhaustively. Although it is possible to use a progressively denser node array, without any local searches, to find an admissible point, such an approach would obviously be laborious if the admissible set is small (it might contain just one single point in the discretised search space, in which case a global search based solely on the node array becomes exhaustive). In contrast, it is easier to use the node array to find a point in the convergence space, especially if this space is larger than the admissible set, whence a local search, such as, for example, a moving boundaries process, converges to an admissible point, in a manner that does not depend on the size of the admissible set.

However, for a given local search method, such as a moving boundaries process, the work required of the local searches is greatly dependent on the intricacy or awkwardness of the shapes of the objective functions. An example of an awkward shape is provided by the Rosenbrock test function (see Satoh, Chapter 6). The more awkward the shapes of the objective functions, the more laborious a local search within that convergence space is likely to be. In contrast, the node array, whether it is regarded as a partial or as a complete search method, is totally impervious to the intricacy or awkwardness of the shapes of the objective functions.

Thus, it is seen that an important characteristic of a good macro search is that its efficiency does not depend on the shapes of the objective function, however intricate or awkward these might be, whilst an important characteristic of a local search is that its efficiency does not depend on the size of the admissible set. The node array method embodies both these characteristics.

It is therefore concluded that, to optimise the node array method, it is necessary to use a local search method giving convergence spaces that are significantly larger than the admissible space but not to allow any local search to labour at very great length and to little effect within the convergence space. This means that some local searches within the convergence space must be terminated if they become too laborious, in accordance with the Stopping Rule 3, defined above.

It is, in principle, possible to devise a local search method with global convergence spaces, for all the objectives in the universal set. A method of this kind would find an admissible point with a single 'local' search starting anywhere in the search space and would, therefore, not necessitate the use of a macro search, such as the node array. However, the efficiency of such a search is greatly dependent on the initial point chosen for the search. An unfortunate choice of initial point, far from the admissible set, would result in a relatively laborious search. An example of a method, almost of this kind, is a probabilistic search technique called *simulated annealing* (see Whidborne, Chapter 7). The method could be used as the local search part of the node array method.

Node sequence: The efficiency of the node array method depends on the way each node sequence is formed. A node sequence can be formed by arbitrary selection of nodes in the incremental node array (a particular arbitrary sequence is defined by the natural sequence of addresses). However, as a search proceeds and gathers more information about the end putative labour of each searched node, it is expected that gains in the efficiency of a global search can be made by the appropriate use of such information. Specifically, the initial elements of the sequence could be chosen for their higher likelihood of resulting in a successful local search.

For example, of all the nodes in the current incremental array, a node nearest (in the sense of the smallest value of the Euclidean distance) to a searched node with the smallest end putative labour might be preferred. A node sequence might therefore start with such a node and proceed to other nodes in the incremental array, in an order of non-decreasing distance from the searched node with the smallest end putative labour.

Other macro methods: An alternative to the node array, as a global search method, for generating a set of points that intersects the convergence space of a local search method, might be provided by the probabilistic global search method called a *genetic algorithm* (Liu et al 1994, 1997). Judging by the test results presented in Part III, this method is, like the global part of the node array method, relatively impervious to the intricacy or awkwardness of the shape of the objective functions. However, although the method appears to be universally globally convergent, its efficiency might be enhanced if it is coupled with a local search method, as is done in the node array method, to make a dual global-local search method. Accordingly, the search would initially set out, with a genetic algorithm, to find a point in the convergence space but would be switched to local searches, at stages in the global search that are analogous to different densities of the node array. An example of a stage, in the progress of a genetic algorithm, might be defined in terms of a number of generations.

A genetic algorithm simulates, to some extent, what is thought to occur in the natural process of biological evolution. Biological evolution can, with some reservations, be viewed as a process of design. Considering the astounding variety and virtuosity of systems created by evolution, it seems reasonable to emulate that process in the design of man-made systems. However, there is one fundamental difference between biological evolution and man-made design. This is that man-made designs usually have to be carried out within a specified period of time (usually fractions of the human life-span), whereas biological evolution and indeed some other natural processes (such as the annealing of metals, upon which the search method called simulated annealing is based) are, apparently, not subject to any such constraint. It might therefore be expected that man-made search methods can be made to be less time consuming or less labour intensive than genetic algorithms or other methods based solely on natural processes. Some support for this hypothesis is found in the results of numerical experiments reported in Part III. This is not to say that parts of genetic algorithms cannot be used to construct efficient methods.

A specific node array method: A node array method (NA) that employs a moving boundaries process (MBP) for its local searches is denoted by NA&MBP(X), where X refers to the particular trial generator used.

The MBP that has been used extensively employs the Rosenbrock (R) trial generator, which is applicable to all objectives in the universal set. This MBP(R) is known, from its extensive use in practice, to have good local convergence properties. To establish its convergence properties more precisely, it might not be difficult to construct various classes of objective vectors that belong to the universal set, for which the convergence space coincides with the finiteness search space, thus demonstrating mathematically some of the convergence properties of the MBP(R). In addition, the convergence spaces of various specific objective functions, possibly having intricate or awkward shape, can be determined by numerical methods so as to give some insight into the efficacy of the method.

For a numerical evaluation of the NA&MBP(R), which is the node array (NA) method employing the MBP(R), see Satoh, Chapter 6. Similar evaluations, using other trial generators, have yet to be carried out.

1.7.6 Traps and Subsidiary Search

A particular situation that might trap the main search of the moving boundaries process is shown in the Venn diagram of Figure 1.5. Suppose that there are two inequalities and the region, in which each is satisfied, is shown by its respective boundary. The intersection of the two regions is the admissible set. A local search with the moving boundaries process, starting in the shaded area labelled local trap, which is in the region of the first inequality, if it attempts to progress towards the region of the second inequality by the shortest route, would be impeded from crossing the boundary of the first region.

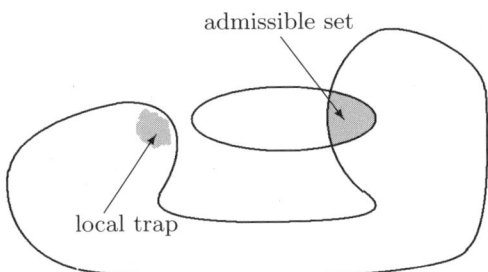

Fig. 1.5. Example of local trap

In this situation, it is relatively easy to devise a subsidiary search that can make progress to an admissible point. Let a subsidiary search proceed, as in the main search, but with the boundary of the first region replaced by a much more inclusive boundary (that is to say, the tolerance associated with the objective function is replaced by a much larger number) that is subsequently reduced gradually to the original boundary during the subsidiary search. Evidently, when the boundary of the first region is expanded, the subsidiary search can progress unhindered to enter the second region. Once inside the second region, the original boundary of the first region is gradually restored and progress towards an admissible point can thereby take place. This process is generalised to more than two inequalities simply by considering each inequality in turn and, if the inequality is satisfied (and perhaps also otherwise), expanding its boundary, within a subsidiary search, and then, when the search has entered the boundary of another inequality (or perhaps sooner), gradually restoring the original boundary. Expanding the boundary of a region is, in effect, the same as partially removing or deactivating its corresponding inequality. Thus only one inequality is deactivated at a time and the other inequalities remain active. This contrasts sharply with the minimax process, where all but one of the inequalities are, in effect, deactivated and the one remaining active is the one most unsatisfied.

Another and more obvious trap that impedes progress occurs at a local minimum of an objective function. Imagine, in Figure 1.5, that a local minimum is situated somewhat far from the admissible set. A search starting close to the local minimum would not progress towards an admissible point. A subsidiary search involving the expansion of boundaries, as outlined in the paragraph above, would again be useful. It would escape from a trap set by a local minimum.

An understanding of the how traps can occur, for example as in Figure 1.5, can be of help when devising ways to avoid them or escape from them. The subsidiary search method outlined here has yet to be programmed for automatic computation or tested in a systematic way, as part of the moving boundaries process. It has, however, been used manually with some success.

1.7.7 Empty Admissible Set

If the admissible set is empty then the node array method can be used to determine another vector of tolerances ε^* that results in a new non-empty admissible set. The problem can be cast as that of finding a new tolerance vector ε^* such that the admissible set is not empty and the difference between the new tolerance vector ε^* and the old is minimised. The designer would then evaluate the new tolerance vector ε^*, to determine if it represents a useful physical or engineering specification of the design problem. Notice that an increase in the tolerances will usually correspond either to a design with more expensive hardware (for example actuators) or a design with diminished performance of the control system.

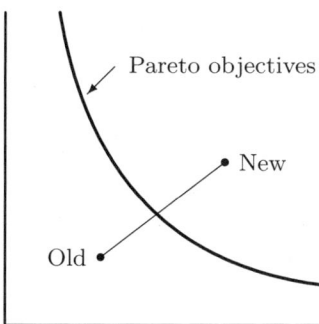

Fig. 1.6. Illustration of new tolerance vector

This is illustrated in Figure 1.6 which shows a region of the plane $\mathbb{R}^2$ and it is assumed that $M = 2$. The points labelled "Old" and "New" show the old and new tolerance vectors respectively and the curve labelled Pareto objectives shows the values of the objective vector corresponding to the set of all Pareto minimisers. That is to say, the curve is the image of the set of all Pareto minimisers, produced by the mapping defined by the objective vector. The length of the line joining Old and New represents the extent of the difference between them. When this is minimised, without making the admissible set empty, the new tolerance ε^* is a point on the curve labelled Pareto objectives. As the figure suggests, it is necessary only to compute the new tolerance ε^*, where the straight line intersects the Pareto objectives curve, and not other parts of the Pareto objectives curve.

To compute a new tolerance vector, it is suggested (Zakian, 1979a) that the new tolerance be expressed by

$$\varepsilon_i^* - \varepsilon_i = xw_i, \quad i \in \widetilde{M} \tag{1.87}$$

Here, w_i is a real non-negative weight that reflects the relative extents to which the old scalar tolerances can be increased and x is a real positive number that cannot be decreased (it has therefore been minimised), without making the admissible set empty. With these new tolerances, the admissible set is minimal and hence every element it contains is a Pareto minimiser (Zakian, 1996; see Section 1.2).

1.7.8 Computation of Objectives

A search for an admissible point usually involves the evaluation of the objectives $\phi_i(c)$ for a long sequence of trial points c. It is therefore important to compute these objectives efficiently. Many objective functions that characterise the dynamical behaviour of the system are functionals of the step or impulse response of the system. Typically, the functionals are the rise-time and overshoot of the step response or the 1-norm or 2-norm of the impulse response or step response of a strictly proper rational transfer function. The impulse response of the transfer function or of the entire system (which may involve delays) can be computed efficiently, if necessary by taking account of any sparsity of any large matrices involved, by using recursions based on I_{MN} approximants (Zakian, 1975, 1979a; Arunsawatwong, 1998). The time response is computed at a sufficiently dense set of points in an interval of time $[0, T]$, where T is much larger than the largest time constant of the system, as determined by its abscissa of stability. If required, the 1-norm or 2-norm is then evaluated by standard methods of numerical integration.

References

Arunsawatwong, S., (1996) Stability of retarded delay differential systems, *Int. J. Contr.*, 65:347-364.

Arunsawatwong, S., (1998) Stability of Zakian I_{MN} recursions for linear delay differential equations, *BIT*, 38:219-233.

Birch, B.J. and Jackson, R., (1959) The behaviour of linear systems with inputs satisfying certain bounding conditions, *J. Electroncs & Control*, 6:366–375.

Bode, H.W., (1960) Feedback — the history of an idea, *Proceedings of the Symposium on Active Networks and Feedback Systems*, 4.

Boyd, S.P. and Barratt, C.H., (1991) Linear Controller Design: Limits of Performance, (Englewood Cliffs, New Jersey, U.S.A.: Prentice Hall).

Desoer, C.A. and Chan, W.S., (1975) The feedback interconnection of linear time-invariant systems, *J. Franklin Inst.*, 300:335-351.

Kuhn, T.S., (1970) *The structure of scientific revolutions* 2nd ed., (Chicago: University Chicago Press).

Lane, P.G., (1995) The principle of matching: a necessary and sufficient condition for inputs restricted in magnitude and rate of change, *Int. J. Contr.*, 62:893 915.

Liu, T.K., Satoh, T., Ishihara, T., and Inooka, H., (1994) An application of genetic algorithms to control system design, *Proceedings of the 1st Asian Control Conference*, Tokyo, 701-704.

Liu, T.K., Ishihara, T., and Inooka, H., (1997) Application of a multiobjective genetic algorithm to control systems design based on the method of inequalities" *Proceedings of the 2nd Asian Control Conference*, Seoul, 289-292.

Maciejowski, J.M., (1989) *Multivariable Feedback Design* (Wokingham, U.K.: Addison-Wesley).

Mayne D.Q., and Sahba M., (1985) An efficient algorithm for solving inequalities, *J. Opt. Theory & Appl.*, 45:407-423.

Maxwell, J.C., (1868) On governors, *Proc. Roy. Soc.*, 16:270-283.

Newton, G.C., Gould, L.A., and Kaiser, J.F., (1957) *Analytical design of linear feedback controls*, (New York: Wiley).

Nyquist, H. (1932) Regeneration theory, *Bell Syst. Tech. J.*, 11:126-147.

Papoulis, A., (1970) Maximum response with input energy constraints and the mathched filter principle, *IEEE Trans. Circuit Theory*, 17:175-182.

Royden, H.L., (1968) *Real Analysis*, 2nd ed. (London: Collier Macmillan).

Rosenbrock, H.H., (1960) An automatic method for finding greatest or least value of a function, *Comput. J.*, 3:175-184.

Rutland, N.K., (1992) Illustration of a new principle of design: vehicle speed control, *Int. J. Contr.*, 55:1319-1334.

Rutland, N.K., (1994a) Illustration of the principle of matching with inputs restricted in magnitude and rate of change: vehicle speed control revisited, *Int. J. Contr.*, 60:395-412.

Rutland, N.K., (1994b) The principle of matching: practical conditions for systems with inputs restricted in magnitude and rate of change, *IEEE Trans. Automatic Control*, AC-39:550-553.

Rutland, N.K. and Lane, P.G., (1995) Computing the 1-norm of the impulse response of linear time-invariant systems, *Syst. Control Lett.*, 26:211-221.

Whidborne, J.F., Postlethwaite, I., and Gu, D.-W., (1994) Robust controller design using H_∞ loop-shaping and the method of inequalities, *IEEE Trans. Control Syst. Technology*, 2(4):455-461.

Zakian, V., (1975) Properties of I_{MN} and J_{MN} approximants and applications to numerical inversion of Laplace transforms and initial-value problems, *J. Math. Anal. & Appl.*, 50:191-222.

Zakian, V., (1979a) New formulation for the method of inequalities, *Proc. IEE*, 126:579-584.

Zakian, V., (1979b) Computation of the abscissa of stability by repeated use of the Routh test, *IEEE Trans. Automatic Control*, AC-24: 604-607.

Zakian, V., (1986a) A performance criterion, *Int. J. Contr.*, 43:921-931.

Zakian, V., (1986b) On performance criteria, *Int. J. Contr.*, 43:1089-1092.

Zakian, V., (1987a) Input spaces and output performance, *Int. J. Contr.*, 46:185-191.

Zakian, V., (1987b) Design formulations, *Int. J. Contr.*, 46: 403-408.

Zakian, V., (1989) Critical systems and tolerable inputs, *Int. J. Contr.*, 49:1285-1289.

Zakian, V., (1991) Well matched systems, *IMA J. Math. Control Inf.*, 8:29-38 (see also Corrigendum, 1992, 9:101).

Zakian, V., (1996) Perspectives on the principle of matching and the method of inequalities, *Int. J. Contr.*, 65: 147-175.

Zakian, V. and Al-Naib, U., (1973) Design of dynamical and control systems by the method of inequalities, *Proc. IEE*, 120:1421-1427.

Part II

Computational Methods (with Numerical Examples)

2 Matching Conditions for Transient Inputs

Paul Geoffrey Lane

Abstract. This chapter provides a method for computing the peak output of a scalar proper rational transfer function, when the possible input set is defined by separate 2-norm bounds on the inputs and their rates of change. The notion of an approximately worst input is considered. A wind-gust/wind-turbine environment/system couple is used to illustrate the use of the method to test matching conditions.

2.1 Introduction

This chapter provides a solution to the problem of evaluating the peak output

$$\hat{v} = \sup\{\|v(f)\|_\infty : f \in \mathcal{P}\} \tag{2.1}$$

of a scalar proper rational transfer function $f \to v(f)$ in the case that the possible set $\mathcal{P}$ is defined by

$$\mathcal{P} = \left\{ f : \|f\|_2 \leqslant M, \|\dot{f}\|_2 \leqslant D \right\} \tag{2.2}$$

where M, D are given positive constants. This set contains transient inputs (see Zakian, 1989).

The output $v(f)$ is given by a convolution integral of the form

$$v(f,t) = \int_{-\infty}^{+\infty} h(t-\tau) f(\tau) \mathrm{d}\tau \quad t \in \mathbb{R} \tag{2.3}$$

where $v(f,t)$ denotes the value of the output $v(f)$ at time t and h denotes the impulse response. The impulse response satisfies $h(t) = 0$ for $t < 0$ and has a proper, rational Laplace transform

$$H(s) = \int_0^\infty h(t) \exp(-st) \mathrm{d}t \tag{2.4}$$

whose poles have negative real parts. This ensures finiteness of the peak output. Because the system is rational, state-space methods can be used to evaluate certain integrals that are required.

An algorithm is presented that enables the peak output to be computed to any desired *pre-specified* degree of accuracy. This enables the matching condition

$$\mathcal{P} \subseteq \mathcal{T} \tag{2.5}$$

to be tested via inequalities of the form (see Zakian, 1989)

$$\hat{v}_i \leqslant \varepsilon_i \quad i = 1, 2, \ldots, m \tag{2.6}$$

A wind-gust/wind-turbine environment-system couple is used to illustrate the use of the algorithm to test the matching condition.

The material is taken largely from the paper by Lane (1995) but includes some new material. The discussion of the stopping criterion for the algorithm has been changed and extended and a formula that is required for the evaluation of a certain lower bound, that was missing in the paper, is provided. An example showing that a certain function is not convex is included that previously was only alluded to. Derivations and other material that would otherwise detract from the narrative are collected in a special section. All results have been calculated anew. There is a discussion of approximately worst inputs and it is shown how they can be constructed. Also the treatment of the numerical example is given a different slant. Here the example is used to illustrate and explain certain features of the algorithm's convergence and to show examples of approximately worst inputs. It is interesting to look at an approximately worst input when a match is not achieved because it may indicate how the system (plant and controller) or its environment (possible set) has to be changed to achieve a match. The idea that a match can sometimes be achieved by modifications to the possible set leads to a discussion of the sensitivity of peak output to changes in the parameters M, D of the possible set $\mathcal{P}$. None of this was discussed in the original paper where the emphasis was on controller design (achievement of a match through changes to the system).

The method used to evaluate peak output relies heavily on facts about convex optimisation problems and especially Lagrangian duality that are to be found in the books by Boyd and Barratt (1991) and Rockafellar (1970).

2.2 Finiteness of Peak Output

The need, for the purposes of design, for conditions that enable an initial design to be found such that

$$\hat{v} < \infty \tag{2.7}$$

is discussed by Zakian (1989).

When the possible set is defined by (2.2), peak output is finite if

$$\alpha < 0 \tag{2.8}$$

where α denotes the largest of the real parts of the poles of the transfer function H. As discussed by Zakian, descent methods may be used to solve an inequality such as (2.8) but cannot solve (2.7) directly.

That (2.8) implies finiteness of the peak output can be seen by the following argument. Let a function w be defined by

$$w(t) = \begin{cases} \exp(-t) & ; t \geq 0 \\ 0 & ; t < 0 \end{cases} \quad (2.9)$$

For proper rational H, (2.8) implies

$$\|h \star w\|_2 < \infty \quad (2.10)$$

where $\star$ denotes convolution. For every input f in the possible set $\mathcal{P}$ of (2.2) there exists $u = f + \dot{f}$ which satisfies $\|u\|_2 \leq M + D$ and is such that $f = w \star u$. So, for every $f \in \mathcal{P}$

$$\begin{aligned} \|v(f)\|_\infty = \|h \star f\|_\infty &= \|h \star w \star u\|_\infty \\ &\leq \|u\|_2 \|h \star w\|_2 \quad \text{by the Schwarz' inequality} \\ &\leq (M+D)\|h \star w\|_2 \\ &< \infty \end{aligned} \quad (2.11)$$

Hence (2.7).

If the bound D on the derivative is absent ($D = \infty$ in (2.2)), the additional condition that the transfer function H is *strictly* proper is required to ensure finiteness of the peak output.

2.3 Evaluation of Peak Output

From (2.1), (2.2), (2.3)

$$\hat{v} = \sup\left\{v(f,0) : \|f\|_2 \leq M, \|\dot{f}\|_2 \leq D\right\} \quad (2.12)$$

Equation (2.12) is a result of time-invariance and linearity of the input-output rule (2.3). Because of time-invariance, one can take it that the peak output occurs at time $t = 0$. Because of linearity and because $f \in \mathcal{P}$ implies $-f \in \mathcal{P}$, the supremum of $|v(f,0)|$ is equal to the supremum of $v(f,0)$.

Problem (2.12) is convex and infinite dimensional. Such problems can be solved by converting the infinite dimensional problem to a sequence of convex problems having finite dimensions through Ritz approximation and by solving the resulting sequence of problems by standard methods for finite dimensional convex optimisation (see Boyd and Barratt, 1991, see also Satoh, Chapter 3). In the present case, however, a Lagrangian dual of the primal problem (2.12) can be constructed that involves the minimisation of a function in just two dimensions. The dual problem is convex and the minimal value of the dual objective is equal to the peak output $\hat{v}$. Thus peak output $\hat{v}$ can be computed by solving the dual of (2.12) by methods for finite-dimensional convex optimisation problems.

A further simplification arises because the two dimensional convex dual can be reduced to a quasi-convex problem in only one dimension. This problem can be solved by a bisection algorithm.

2.3.1 Solution via the Lagrangian Dual

The required dual of the primal problem (2.12) is obtained as follows. A Lagrange multiplier $\lambda = (\lambda_1, \lambda_2) \in \mathbb{R}^2$ is introduced and the Lagrangian function

$$L(f, \lambda) = v(f, 0) + \frac{\lambda_1}{2}(M^2 - \|f\|_2^2) + \frac{\lambda_2}{2}(D^2 - \|\dot{f}\|_2^2) \qquad (2.13)$$

is considered. The dual objective function ψ is defined by

$$\psi(\lambda) = \sup \left\{ L(f, \lambda) : \|f\|_2 < \infty, \|\dot{f}\|_2 < \infty \right\} \qquad (2.14)$$

This function is convex because it is the pointwise supremum of affine functions of λ. The required dual of the primal problem (2.12) is to evaluate

$$\hat{\psi} = \inf \{\psi(\lambda) : \lambda \geqslant 0\} \qquad (2.15)$$

where $\lambda \geqslant 0$ means $\lambda_1 \geqslant 0$, $\lambda_2 \geqslant 0$.

The terms added to the primal objective $v(f, 0)$ to form the Lagrangian of equation (2.13) are non-negative whenever the input f satisfies the constraints of the primal problem and $\lambda \geqslant 0$. Therefore, for all $\lambda \geqslant 0$,

$$\begin{aligned}
\hat{v} &= \sup \left\{ v(f, 0) : \|f\|_2 \leqslant M, \|\dot{f}\|_2 \leqslant D \right\} \\
&\leqslant \sup \left\{ L(f, \lambda) : \|f\|_2 \leqslant M, \|\dot{f}\|_2 \leqslant D \right\} \\
&\leqslant \sup \left\{ L(f, \lambda) : \|f\|_2 < \infty, \|\dot{f}\|_2 < \infty \right\} = \psi(\lambda)
\end{aligned} \qquad (2.16)$$

This chain of inequalities and equalities holds of all $\lambda \geqslant 0$. Therefore

$$\hat{v} \leqslant \hat{\psi} \qquad (2.17)$$

Because the primal problem (2.12) is convex, one conjectures that inequality (2.17) actually holds with equality. Lane (1995) shows that this is the case and, further, that a minimiser $\bar{\lambda} \geqslant 0$ exists such that $\hat{\psi} = \psi(\bar{\lambda})$. Thus, peak output $\hat{v}$ can be computed by minimising $\psi(\lambda)$ subject to the constraint $\lambda \geqslant 0$.

2 Matching Conditions for Transient Inputs

This observation is useful because the dual objective function ψ turns out to be easy to evaluate. The evaluation of ψ from (2.14) requires a supremem operation over inputs f but this problem can be solved analytically - see Section 2.5.1. The solution involves the construction of a maximising input $f_0(\lambda)$ (a different input for each different λ) such that

$$\psi(\lambda) = L(f_0(\lambda), \lambda) \tag{2.18}$$

An alternative formula for the evaluation of ψ is (Section 2.5.1)

$$\psi(\lambda) = \frac{1}{2}\left(\lambda_1 \|f_0(\lambda)\|_2^2 + \lambda_2 \|\dot{f}_0(\lambda)\|_2^2 + \lambda_1 M^2 + \lambda_2 D^2\right) \tag{2.19}$$

This formula is useful because $\|f_0(\lambda)\|_2$, $\|\dot{f}_0(\lambda)\|_2$ can be computed conveniently by state-space methods - see Section 2.5.4.

In addition, one readily obtains a subgradient of ψ at any given point $\lambda > 0$. A subgradient of ψ at λ is, by definition, any vector $g \in \mathbb{R}^2$ such that

$$\psi(\lambda') \geq \psi(\lambda) + g^T(\lambda' - \lambda) \text{ for all } \lambda' \geq 0 \tag{2.20}$$

From the definitions (2.13), (2.14) and using (2.18) a subgradient of ψ at $\lambda > 0$ is

$$g = \begin{pmatrix} M^2 - \|f_0(\lambda)\|_2^2 \\ D^2 - \|\dot{f}_0(\lambda)\|_2^2 \end{pmatrix} \tag{2.21}$$

One can easily determine a bounded region in which a minimiser must lie. Let $\bar{\lambda} \geq$ denote a minimiser and let ψ be evaluated at an arbitrary point $\lambda' > 0$. Then, $\infty > \psi(\lambda') \geq \psi(\bar{\lambda})$ and from (2.14), $\psi(\bar{\lambda}) \geq L(0, \bar{\lambda})$. Therefore

$$\bar{\lambda}_1 M^2 + \bar{\lambda}_2 D^2 \leq 2\psi(\lambda') \tag{2.22}$$

which, together with $\bar{\lambda} \geq 0$, defines a bounded triangular region in which the minimiser must lie.

A bounded region containing a minimiser and the means to evaluate ψ and a subgradient of it are all that is required to solve the dual problem using an ellipsoid method such as described by Boyd and Barratt (1991) on page 328. A standard stopping criterion for the ellipsoid algorithm (based on a lower bound computed using the subgradient) enables the minimal value of ψ to be computed to any desired pre-specified accuracy.

An alternative stopping criterion, which has two important advantages and will be useful in the next section, is obtained as follows. Lane (1995) proves that inequality (2.17) holds with equality by showing that, when $\bar{\lambda} \geq 0$ is a minimiser for the dual problem, the corresponding input $f_0(\bar{\lambda})$ satisfies the constraints of the primal problem, *i.e.*,

$$\begin{aligned} \|f_0(\bar{\lambda})\|_2 &\leq M \\ \|\dot{f}_0(\bar{\lambda})\|_2 &\leq D \end{aligned} \tag{2.23}$$

and, that

$$\psi(\bar{\lambda}) = v(f_0(\bar{\lambda}), 0) \tag{2.24}$$

It then follows from (2.12) that

$$\hat{v} \geq \psi(\bar{\lambda}) = \hat{\psi} \tag{2.25}$$

which, together with inequality (2.17), implies that (2.17) holds with equality. This argument suggests that a stopping criterion for an algorithm to solve (2.15) be constructed as follows. Let $\lambda^1, \lambda^2, \ldots \geq 0$ denote a sequence of points generated by the algorithm. It is assumed that the sequence converges to a minimiser $\bar{\lambda}$ or contains a convergent subsequence. At the kth iteration, an upper bound for $\hat{\psi}$ is

$$U^k = \min\{\psi(\lambda^i) : i = 1, 2, \ldots, k\} \tag{2.26}$$

A lower bound for $\hat{v}$ is

$$L^k = \max\{\gamma(\lambda^i) : i = 1, 2, \ldots, k\} \tag{2.27}$$

where

$$\gamma(\lambda) = v(\rho(\lambda) f_0(\lambda), 0) \tag{2.28}$$

and

$$\rho(\lambda) = \min\left\{\frac{M}{\|f_0(\lambda)\|_2}, \frac{D}{\|\dot{f}_0(\lambda)\|_2}\right\} \tag{2.29}$$

is the largest number such that the input $\rho(\lambda) f_0(\lambda)$ satisfies the constraints of the primal problem (2.12).

The quantity $\gamma(\lambda)$ is a lower bound for $\hat{v}$ because the input $\rho(\lambda) f_0(\lambda)$ satisfies the constraints of the primal problem. In view of (2.23), (2.24), the upper and lower bounds are expected to converge. If the algorithm is stopped when

$$\frac{U^k - L^k}{U^k + L^k} \leq \text{tol} \tag{2.30}$$

then, $\hat{v}$, $\hat{\psi}$ differ from the value $(U^k + L^k)/2$ by no more than tol $\times$ 100 percent.

This stopping criterion has the following advantages. First, the latter statement is a result of the inequalities $L^k \leq \hat{v}$ and $\hat{\psi} \leq U^k$, (which follow directly from the definitions (2.12), (2.15) of the primal and dual problems) and inequality (2.17). This means that when the criterion (2.30) is satisfied, one is assured that peak output $\hat{v}$ has been computed to the desired accuracy and one does not need a proof that equality holds in (2.17) to know this.

This is not the case when the standard stopping criterion for the ellipsoid algorithm based on the subgradient is used. This assures that $\hat{\psi}$ is computed to the desired accuracy but says nothing about $\hat{v}$.

Second, it provides an approximately worst input - that is, a possible input that produces approximately the peak output. Suppose that the algorithm halts according to the criterion (2.30) at iteration k and let $j \leqslant k$ denote the iteration at which the best lower bound L^k of (2.27) was produced. Then the input $\rho(\lambda^j) f_0(\lambda^j)$ is possible (satisfies the constraints of the primal problem) and produces an output $v(\rho(\lambda^j) f_0(\lambda^j), 0)$ that is precisely tol $\times$ 100 percent less than the computed value of peak output and not more than two times tol $\times$ 100 percent (of the computed value) less than the true value of the peak output.

The quantity $\gamma(\lambda)$ of (2.28) can be evaluated using (by linearity) $v(\rho f_0(\lambda), 0) = \rho v(f_0(\lambda), 0)$ and (by Section 2.5.1)

$$v(f_0(\lambda), 0) = \lambda_1 \|f_0(\lambda)\|_2^2 + \lambda_2 \|\dot{f}_0(\lambda)\|_2^2 \tag{2.31}$$

2.3.2 Reduction to a One-dimensional Problem

Let the point $(1 - \mu, \mu) \in \mathbb{R}^2$ be denoted by λ_μ (not to be confused with the μth component of λ). Then the dual problem (2.15) can be written

$$\hat{\psi} = \inf \{\psi(r\lambda_\mu) : r \geqslant 0,\ 0 \leqslant \mu \leqslant 1\} \tag{2.32}$$

Let

$$\phi(\mu) = \inf \{\psi(r\lambda_\mu) : r \geqslant 0\} \tag{2.33}$$

Then

$$\hat{\psi} = \inf \{\phi(\mu) : 0 \leqslant \mu \leqslant 1\} \tag{2.34}$$

The infimum operation over r in (2.33) can be carried out analytically. The infimum is (Section 2.5.2)

$$\phi(\mu) = \left((1-\mu)\|f_0(\lambda_\mu)\|_2^2 + \mu\|\dot{f}_0(\lambda_\mu)\|_2^2\right)^{\frac{1}{2}} \left((1-\mu)M^2 + \mu D^2\right)^{\frac{1}{2}} \tag{2.35}$$

and occurs at

$$\bar{r} = \operatorname{argmin}\{\psi(r\lambda_\mu) : r \geqslant 0\}$$
$$= \left((1-\mu)\|f_0(\lambda_\mu)\|_2^2 + \mu\|\dot{f}_0(\lambda_\mu)\|_2^2\right)^{\frac{1}{2}} \Big/ \left((1-\mu)M^2 + \mu D^2\right)^{\frac{1}{2}} \tag{2.36}$$

A geometric argument shows (see Figure 2.1) that the α-level sets of ϕ (sets of the form $\{\mu \in [0,1] : \phi(\mu) \leqslant \alpha\}$) are convex. Therefore ϕ is quasi-convex.

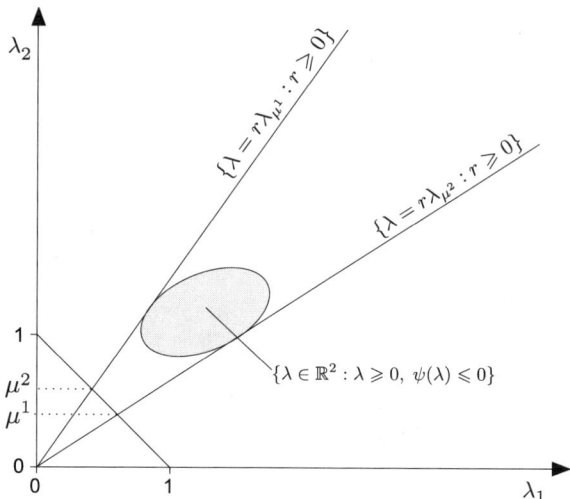

Fig. 2.1. The α-level set of ψ is $\{\lambda \in \mathbb{R}^2 : \lambda \geq 0,\ \psi(\lambda) \leq \alpha\}$ and is convex because ψ is convex. If the ray $\{\lambda = r\lambda_\mu : r \geq 0\}$ intersects this set, then $\phi(\mu) \leq \alpha$. Otherwise $\phi(\mu) > \alpha$. Thus, in the instance shown above, the α-level set $\{\mu \in [0,1] : \phi(\mu) \leq \alpha\}$ of ϕ is equal to the set $[\mu^1, \mu^2]$. This set is convex because the α-level set $\{\lambda \in \mathbb{R}^2 : \lambda \leq 0,\ \psi(\lambda) \leq \alpha\}$ of ψ is convex

The following bisection algorithm solves (2.34).

```
a=0;b=1;
k=0;
repeat
   k=k+1;
   μᵏ=(a+b)/2;
   evaluate the gradient ∇ϕ(μᵏ);
   if ∇ϕ(μᵏ) ≥ 0
      b=μᵏ;
   else
      a=μᵏ;
   end
until(stopping criterion)
```

The algorithm maintains an interval $[a, b]$ of ever decreasing length in which a minimiser is known to lie. The interval is halved at each iteration and the gradient $\nabla\phi$ determines which half is to be discarded. This works because ϕ is quasi-convex. The gradient $\nabla\phi(\mu)$ is given by (Section 2.5.2)

$$\nabla\phi(\mu) = \left(D^2 \|f_0(\lambda_\mu)\|_2^2 - M^2 \|\dot{f}_0(\lambda_\mu)\|_2^2\right) \big/ 2\phi(\mu) \tag{2.37}$$

To obtain a stopping criterion one considers that the above algorithm generates a sequence of points $\lambda^1, \lambda^2, \ldots \in \mathbb{R}^2$ that converges to a minimiser

$\bar{\lambda}$ of the two dimensional problem (2.15) where

$$\lambda^i = \bar{r}^i \lambda_{\mu^i} \quad i = 1, 2, \ldots \tag{2.38}$$

and

$$\bar{r}^i = \operatorname{argmin}\{\psi(r\lambda_{\mu^i}) : r \geqslant 0\} \quad i = 1, 2, \ldots \tag{2.39}$$

and $\mu^1, \mu^2, \ldots$ are the values of μ visited by the algorithm. The stopping criterion introduced at the end of Section 2.3.1 can then be used. Inserting (2.38) in (2.26) yields the obvious upper bound

$$U^k = \min\{\phi(\mu^i) : i = 1, 2, \ldots, k\} \tag{2.40}$$

Inserting (2.38) in (2.27) yields

$$L^k = \max\{\gamma(r^i \lambda_{\mu^i}) : i = 1, 2, \ldots, k\} \tag{2.41}$$

which simplifies (Section 2.5.2)

$$L^k = \max\{\gamma(\lambda_{\mu^i}) : i = 1, 2, \ldots, k\} \tag{2.42}$$

The stopping criterion

$$\frac{U_k - L_k}{U_k + L_k} \leqslant \text{tol} \tag{2.43}$$

then ensures that peak output $\hat{v}$ differs from the value $(U^k + L^k)/2$ by no more than tol × 100 percent when the algorithm halts.

The comments regarding the stopping criterion at the end of Section 2.3.1 apply also to the stopping criterion for the one dimensional problem. Thus, one does not need a proof of equality in (2.17) to be assured that the true value of peak output lies between the upper and lower bounds given by (2.40), (2.42). If the algorithm halts at iteration k in accordance with the stopping criterion (2.43) and if the best lower bound L^k of (2.42) was produced at iteration $j \leqslant k$ then, the input $\rho(\lambda^j) f_0(\lambda^j)$ is possible and produces approximately the peak output (to the tolerance tol).

A stopping criterion based on the subgradient cannot be used for the one dimensional problem because although ϕ is quasi-convex it is not generally convex - see Figure 2.2.

The reader is reminded that state-space methods for the evaluation of the quantities $\|f_0(\lambda)\|_2$, $\|\dot{f}_0(\lambda)\|_2$ are described in Section 2.5.4. Thus, $\phi(\mu)$ and the gradient $\nabla\phi(\mu)$ can be evaluated using formulae (2.35), (2.37) and $\gamma(\lambda_\mu)$ (required for the lower bound (2.42)) can be obtained from (2.28) using (2.29) and (2.31).

Section 2.5.3 shows how to calculate an input $f_0(\lambda)$ as a function of time. This is not necessary to compute peak output but it is necessary if one wishes to inspect a graph of an approximately worst input.

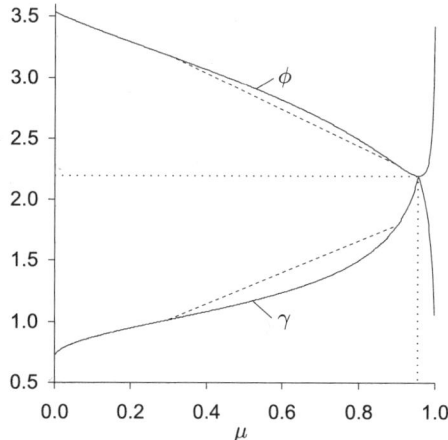

Fig. 2.2. The quantities $\phi(\mu)$ and $\gamma(\lambda_\mu)$ are shown plotted against μ for the system with transfer function $H(s) = 1/(s^2 + s + 1)$ and the possible set $\mathcal{P} = \left\{ f : \|f\|_2 \leqslant 5.0, \|\dot{f}\|_2 \leqslant 1.0 \right\}$. The dashed lines show that ϕ is not convex and (incidentally) that γ is not concave. Function ϕ is, however, quasi-concave. No claim is made regarding γ. Peak output $\hat{v}$ does not exceed $\phi(\mu)$ and is not less than $\gamma(\lambda_\mu)$ for any μ. Thus, it is evident from the graphs that peak output $\hat{v}$ is approximately 2.2 in this case

2.4 Example

In this section the algorithm of Section 2.3.2 is used to test matching conditions for a wind-gust/wind-turbine environment-system couple. The system consists of a wind turbine supported on a flexible tower and its controller. The system input is the wind speed f.

Five system outputs are used to define a tolerable wind gust. These are the power output, v_1; the gearbox and drive-train torsion angle, v_2; the tower deflection, v_3; the turbine blade pitch angle, v_4; and, the rate of change of the blade pitch angle, v_5. A wind gust is tolerable if and only if

$$\|v_1(f)\|_\infty \leqslant \varepsilon_1 = 1.2 \text{ MW}$$
$$\|v_2(f)\|_\infty \leqslant \varepsilon_2 = 0.06 \text{ rad}$$
$$\|v_3(f)\|_\infty \leqslant \varepsilon_3 = 0.4 \text{ m} \qquad (2.44)$$
$$\|v_4(f)\|_\infty \leqslant \varepsilon_4 = 0.2 \text{ rad}$$
$$\|v_5(f)\|_\infty \leqslant \varepsilon_5 = 0.1 \text{ rad s}^{-1}$$

The first inequality expresses a mild power output regulation requirement. A gust that does not satisfy the second and third inequalities causes structural damage. (An excessively large strain in the gearbox and drive-train or in the tower results in a permanent deformation.) Therefore, these are critical design requirements (Zakian, 1989). A gust that does not satisfy the remaining inequalities causes the limitations of the blade pitch actuator to be exceeded and/or results in excessive stress in the turbine blade roots. The tolerable set of wind gusts is

$$\mathcal{T} = \{f : \|v_i(f)\|_\infty \leqslant \varepsilon_i,\ i = 1, 2, \ldots, 5\} \tag{2.45}$$

The possible set $\mathcal{P}$ of wind gusts is of the form (2.2) with the following parameters' values.

$$\begin{aligned} M &= 7.6\text{ m s}^{-1/2} \\ D &= 4.8\text{ m s}^{-3/2} \end{aligned} \tag{2.46}$$

The wind turbine system with controller is described by a linear and time-invariant state-space system. The details are provided in Section 2.5.5, which also contains a note on the state-space realisations required for the evaluation of peak output.

The matching condition $\mathcal{P} \subseteq \mathcal{T}$ is satisfied if and only if

$$\hat{v}_i \leqslant \varepsilon_i \quad i = 1, 2, \ldots, 5 \tag{2.47}$$

Peak output was calculated to a tolerance of 0.01% for each output $v_1, \ldots, v_5$ in turn using the algorithm of Section 2.3.2. The results are

$$\begin{aligned} \hat{v}_1 &= 1.1834\text{ MW} \\ \hat{v}_2 &= 0.054523\text{ rad} \\ \hat{v}_3 &= 0.37324\text{ m} \\ \hat{v}_4 &= 0.11864\text{ rad} \\ \hat{v}_5 &= 0.12924\text{ rad s}^{-1} \end{aligned} \tag{2.48}$$

The progress of the algorithm during the evaluation of $\hat{v}_1, \ldots, \hat{v}_5$ is shown in Figure 2.3. Certain features of the convergence of the algorithm are explained in Figure 2.4. Comparing the results in (2.48) with the bounds that define a tolerable input in (2.44), taking into account the tolerance to which the results were calculated, one can see that the first four matching conditions of (2.47) are definitely satisfied while the fifth condition is definitely not satisfied. This shows that the matching condition $\mathcal{P} \subseteq \mathcal{T}$ is *not* satisfied.

Figure 2.5 shows approximately worst inputs. As discussed at the end of Section 2.3.1 and again at the end of Section 2.3.2, an approximately worst input is a possible input that produces approximately the peak output. There is a different worst input for each output. A worst input is constructed by determining the response of the output in question to a pulse (Secton 2.5.3). The worst input is the response reversed in time. The worst inputs for the outputs $v_1, v_2, \ldots, v_5$ are shown on the left of Figure 2.5. The pulses that were used to construct them are shown inset. The responses of the outputs $v_1, v_2, \ldots, v_5$ to their respective worst inputs are shown on the right. The responses to the worst inputs are symmetrical about $t = 0$ (Section 2.5.3) which is also the time at which the peak output occurs. It was verified numerically that the worst inputs are possible and produce (to the tolerance tol) the computed peak output. The computed value of the peak output is marked by the dashed horizontal lines in the graphs on the right. The peak output may *seem* not to be attained in some of the graphs on the right but this is due to the narrowness of the peaks on the time scale shown. Similarly, the worst input for output v_5 seems to be discontinuous at time $t = 0$ on the time scale shown but closer inspection shows that this is not the case. The outputs v_1, v_2, v_5 appear to exhibit approximately the characteristics of a differentiator. The worst inputs for these outputs are equal (approximately and to within a scaling factor that may be negative) to the derivative of the pulses which were used to generate them. The worst inputs for outputs v_1, v_2 are, in fact, almost identical. This is because the outputs themselves are nearly proportional. This has a physical explanation, namely, that power output and shaft strain are both proportional to shaft torque when shaft dynamics are ignored. The outputs v_3, v_4 exhibit roughly (inverse) proportional behaviour. Evidently, resonance (associated with the flexible tower) plays a significant rôle in determining the worst input in the case of output v_3.

When matching conditions are not met, the designer can consider a number of options. For example, at the expense of installing a more powerful blade pitch actuator and/or of strengthening the blade roots, the designer could arrange for the bound ε_5 to be relaxed and thereby achieve a match. Alternatively, modifications could be made that allow one or more of the bounds $\varepsilon_1, \ldots, \varepsilon_4$ to be relaxed. It may then be possible to redesign the controller so that performance in respect of one or more of these outputs is traded for improved performance in respect of the output v_5. Or a match could be achieved simply by redesigning the controller - possibly by searching in a larger class of controllers than has hitherto been considered. A further possibility is to modify the possible set. In practice a wind turbine is shut down and driven to a safe position in bad weather. Thus, although one cannot influence the weather, one can influence the inputs that the turbine is subjected to through this action.

Table 2.1 shows the sensitivity of peak output to changes in the parameters M and D of the possible set for the outputs $v_1, v_2, \ldots, v_5$. One does not need the algorithm of Section 2.3.2 to show that, when M, D are simultaneously reduced by a given proportion, the peak output is reduced by the same proportion. This is essentially a result of linearity of the input-output rule. But it is reassuring that the results obtained using the algorithm confirm that this is the case for a simultaneous reduction of 25%. (Third column in the table). The reductions in peak output that result from individual 25% reductions in the bounds M, D are shown in the first and second columns of the table. In the case of outputs v_3, v_5 the change in M barely influences the peak output. When one of the bounds M, D has nearly no influence, the lin-

Table 2.1. Changes in peak output due to changes in the parameters M, D of the possible set for the outputs $v_1, v_2, \ldots, v_5$ of the wind turbine system. The first three columns show the percentage reductions in peak output produced by 25% reductions in the bounds M, D (individually and together). The fourth column shows the relative importance of the bounds M and D judged by dividing the results in the first column by those in the second. Values greater than one indicate that M is more important and values less than one indicate that D is more important. Peak output was calculated to a tolerance of 0.01% using the algorithm of Section 2.3.2. The percentages are shown rounded to one decimal place. Importance is rounded to two decimal places.

	25% reduction in M	25% reduction in D	25% reduction in M, D	Importance $M:D$
v_1	10.4%	18.3%	25.0%	0.57
v_2	12.1%	17.1%	25.0%	0.71
v_3	1.5%	23.7%	25.0%	0.06
v_4	15.0%	11.9%	25.0%	1.26
v_5	0.4%	24.8%	25.0%	0.02

earity argument implies that peak output is proportional to the other bound. Thus in the case of outputs v_3, v_5 one expects $\hat{v} \propto D$ which is evidently very nearly the case. (The changes in peak output shown in column 2 for outputs v_3, v_5 are nearly equal to the 25% change in D.) The relative importance of the bounds M, D as judged by the ratio of the data in columns 1 and 2 is shown in the last column in the table. A value greater than one indicates that M is important and a value less than one indicates that D is important. The overwhelming importance of D in the case of output v_5 is very likely due to the (approximate) behaviour of this output as a differentiator - see Figure 2.5 and the discussion of worst inputs above. In the case of output v_3 the importance of D can probably be attributed largely to its rôle in limiting the extent to which the resonance is excited. The importance of the bounds M, D is more finely balanced in the case of outputs v_1, v_2, v_4. It is tempting to suggest that the importance M is greater in the case of output v_4 than it is in the case of outputs v_1, v_2 because output v_4 exhibits approximately proportional behaviour whereas outputs v_1, v_2 exhibit more nearly differentiator behaviour.

The bound ε_5 in (2.44) is around 23% less than the value of peak output $\hat{v}_5$ calculated for the nominal values (2.46) of the bounds M, D. Thus, in the light of the results above, a match could be achieved at the expense of a (roughly) 23% reduction in the bound D. In practice, this might mean that the wind turbine would have to be shut down in milder weather conditions than would otherwise have been the case. A reduction in the bound M is evidently not helpful if the system is not changed. But a reduction in M may be helpful if an improvement in performance of outputs other than v_5 could be traded for improvement in respect of v_5 by redesigning the controller.

Table 2.2. *Estimated* changes in peak output due to non-vanishing changes in the parameters M, D of the possible set calculated using sensitivity theory. The first three columns show the *estimated* percentage reductions in peak output produced by 25% reductions in the bounds M, D (individually and together). The fourth column shows the estimated relative importance of the bounds M and D judged by dividing the results in the first column by those in the second. Change in peak output was estimated using a linear approximation based on local sensitivities (equation (2.49)). The percentages are rounded to one decimal place. Importance is rounded to two decimal places.

	25% reduction in M	25% reduction in D	25% reduction in M, D	Importance $M : D$
v_1	8.4%	16.6%	25.0%	0.51
v_2	9.9%	15.1%	25.0%	0.65
v_3	1.4%	23.6%	25.0%	0.06
v_4	14.0%	11.0%	25.0%	1.27
v_5	0.3%	24.7%	25.0%	0.01

It is noted in passing that there is a sensitivity theory for convex optimisation problems that enables (amongst other things) local sensitivities of the primal optimum to be calculated from the dual optimiser (See ??). In the present case (assuming that the derivatives exist and ignoring technical details)

$$\frac{\partial \hat{v}}{\partial M} = \bar{\lambda}_1 M$$
$$\frac{\partial \hat{v}}{\partial D} = \bar{\lambda}_2 D$$
(2.49)

where $\bar{\lambda} = (\bar{\lambda}_1, \bar{\lambda}_2)$ is the minimiser for the two dimensional dual problem (2.15). The algorithm of Section 2.3.2 does not provide an exact minimiser but one obtains an approximate minimiser by taking the last point visited by the algorithm. The results in Table 2.2 were calculated using the results from the evaluations of peak output at the nominal values of M, D and a linear approximation based on the local sensitivities (2.49). The linear approximation is exact when the bounds M, D are changed in the same proportion (as it should be) and generally performs remarkably well - *cp.* Table 2.1.

2.5 Miscellaneous Results

2.5.1 Formulae for Section 2.3.1

By Parseval's theorem, the Lagrangian $L(f,\lambda)$ of equation (2.13) can be written

$$L(f,\lambda) = \frac{1}{2\pi}\int_{-\infty}^{+\infty} H(j\omega)F(j\omega)\mathrm{d}\omega + \frac{\lambda_1}{2}\left(M^2 - \frac{1}{2\pi}\int_{-\infty}^{+\infty}|F(j\omega)|^2\mathrm{d}\omega\right)$$
$$+ \frac{\lambda_2}{2}\left(D^2 - \frac{1}{2\pi}\int_{-\infty}^{+\infty}|\omega F(j\omega)|^2\mathrm{d}\omega\right) \quad (2.50)$$

where F, H denote the Fourier transforms of the input f and the impulse response h and are defined by

$$F(j\omega) = \int_{-\infty}^{+\infty} f(t)\exp(-j\omega t)\mathrm{d}t \quad (2.51)$$

$$H(j\omega) = \int_{-\infty}^{+\infty} h(t)\exp(-j\omega t)\mathrm{d}t \quad (2.52)$$

Rearranging (2.50) and completing the square yields

$$L(f,\lambda) = \frac{1}{4\pi}\int_{-\infty}^{+\infty}\frac{|H(j\omega)|^2}{\lambda_1 + \lambda_2\omega^2}\mathrm{d}\omega + \frac{\lambda_1}{2}M^2 + \frac{\lambda_2}{2}D^2$$
$$- \frac{1}{4\pi}\int_{-\infty}^{+\infty}\left|\sqrt{\lambda_1 + \lambda_2\omega^2}\,F(j\omega) - \frac{H^*(j\omega)}{\sqrt{\lambda_1 + \lambda_2\omega^2}}\right|^2\mathrm{d}\omega \quad (2.53)$$

where * denotes complex conjugate. Evidently, the Lagrangian $L(f,\lambda)$ is maximised when the input f has Fourier transform $F_0(\lambda)$ given by

$$F_0(\lambda, j\omega) = \frac{H^*(j\omega)}{\lambda_1 + \lambda_2\omega^2} \quad (2.54)$$

and then

$$\psi(\lambda) = L(f_0(\lambda),\lambda)$$
$$= \frac{1}{2}\left(\frac{1}{2\pi}\int_{-\infty}^{+\infty}\frac{|H(j\omega)|^2}{\lambda_1 + \lambda_2\omega^2}\mathrm{d}\omega + \lambda_1 M^2 + \lambda_2 D^2\right) \quad (2.55)$$

where $f_0(\lambda)$ denotes the input whose Fourier transform is given by (2.54).

There may be points on the axes $\lambda_1 = 0$ or $\lambda_2 = 0$ for which, strictly speaking, the above derivation does not work (see Lane (1995)). But this is not important in practice, because it turns out that it is not necessary to evaluate ψ on the axes. This can be avoided in the solution of the dual problem of Section 2.3.1 by the ellipsoid method by arranging for a so called *constraint iteration* to be carried out on the axes. This means that a minimiser that happens to lie on one of the axes will be approached from the interior of the region $\{\lambda \in \mathbb{R}^2 : \lambda \geqslant 0\}$. In the case of the bisection method for the solution of the one-dimensional problem of Section 2.3.2, the algorithm only ever calls for evaluations of the function ϕ of equation (2.33) in the interior of the interval $[0, 1]$ - which corresponds to the evaluation of ψ in the interior of $\{\lambda \in \mathbb{R}^2 : \lambda \geqslant 0\}$.

Formula (2.19) is obtained by writing the integrand in equation (2.55) in the form $\lambda_1|F_0(\lambda, j\omega)|^2 + \lambda_2|j\omega F_0(\lambda, j\omega)|^2$ and then using Parseval's theorem. Formula (2.31) is obtained by noting that the same integral (including the factor $1/2\pi$) is equal to $v(f_0(\lambda), 0)$ (also by Parseval).

2.5.2 Formulae for Section 2.3.2

From (2.13), (2.14),

$$\psi(\lambda) = \sup\left\{v(f, 0) : \|f\|_2 < \infty, \|\dot{f}\|_2 < \infty\right\} = \infty \tag{2.56}$$

(unless $h = 0$). Therefore, to calculate the infimum in (2.33) it is only necessary to consider $r > 0$. From (2.55),

$$\psi(r\lambda_\mu) = \frac{1}{2r}\left(\frac{1}{2\pi}\int_{-\infty}^{+\infty}\frac{|H(j\omega)|^2}{1 - \mu + \mu\omega^2}d\omega\right) + \frac{r}{2}\left((1 - \mu)M^2 + \mu D^2\right) \tag{2.57}$$

The infimum can be determined using elementary calculus. It is

$$\phi(\mu) = \left(\frac{1}{2\pi}\int_{-\infty}^{+\infty}\frac{|H(j\omega)|^2}{1 - \mu + \mu\omega^2}d\omega\right)^{\frac{1}{2}}\left((1 - \mu)M^2 + \mu D^2\right)^{\frac{1}{2}} \tag{2.58}$$

and is attained at

$$\bar{r} = \left(\frac{1}{2\pi}\int_{-\infty}^{+\infty}\frac{|H(j\omega)|^2}{1 - \mu + \mu\omega^2}d\omega\right)^{\frac{1}{2}}\bigg/\left((1 - \mu)M^2 + \mu D^2\right)^{\frac{1}{2}} \tag{2.59}$$

Formulae (2.35), (2.36) are obtained from the above formulae by rewriting the integral as described at the end of Section 2.5.1. The formula for the gradient (2.37) can be obtained by differentiation of (2.58) and rewriting integrals appropriately.

From (2.54), $f_0(r\lambda) = f_0(\lambda)/r$ for all $r > 0$ and all $\lambda \in \mathbb{R}^2$, $\lambda > 0$. Therefore, (under the same conditions on r and λ) $\rho(r\lambda)f_0(r\lambda) = \rho(\lambda)f_0(\lambda)$ and $\gamma(r\lambda) = \gamma(\lambda)$ where $\rho(\lambda)$, $\gamma(\lambda)$ are defined by (2.29), (2.28) respectively. Thus, the lower bound of equation (2.41) simplifies to that of equation (2.42).

2.5.3 Construction of Approximately Worst Inputs

An approximately worst input, as described at the end of Section 2.3.1 and again at the end of 2.3.2, is proportional to the input $f_0(\lambda)$ (when the value of λ is properly chosen).

The Fourier transform $F_0(\lambda)$ of the input $f_0(\lambda)$ is given by (2.54) and can be inverted numerically as follows. The input $f_0(\lambda)$, reversed in time, has Fourier transform equal to the complex conjugate $F_0^*(\lambda)$ of $F_0(\lambda)$ given by

$$F_0^*(\lambda, j\omega) = \frac{H(j\omega)}{\lambda_1 + \lambda_2 \omega^2} \tag{2.60}$$

This can be considered to be the product of the terms $H(j\omega)$ and $1/(\lambda_1 + \lambda_2 \omega^2)$. The inverse Fourier transform of the second term is equal to the pulse p defined by

$$p(t) = \frac{1}{2\sqrt{\lambda_1 \lambda_2}} \exp\left(-\sqrt{\lambda_1/\lambda_2}\, |t|\right) \quad \in \mathbb{R} \tag{2.61}$$

From (2.60), by the convolution theorem, the input $f_0(\lambda)$ (reversed in time) is equal to the output of the system with impulse response h and input p. Since the system is assumed to be rational, the response can be determined by simulating the response of a state-space realisation of the system. The MATLAB Control System Toolbox (The MathWorks, 1998) contains a function lsim that simulates linear time-invariant systems.

The Fourier transform of the response to the input $f_0(\lambda)$ is real. Therefore the response to an approximately worst input is symmetrical about time $t = 0$.

2.5.4 State-space Methods

Let the transfer function G be defined by

$$G(s) = \frac{H(s)}{\left(\sqrt{\lambda_1} + \sqrt{\lambda_2} s\right)^2} \tag{2.62}$$

Then,

$$\|f_0(\lambda)\|_2^2 = \frac{1}{2\pi} \int_{-\infty}^{+\infty} |F_0(\lambda, j\omega)|^2 d\omega = \frac{1}{2\pi} \int_{-\infty}^{+\infty} |G(j\omega)|^2 d\omega = \|g\|_2^2 \tag{2.63}$$

and

$$\|\dot{f}_0(\lambda)\|_2^2 = \frac{1}{2\pi} \int_{-\infty}^{+\infty} \omega^2 |F_0(\lambda, j\omega)|^2 d\omega = \frac{1}{2\pi} \int_{-\infty}^{+\infty} \omega^2 |G(j\omega)|^2 d\omega = \|\dot{g}\|_2^2 \tag{2.64}$$

where g denotes the impulse response of the system with transfer function G.

Let A, b, c, d ($d = 0$) denote the matrices of a state-space realisation of the transfer function $G(s)$. The realisation exists because $H(s)$ is assumed rational. The matrix b is a column vector and c is a row vector. The scalar d is zero because $G(s)$ is strictly proper. The integral on the right in (2.63) is given by (Doyle et al, 1992)

$$\|g\|_2^2 = cWc' \tag{2.65}$$

where $'$ denotes transpose and W is the controllability grammian and satisfies the Lyapunov equation

$$AW + WA + bb' = 0 \tag{2.66}$$

If the system $(A, b, c, 0)$ realises $G(s)$ then, the system $(A, b, cA, 0)$ realises the transfer function $sG(s)$, so the integral on the right in (2.64) is given by

$$\|\dot{g}\|_2^2 = (cA)W(cA)' \tag{2.67}$$

The Lyapunov equation (2.66) can be solved numerically. For example, the MATLAB Control System Toolbox (The MathWorks, 1998) contains a function that computes grammians. Given the grammian W, the quantities $\|f_0(\lambda)\|_2^2$, $\|\dot{f}_0(\lambda)\|_2^2$ can be obtained from (2.63), (2.64) using (2.65), (2.67).

2.5.5 Wind Turbine and Controller System Matrices

The state-space system matrices (for the system comprising wind turbine and controller) that were used in the example of Section 2.4 are as follows.

```
A=
[            0  1.0000e+000              0              0              0              0
   -7.8083e+000 -3.1912e+000              0  -5.4902e-002   5.8403e-001  -7.7338e-001
              0             0              0   1.0000e+000              0              0
   -2.7533e+001 -1.0942e+001  -4.4762e+000  -2.2857e-001   2.5531e+000  -3.3808e+000
    2.0159e+001  7.8540e+000              0              0  -3.4843e+000              0
              0             0              0              0   2.5000e-001             0]


B=
[           0
   4.7059e-002
             0
   1.4286e-001
             0
             0]


C=
[ 2.0159e+001  7.8540e+000              0              0              0              0
   1.0000e+000             0              0              0              0              0
             0             0   1.0000e+000              0              0              0
  -3.2122e+000 -1.2515e+000              0              0   2.9786e-001  -3.9443e-001
   1.5777e+001  3.1210e+000              0   6.8711e-002  -1.8674e+000   9.6790e-001]
```

```
D=
[          0
           0
           0
           0
 -5.8895e-002]
```

whereby, the units of the system input are m s^{-1} and the outputs have units MW, rad, m, rad, rad s^{-1}.

The above closed loop system matrices were obtained by combining a model of the wind turbine with a state-space realisation of a PID controller using parameters' values that are to be found in Lane (1995). The wind turbine model came originally from Mattson (1984).

2.6 Conclusion

This chapter provides a method to compute the peak output of a scalar proper rational transfer function in the case that the possible set is of the form (2.2). The problem is formulated as an infinite-dimensional convex optimisation problem. This problem is solved by constructing a two-dimensional dual problem by a standard Lagrangian construction. The two-dimensional dual problem may be solved to a pre-specified accuracy by standard methods for finite-dimensional convex optimisation problems, or it may be reduced to a quasi-convex problem in only one dimesion. A special stopping criterion enables the latter problem to be solved to pre-specified accuracy by a bisection algorithm.

A wind-turbine/wind-gust system-environment couple illustrates the use of the method to test the matching condition $\mathcal{P} \subseteq \mathcal{T}$ in the case that the tolerable set $\mathcal{T}$ is defined by criteria of the form (2.44). Examples of approximately worst inputs are shown. These are possible inputs that produce approximately the peak output. Examination of an approximately worst input may indicate how the system or the environment (or both) have to be changed in order to achieve or improve a match.

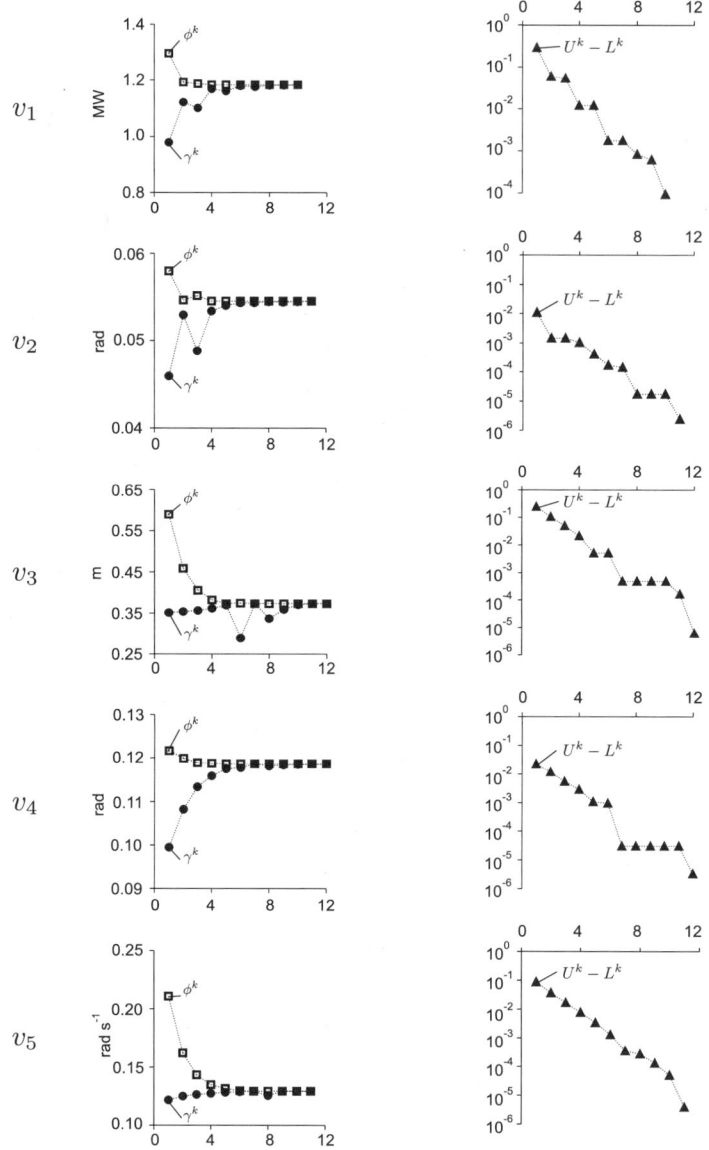

Fig. 2.3. Evaluation of peak output to a tolerance of 0.01% for the outputs $v_1, v_2, \ldots, v_5$ of the wind turbine system by the algorithm of Section 2.3.2. The graphs on the left show the upper bound $\phi(\mu^k)$ and the lower bound $\gamma(\lambda_{\mu^k})$ plotted against iteration number k. The upper and lower bounds converge in the long run although it is evident that the bounds do not always improve with increasing iteration number. This is why the best bounds U^k, L^k of equations (2.40), (2.42) are used for the stopping criterion. The stopping criterion was met after between ten and twelve iterations in each case. The difference $U^k - L^k$ is plotted on a logarithmic scale against iteration number in the graphs on the right

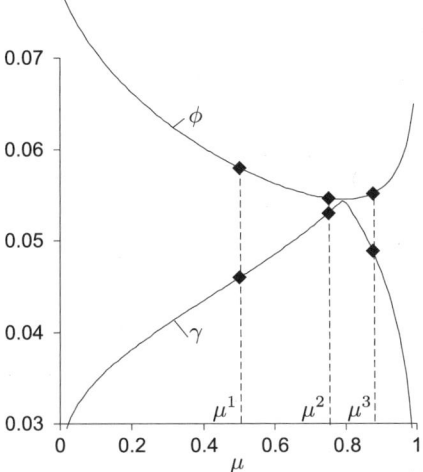

Fig. 2.4. This figure explains why the bounds $\phi(\mu^k)$, $\gamma(\lambda_{\mu^k})$ do not always improve with increasing iteration number k. The evaluation of the peak output for output v_2 of the wind turbine system is taken as an example. Both bounds improve at iteration $k = 2$, but are worse at iteration $k = 3$ - see Figure 2.3, second graph from the top, on the left. The bounds $\phi(\mu)$ and $\gamma(\lambda_\mu)$ for this problem are plotted against μ above. Also shown are the first three values μ^1, μ^2, μ^3 of μ visited by the algorithm of Section 2.3.2 during the evaluation of $\hat{v}_2$. The bounds calculated at μ^2 improve on those calculated at μ^1 because μ^2 lies on the same side, and closer to, the minimiser of ϕ (maximiser of γ). But when the algorithm jumps from one side of the minimiser to the other, as happens at μ^3, the bounds can (as in this case) worsen. Similar considerations explain why the difference $U^k - L^k$ between the best upper and lower bounds sometimes does not improve over a number of consecutive iterations. This is especially noticeable in the case of the evaluation of $\hat{v}_4$ - Figure 2.3, second graph from the bottom, on the right. But in all cases the bounds improve in the long run as the algorithm visits points closer and closer to the minimiser

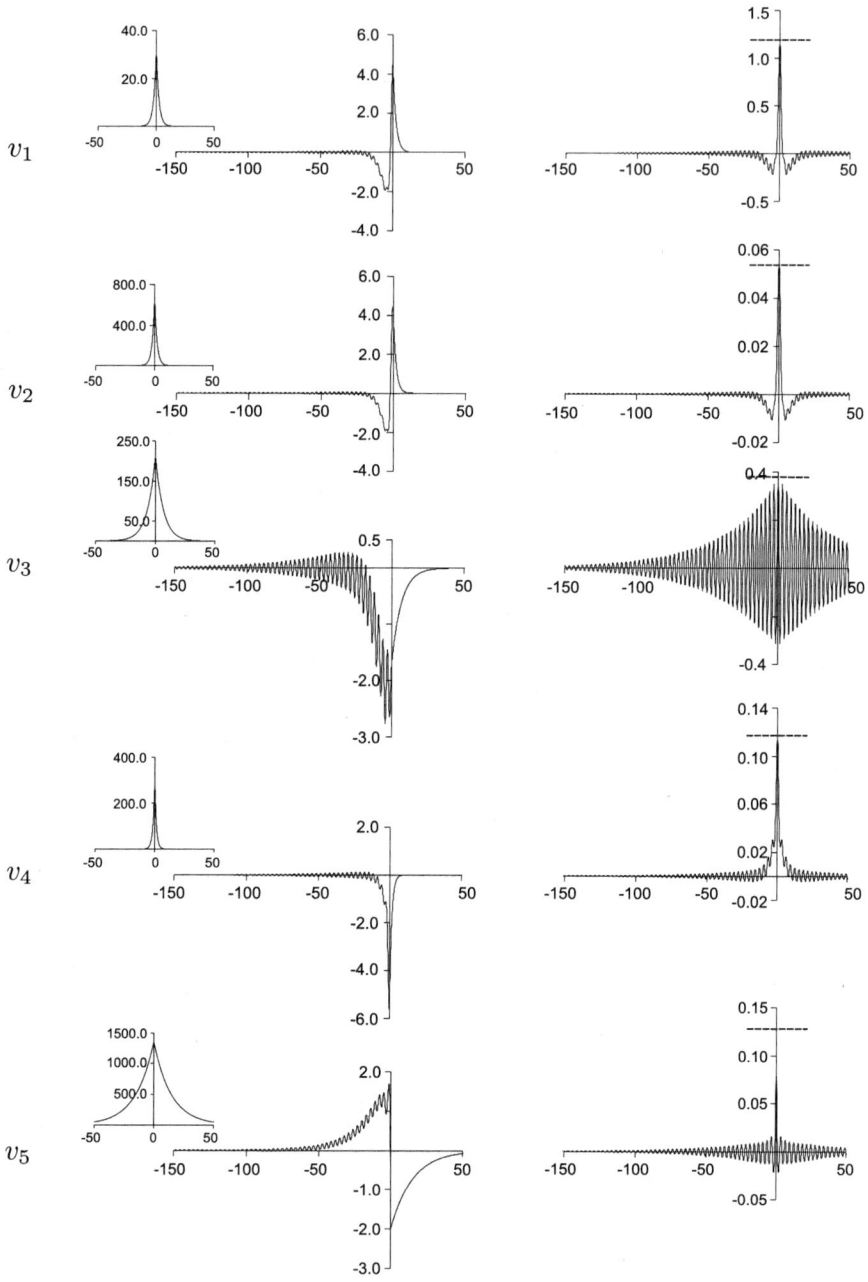

Fig. 2.5. Approximately worst inputs and the corresponding responses of the outputs $v_1, v_2, \ldots, v_5$ of the wind turbine system plotted against time in seconds. The inputs (in m s^{-1}) and the impulse-like functions used to generate them are shown on the left. The responses are shown on the right. The units of the responses are respectively MW, rad, m, rad, rad s^{-1}. The dashed horizontal lines mark the computed values of the peak output

References

Boyd, S.P. and Barratt, C.H. (1991). *Linear Controller Design: Limits of Performance*, Prentice-Hall, Englewood Cliffs, NJ.

Doyle, J.C., Francis, B.A.and Tannenbaum, A.R. (1992). *Feedback Control Theory*, Macmillan, New York, NY.

Lane, P.G. (1995). The principle of matching: a necessary and sufficient condition for inputs restricted in magnitude and rate of change, *International Journal of Control*, 62(4):893–915.

Mattson, S.E. (1984). *Modelling and control of large horizontal axis wind power plant*, PhD thesis, Lund University.

Rockafellar, R.T. (1970). *Convex Analysis*, Princeton University Press, Princeton, NJ.

The MathWorks, Inc (1998). *Control System Toolbox User's Guide Version 4*, The MathWorks, Natick, MA.

Zakian, V. (1989). Critical systems and tolerable inputs, *International Journal of Control*, 49(4):1285-1289.

3 Matching to Environment Generating Persistent Disturbances

Toshiyuki Satoh

Abstract. This chapter considers the design of matched systems that are subject to persistent disturbances limited in both magnitude and rate of change. A computational method, that uses convex optimisation, for computing the approximate supremum absolute value of each output is presented. Together with the Method of Inequalities, the method is utilised in a numerical example to illustrate how a match can be achieved.

3.1 Introduction

This chapter considers the principle of matching (Zakian, 1991; Zakian, 1996) and focuses on the design of systems that are required to be matched to an environment that subjects the system to persistent disturbances. Such persistent input continues or repeats indefinitely. It is a form of disturbance that appears frequently in design problems. For example, persistent random noise or sinusoidal input are persistent input.

The principle of matching requires the designer to model the environment by a set, called the *possible set*, that comprises all inputs that can happen or are likely to happen. The possible set adopted here comprises all persistent functions characterised by separate uniform bounds on magnitude and rate of change. It has been shown that both these restrictions are necessary in general (Zakian, 1996).

In order to achieve a match between the control system and its environment, it is necessary to compute the supremum of the absolute response, called the peak output, at each output port with respect to the possible set of inputs generated by the environment. Unfortunately, it seems that an analytical expression for the peak output cannot be obtained for the possible set considered in this chapter. Hence, a numerical method for the computation of the peak output is developed, and the method is illustrated by means of a design example.

The problem of determining the peak output is not new and has been discussed by a number of researchers.

Birch and Jackson (1959) seem to have been the first to tackle this problem. Their approach is to construct an input that gives rise to the greatest value of the output at some specified time. Such an input is called a *maximal input*. Birch and Jackson gave a set of conditions that a maximal input must satisfy. However, the procedure for constructing maximal inputs is complicated and unclear.

Chang (1962) obtained necessary conditions for an input to be maximal in terms of the concepts of the time-optimal control of linear systems.

Horowitz (1963) wrote down a set of rules for the construction of maximal inputs.

Bongiorno (1967) proved that the results obtained by Chang (1962) are also sufficient conditions for an input to be maximal and derived conditions under which the maximal input is unique. However, Bongiorno did not give a procedure for constructing maximal inputs.

Recently, Reinelt (2000) presented an algorithm that constructs the maximal input and calculates the maximum output magnitude on the basis of linear programming techniques. The algorithm is utilised for a design of optimal controllers using Youla parametrisation and nonlinear optimisation.

It may be noted here that the above researches were done outside the framework of this book and did not aim at designing matched systems.

In contrast, Lane (1992), who worked within the framework of this book, gave a general solution to the problem of finding maximal inputs. The solution is obtained in terms of a *switching function*. This is a function of time and is obtained from the step response according to simple rules. Maximal inputs can be obtained from the graph of the switching function. Lane's switching function method always succeeds in constructing a maximal input.

In an effort to derive a simple expression, Zakian (1979) obtained an upper bound for the peak output. The upper bound is easier to compute than the exact value. This gives a simple sufficient condition for a match, which is useful when there exists a solution of the design problem that satisfies the condition.

The purpose of this chapter is to give a new method for computing the peak output via convex optimisation technique. The advantage of this method over previous ones is that it is easy to implement as a computer program since, unlike the maximal input construction methods, it is not a rule-based approach, and well-developed algorithms for convex optimisation problem can be utilised.

The organisation of this chapter is as follows. Section 3.3.1 explains that the problem of finding the supremum absolute value of each response can be cast as a finite-dimensional convex optimisation problem. Section 3.3.2 shows subgradients of both objective and constraint functions, which are necessary to solve the problem by using a special algorithm for convex optimisation. Section 3.4 explains the numerical algorithm for the computation of the peak output. A design example of a matched system using the computational method for the peak output is given in Section 3.5. Conclusions are stated in Section 3.6.

3.2 Preliminaries

3.2.1 Notations and Definitions

Let $\mathbb{R}$ and $\mathbb{R}_+$ be the set of all real numbers and the set of all nonnegative real numbers, respectively. For a piecewise continuous signal $x : \mathbb{R}_+ \to \mathbb{R}$, the ∞-*norm* of x is defined as

$$\|x\|_\infty := \sup\{|x(t)| : t \geqslant 0\} \tag{3.1}$$

and the 1-*norm* of x is defined as

$$\|x\|_1 := \int_0^\infty |x(t)|\,\mathrm{d}t \tag{3.2}$$

3.2.2 Principle of Matching

The principle of matching (PoM) is one of the three basic concepts of the new foundation of control systems design (Zakian, 1991; Zakian, 1996).

Let w and $z_i(w)$ $(i = 1, 2, \ldots, m)$ denote, respectively, any input generated by the environment and the corresponding output caused by the input. Let $\mathcal{P}$ and $\mathcal{T}$ denote the possible set and the tolerable set, respectively.

For the present purpose, the tolerable set $\mathcal{T}$ is defined as

$$\mathcal{T} := \{w \in \mathcal{F} : \|z_i(w)\|_\infty \leqslant \varepsilon_i,\ i = 1, 2, \ldots, m\} \tag{3.3}$$

where $\mathcal{F}$ denotes the linear space of continuous functions $f(t)$ on the real interval $(-\infty, \infty)$ that map the real line into itself. Furthermore, let $\hat{z}_i(\mathcal{P})$ denote the peak output corresponding to $w \in \mathcal{P}$ defined by

$$\hat{z}_i(\mathcal{P}) := \sup\{\|z_i(w)\|_\infty : w \in \mathcal{P}\} \tag{3.4}$$

A necessary and sufficient condition for matching is given as follows: (Zakian, 1996)

$$\mathcal{P} \subseteq \mathcal{T} \iff \hat{z}_i(\mathcal{P}) \leqslant \varepsilon_i \text{ for all } i \in \{1, 2, \ldots, m\} \tag{3.5}$$

The possible set considered in this chapter is defined as

$$\mathcal{F}_\infty(M, D) := \left\{ w \in \mathcal{F} : \begin{array}{l} \|w\|_\infty \leqslant M, \\ \|\dot{w}\|_\infty \leqslant D, \\ \dot{w} \in C(-\infty, \infty) \end{array} \right\} \tag{3.6}$$

where $\dot{w}$ is the first derivative of w with respect to time, $M \in \mathbb{R}_+$ and $D \in \mathbb{R}_+$ represent the bounds on the magnitude and rate of change of the input respectively, and $C(-\infty, \infty)$ is the set of all real-valued continuous functions on the real interval $(-\infty, \infty)$. $\dot{w}$ is assumed to be continuous for the reason of a mathematical technique used in Section 3.3. The possible set

$\mathcal{F}_\infty(M, D)$ is defined by separate restrictions on the magnitude of the input (i.e., $\|w\|_\infty \leqslant M$) and the rate of change of the input (i.e., $\|\dot{w}\|_\infty \leqslant D$). Therefore, $\mathcal{F}_\infty(M, D)$ comprises transient inputs as well as persistent inputs. However, the set $\mathcal{F}_\infty(M, D)$ is especially suitable for characterising persistent inputs.

Unlike the input set characterised by a single bound on the magnitude or rate of change of inputs, for which a simple expression for the peak output can be derived, it seems that there is no simple formula for the peak output (3.4) when the possible set $\mathcal{P}$ is equal to $\mathcal{F}_\infty(M, D)$. Hence, some kind of numerical method is requried to compute the peak output $\hat{z}_i(\mathcal{F}_\infty(M, D))$.

3.3 Computation of Peak Output via Convex Optimisation

3.3.1 Approximation to a Finite-dimensional Problem

It can be seen from (3.5) and (3.6) that it is necessary to evaluate the peak output defined by

$$\hat{z}_i(\mathcal{F}_\infty(M, D)) := \sup \{\|z_i(w)\|_\infty : w \in \mathcal{F}_\infty(M, D)\} \tag{3.7}$$

Assume that every input-output relation of the control system is characterised by the following convolution integral of the form

$$\begin{aligned} z_i(t, w) &= \int_{-\infty}^{t} z_i(t - \tau, \delta) w(\tau) \mathrm{d}\tau \\ &= \int_{0}^{\infty} z_i(\tau, \delta) w(t - \tau) \mathrm{d}\tau, \quad t > 0, \quad i = 1, 2, \ldots, n_z \end{aligned} \tag{3.8}$$

Evaluation of the peak output defined in (3.7) generally requires a supremum operation both over all time t and over inputs that belong to $\mathcal{F}_\infty(M, D)$. However, if the input w belongs to the possible set $\mathcal{F}_\infty(M, D)$ then the time-shifted input w_d defined by $w_d(t) := w(t - d)$ ($d \in \mathbb{R}$) also belongs to $\mathcal{F}_\infty(M, D)$, and the response $z_i(t, w_d)$ is equal to the time-shifted response $z_i(t - d, w)$ since the input-output relation given in (3.8) is time-invariant. Hence, considering one particular time t is adequate to determine the peak output $\hat{z}_i(\mathcal{F}_\infty(M, D))$. Here, $t = 0$ is taken for the sake of convenience. In addition, since the input-output relation is linear and $-w$ also belongs to $\mathcal{F}_\infty(M, D)$ when w is a member of $\mathcal{F}_\infty(M, D)$, the peak output over inputs in $\mathcal{F}_\infty(M, D)$ is equal to the supremum response over inputs in $\mathcal{F}_\infty(M, D)$;

3 Matching to Environment Generating Persistent Disturbances

that is,

$$\left| \int_0^\infty z_i(\tau,\delta) w_{\text{opt}}(-\tau) d\tau \right|$$
$$= \begin{cases} \int_0^\infty z_i(\tau,\delta) w_{\text{opt}}(-\tau) d\tau, & \text{if } \int_0^\infty z_i(\tau,\delta) w_{\text{opt}}(-\tau) d\tau \geq 0 \\ \int_0^\infty z_i(\tau,\delta) \left(-w_{\text{opt}}(-\tau)\right) d\tau, & \text{otherwise} \end{cases}$$
(3.9)

where $w_{\text{opt}} \in \mathcal{F}_\infty(M,D)$ denotes an input that causes the peak output. The negative input $-w_{\text{opt}}$ also belongs to $\mathcal{F}_\infty(M,D)$, so that the peak output can be determined only by the supremum operation on the integration (3.8) over $w \in \mathcal{F}_\infty(M,D)$. Therefore, the problem of finding the peak output $\hat{z}_i(\mathcal{F}_\infty(M,D))$ $(i = 1, 2, \ldots, m)$ is equivalent to the following infinite-dimensional convex optimisation problem:

$$\begin{aligned} \text{infimum} \quad & -\int_0^\infty z_i(\tau,\delta) f(\tau) d\tau \\ \text{subject to} \quad & \|f\|_\infty - M \leqslant 0 \\ & \|\dot{f}\|_\infty - D \leqslant 0 \end{aligned}$$
(3.10)

where $f \in C[0,\infty)$ and $\dot{f} \in C[0,\infty)$, and $C[0,\infty)$ is the set of all real-valued continuous functions on the real interval $[0,\infty)$. Notice that the peak output $\hat{z}_i(\mathcal{F}_\infty(M,D))$ is obtained by multiplying the optimal objective value of (3.10) by -1. It is noted here that Boyd and Barratt (1991) gave the same formulation (3.10) and made a comment that (3.10) can be solved via convex optimisation. However, they did not show a detailed procedure for the computation of $\hat{z}_i(\mathcal{F}_\infty(M,D))$.

Since it is difficult to solve directly the above infinite-dimensional convex optimisation problem (3.10), a finite-dimensional approximation of f is used and (3.10) is solved as a finite-dimensional convex optimisation problem. That is, $f(t)$ in (3.10) is replaced with $f_N(t,x)$ defined as follows:

$$f_N(t,x) := R_0(t) + \sum_{i=1}^N x_i R_i(t)$$
(3.11)

where $x := [x_1, x_2, \ldots, x_N]^T \in \mathbb{R}^N$, and each $R_i \in C[0,\infty)$ $(i = 0, 1, \ldots, N)$ is a given continuous function. (3.11) is also refered to as a *Ritz approximation* (see, for example, Boyd and Barratt, 1991). The original problem (3.10) is then reduced to the following finite-dimensional convex optimisation problem:

$$\text{infimum} \quad -\int_0^\infty z_i(\tau,\delta)f_N(\tau,x)d\tau$$
$$\text{subject to} \quad \|f_N(x)\|_\infty - M \leqslant 0 \qquad (3.12)$$
$$\left\|\dot{f}_N(x)\right\|_\infty - D \leqslant 0$$

where $x \in \mathbb{R}^N$ is the vector of variables to be optimised.

In (3.11), each continuous function $R_i \in C[0,\infty)$ ($i=0,1,\ldots,N$) can be freely chosen. However, $f_N(t,x)$ should be a good approximation of $f(t)$ when N is large enough. Therefore, it is preferable that each $R_i \in C[0,\infty)$ ($i=0,1,\ldots,N$) should be taken such that a sequence of continuous functions $R_1, R_2, \ldots, R_N$ constitutes a basis which spans the infinite-dimensional vector space to which f belongs as $N \to \infty$. Hence, without loss of generality, it is assumed that f is a member of $C_P[0,2L]$ ($0 < L < \infty$) where $C_P[0,T]$ is the set of all real-valued periodic functions on the closed real interval $[0,T]$ defined by

$$C_P[0,T] := \{f \in C[0,T] : f(0) = f(T)\} \qquad (3.13)$$

and $f_N(t,x)$ is defined as the following truncated Fourier series expansion:

$$f_N(t,x) := \frac{1}{2}x_1 + \sum_{k=1}^{(N-1)/2}\left(x_{2k}\cos\frac{k\pi t}{L} + x_{2k+1}\sin\frac{k\pi t}{L}\right), \quad N = 3, 5, \ldots \qquad (3.14)$$

The reason for the above-mentioned assumption on f and on the definition of $f_N(t,x)$ is explained in what follows. When $f_N(t,x)$ is defined as (3.14), the optimised variables $x_1, x_2, \ldots, x_N$ are the so-called Fourier coefficients. Hence, the convex optimisation problem (3.12) results in the problem of finding appropriate Fourier coefficients.

Let $f_N(t,x^\star)$ be the optimal solution of the finite-dimensional convex optimisation problem defined in (3.12). In general, there is no guarantee that $f_N(t,x^\star)$ is a continuous periodic function on a closed real interval. If $f_N(t,x^\star)$ is not a continuous periodic function, it does not make sense to restrict $f_N(t,x)$ to the Fourier series expansion in (3.14). However, it is actually possible to regard $f_N(t,x)$ as a continuous periodic function on a closed real interval. The validity of this restriction on $f_N(t,x)$ is supported by the undermentioned observation.

Suppose that the internal stability of the closed-loop system is ensured in the following. If the closed-loop system is stable then the impulse response $z_i(t,\delta) \to 0$ as $t \to \infty$. Hence, in the objective function of the finite-dimensional convex optimisation problem (3.12), $z_i(t,\delta)f_N(t,x) \to 0$ as $t \to \infty$ as long as $f_N(t,x)$ is a finite real-valued function. $f_N(t,x)$ is

3 Matching to Environment Generating Persistent Disturbances

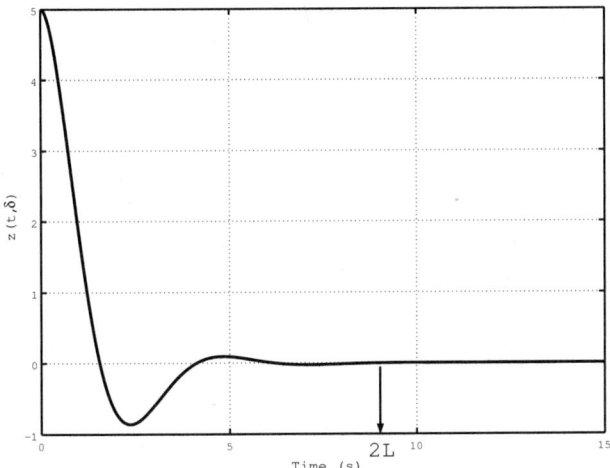

Fig. 3.1. The impulse response of a stable transfer function is approximately zero after the time $2L$

bounded in magnitude from the assumption and is hence finite. Therefore, if it is possible to estimate such time $2L > 0$ as $z_i(t, \delta) \simeq 0$ for all $t \geqslant 2L$ (see Figure 3.1), then the value of $f_N(t, x)$ for $t \geqslant 2L$ may be arbitrarily decided. In many cases, such time $2L > 0$ can be found through a simulation of the impulse response. The above discussion naturally leads us to the conclusion that $f_N(t, x)$ may be approximately regarded as a continuous periodic function on a closed real interval $[0, 2L]$.

To solve the original infinite-dimensional convex optimisation problem (3.10) via the finite-dimensional problem (3.12), it is necessary to show that the solution of (3.12) approaches that of (3.10) as $N \to \infty$. The following proposition ensures this convergence.

Proposition 3.1. *Suppose that a finite-dimensional approximation of $f \in C_P[0, 2L]$ is given in (3.14). Let $\phi_i(f) : C_P[0, 2L] \to \mathbb{R}$ ($i = 1, 2, 3$) be the functionals defined as follows:*

$$\phi_1(f) := -\int_0^\infty z_i(\tau, \delta) f(\tau) \mathrm{d}\tau \tag{3.15}$$

$$\phi_2(f) := \|f\|_\infty \tag{3.16}$$

$$\phi_3(f) := \|\dot{f}\|_\infty \tag{3.17}$$

In addition, let $\mathcal{A}$ and $\mathcal{A}_N$ denote the following achievable specifications for the infinite-dimensional convex optimisation problem (3.10) and for the finite-

dimensional convex optimisation problem (3.12), *respectively:*

$$\mathcal{A} := \left\{ a \in \mathbb{R}^3 : \begin{array}{l} \phi_i(f) \leqslant a_i, \\ \dot{f} \in C_P[0, 2L], \\ i = 1, 2, 3 \end{array} \right\} \tag{3.18}$$

$$\mathcal{A}_N := \left\{ a^{(N)} \in \mathbb{R}^3 : \begin{array}{l} \phi_i(f_N) \leqslant a_i^{(N)}, \\ \dot{f}_N(x) \in C_P[0, 2L], \\ i = 1, 2, 3 \end{array} \right\} \tag{3.19}$$

Then the set $\mathcal{A}_N$ approaches the set $\mathcal{A}$ as $N \to \infty$.

Proof. The convergence $\mathcal{A}_N \to \mathcal{A}$ ($N \to \infty$) means that, for any vector $a \in \mathcal{A}$, there exists a sequence $\{a^{(N)}\}_{N=1}^{\infty} \subset \bigcup_{N=1}^{\infty} \mathcal{A}_N$ that converges to $a \in \mathcal{A}$. In other words, every element of $\mathcal{A}$ is the limit of a sequence from $\bigcup_{N=1}^{\infty} \mathcal{A}_N$. Therefore, our task is to show that $\bigcup_{N=1}^{\infty} \mathcal{A}_N$ is *dense* in $\mathcal{A}$; that is, $\bigcup_{N=1}^{\infty} \mathcal{A}_N \subseteq \mathcal{A} \subseteq \overline{\bigcup_{N=1}^{\infty} \mathcal{A}_N}$ (see, for example, Megginson, 1991) where $\overline{\bigcup_{N=1}^{\infty} \mathcal{A}_N}$ denotes the *closure* of $\bigcup_{N=1}^{\infty} \mathcal{A}_N$.

First, $\bigcup_{N=1}^{\infty} \mathcal{A}_N \subseteq \mathcal{A}$ is established. Since the approximations in the form of (3.11) yields the *inner* approximations of $\mathcal{A}$ (Boyd and Barratt, 1991); that is, $\mathcal{A}_1 \subseteq \mathcal{A}_2 \cdots \subseteq \mathcal{A}_N \cdots \subseteq \mathcal{A}$. It follows that $\bigcup_{N=1}^{\infty} \mathcal{A}_N \subseteq \mathcal{A}$.

Next, $\mathcal{A} \subseteq \overline{\bigcup_{N=1}^{\infty} \mathcal{A}_N}$ is shown. Suppose that $a \in \mathcal{A}$. Then there exists a periodic function $f \in C_P[0, 2L]$ that satisfies $\phi_i(f) \leqslant a_i$ ($i = 1, 2, 3$). It can be shown from (3.15), (3.16) and (3.17) that the following inequalities hold:

$$\phi_1(f_N) \leqslant a_1 + \|z_i(\delta)\|_1 \|f - f_N\|_\infty \tag{3.20}$$

$$\phi_2(f_N) \leqslant a_2 + \|f - f_N\|_\infty \tag{3.21}$$

$$\phi_3(f_N) \leqslant a_3 + \left\|\dot{f} - \dot{f}_N\right\|_\infty \tag{3.22}$$

Since $f \in C_P[0, 2L]$, it follows from the *Weierstrass trigonometric polynomial approximation theorem* (see, for example, Zeidler, 1991) that, for each $\varepsilon \in \mathbb{R}_+$, there exists a trigonometric polynomial

$$f_N(t) = \frac{1}{2}x_1 + \sum_{k=1}^{(N-1)/2} \left(x_{2k} \cos \frac{k\pi t}{L} + x_{2k+1} \sin \frac{k\pi t}{L} \right) \tag{3.23}$$

such that $\|f - f_N\|_\infty < \varepsilon$ where $x_i \in \mathbb{Q}$ ($i = 1, 2, \ldots, N$), and $\mathbb{Q}$ is the set of all rational numbers.

For any $h \in \mathbb{R}_+$ and $\eta \in \mathbb{R}_+$, λ is defined as $\lambda := \eta h$. Then $\lambda \to 0$ as $\eta \to 0$. Now the trigonometric polynomial $f_N(t)$ on $[0, 2L]$ is chosen such that

$$\|f - f_N\|_\infty < \lambda \tag{3.24}$$

Since $\dot{f} \in C_P[0, 2L]$ and $\dot{f}_N \in C_P[0, 2L]$, both f and f_N are differentiable on the closed interval $[0, 2L]$. Hence, the following inequality holds:

$$\left|\dot{f}(t) - \dot{f}_N(t)\right| \leqslant \lim_{h \to 0} \frac{|f(t+h) - f_N(t+h)| + |f(t) - f_N(t)|}{h} \tag{3.25}$$

Using the inequalities (3.24) and (3.25), the next inequality is obtained:

$$\left\|\dot{f} - \dot{f}_N\right\|_\infty \leqslant \lim_{h \to 0} \frac{2\left\|f - f_N\right\|_\infty}{h}$$
$$< 2\eta \tag{3.26}$$

It can be seen from (3.24) and (3.26) that $f_N(t) \to f(t)$ and $\dot{f}_N(t) \to \dot{f}(t)$ on the closed interval $[0, 2L]$ as $\eta \to 0$. Therefore, for each $\zeta \in \mathbb{R}_+$, a trigonometric polynomial $f_N(t) \in C_P[0, 2L]$ can be chosen such that $\|f - f_N\|_\infty < \zeta$ and $\left\|\dot{f} - \dot{f}_N\right\|_\infty < \zeta$ simultaneously hold.

Hence, defining $\varepsilon^{(N)} \in \mathbb{R}_+$ as

$$\varepsilon^{(N)} := \begin{cases} \zeta \|z_i(\delta)\|_1, & \text{if } \|z_i(\delta)\|_1 \geqslant 1 \\ \zeta, & \text{otherwise} \end{cases} \tag{3.27}$$

the following inequalities from (3.20), (3.21) and (3.22) are obtained:

$$\phi_i(f_N) < a_i + \varepsilon^{(N)}, \ i = 1, 2, 3 \tag{3.28}$$

It can be seen from (3.18), (3.19) and (3.28) that $\mathcal{A} \subseteq \overline{\bigcup_{N=1}^\infty \mathcal{A}_N}$.

Consequently, the union $\bigcup_{N=1}^\infty \mathcal{A}_N$ is dense in the set $\mathcal{A}$, and it follows that $\mathcal{A}_N \to \mathcal{A}$ as $N \to \infty$. □

3.3.2 Subgradients of Objective and Constraint Functions

In order to solve the finite-dimensional convex optimisation problem (3.12), special algorithms such as *cutting-plane* or *ellipsoid* algorithm (see, for example, Boyd and Barratt, 1991) can be employed. These algorithms require the evaluation of any one subgradient in addition to the function value. Hence, in this subsection, subgradients for both the objective and the constraint functions of the finite-dimensional convex optimisation problem (3.12) are derived.

A vector $g \in \mathbb{R}^n$ is called a *subgradient* of a convex function $f : \mathbb{R}^n \to \mathbb{R}$ at a point $x \in \mathbb{R}^n$ if (see, for example, Rockafellar, 1970)

$$f(z) \geqslant f(x) + g^T(z - x) \text{ for all } z \in \mathbb{R}^n \tag{3.29}$$

Every convex function always has at least one subgradient g at every point x. Here, f does not need to be differentiable. For example, a convex function on $\mathbb{R}$ shown in Figure 3.2 is not differentiable at x_0. However, this convex function has two subgradients at x_0, which are indicated by two different tangent lines. It is readily understood from (3.29) that the gradient of a differentiable convex function is equal to the subgradient. In the case of a

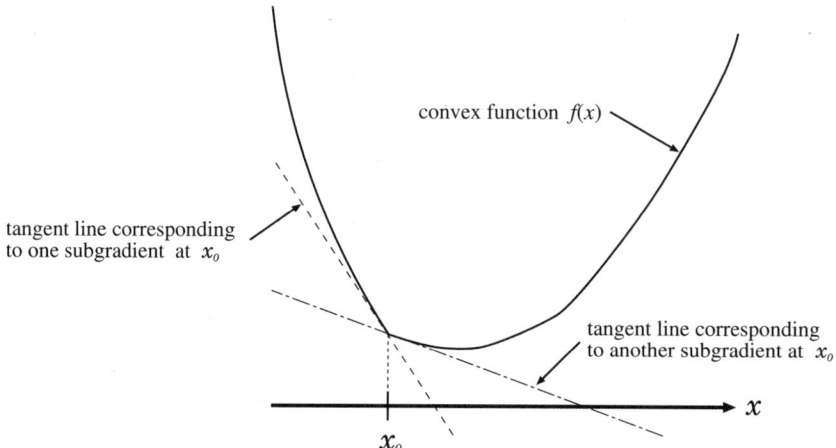

Fig. 3.2. A convex function $f(x)$ is not differentiable at x_0. However, there are two subgradients at x_0, which are corresponding to the two tangent lines at x_0

convex functional on a linear space $\mathcal{S}$, a linear functional $f^{\text{sg}} \in \mathcal{S}$ is called a subgradient for f at $x \in \mathcal{S}$ if

$$f(z) \geqslant f(x) + f^{\text{sg}}(z - x) \text{ for all } z \in \mathcal{S} \tag{3.30}$$

This definition of a subgradient is a natural extension of g in (3.29).

Let $\phi(f) : C_P[0, 2L] \to \mathbb{R}$ and $\psi_i(f) : C_P[0, 2L] \to \mathbb{R}$ $(i = 1, 2)$ be the objective functional and the constraint functionals defined as follows:

$$\phi(f) := -\int_0^\infty z_i(\tau, \delta) f(\tau) d\tau \tag{3.31}$$

$$\psi_1(f) := \|f\|_\infty - M \tag{3.32}$$

$$\psi_2(f) := \|\dot{f}\|_\infty - D \tag{3.33}$$

Then the infinite-dimensional convex optimisation problem (3.10) is expressed as

$$\begin{aligned}\text{infimum} \quad & \phi(f) \\ \text{subject to} \quad & \psi_1(f) \leqslant 0 \\ & \psi_2(f) \leqslant 0\end{aligned} \tag{3.34}$$

Let $\tilde{f}$ be any function in $C_P[0, 2L]$. Then it can be seen that the following $\phi^{\text{sg}}(f)$ and $\psi_i^{\text{sg}}(f)$ $(i = 1, 2)$ are subgradients of the objective functional $\phi(f)$ and the constraint functionals $\psi_i(f)$ at the function $\tilde{f} \in C_P[0, 2L]$, respectively:

$$\phi^{\text{sg}}(f) = -\int_0^\infty z_i(\tau,\delta) f(\tau) d\tau \tag{3.35}$$

$$\psi_1^{\text{sg}}(f) = \text{sgn}\left(\tilde{f}(t_0)\right) f(t_0) \tag{3.36}$$

$$\psi_2^{\text{sg}}(f) = \text{sgn}\left(\dot{\tilde{f}}(t_1)\right) \dot{f}(t_1) \tag{3.37}$$

where $t_0 \in \mathbb{R}_+$ and $t_1 \in \mathbb{R}_+$ denote any time such that $\left\|\tilde{f}\right\|_\infty$ and $\left\|\dot{\tilde{f}}\right\|_\infty$ are achieved respectively, and $\text{sgn}(f)$ denotes the sign function defined as

$$\text{sgn}(f) := \begin{cases} 1, & \text{if } f > 0 \\ 0, & \text{if } f = 0 \\ -1, & \text{if } f < 0 \end{cases} \tag{3.38}$$

The subgradients (3.35), (3.36) and (3.37) are linear functionals on the infinite-dimensional space of continuous periodic functions. The actual optimisation is carried out on the finite-dimensional space. Hence, it is necessary to compute subgradients of both the objective and the constraint functions on the finite-dimensional subspace of continuous periodic functions.

Now the finite-dimensional counterparts of (3.31), (3.32) and (3.33) are defined. That is, let $\zeta(x) : \mathbb{R}^N \to \mathbb{R}$ and $\xi_i(x) : \mathbb{R}^N \to \mathbb{R}$ $(i = 1, 2)$ be the objective function and the constraint functions for the finite-dimensional problem defined below:

$$\zeta(x) := -\int_0^\infty z_i(\tau,\delta) f_N(\tau,x) d\tau \tag{3.39}$$

$$\xi_1(x) := \|f_N(x)\|_\infty - M \tag{3.40}$$

$$\xi_2(x) := \left\|\dot{f}_N(x)\right\|_\infty - D \tag{3.41}$$

Then the following proposition gives subgradients on the finite-dimensional subspace.

Proposition 3.2. *Suppose that the truncated Fourier expansion of* $f \in C_P[0, 2L]$ *is given in* (3.14). *Let* $\tilde{x} \in \mathbb{R}^N$ *be any point in the N-dimensional linear space* $(N = 3, 5, \ldots)$, *and let* g, h_1 *and* h_2 *be subgradients of the objective function* $\zeta(x)$, *the constraint functions* $\xi_1(x)$ *and* $\xi_2(x)$ *that satisfy*

$$\zeta(x) \geqslant \zeta(\tilde{x}) + g^T(x - \tilde{x}) \tag{3.42}$$

$$\xi_1(x) \geqslant \xi_1(\tilde{x}) + h_1^T(x - \tilde{x}) \tag{3.43}$$

$$\xi_2(x) \geqslant \xi_2(\tilde{x}) + h_2^T(x - \tilde{x}) \tag{3.44}$$

respectively. Then g, h_1 and h_2 can be chosen as follows:

$$g = \begin{pmatrix} -\int_0^\infty z_i(\tau,\delta)/2 \, d\tau \\ -\int_0^\infty z_i(\tau,\delta) \cos \dfrac{\pi\tau}{L} d\tau \\ -\int_0^\infty z_i(\tau,\delta) \sin \dfrac{\pi\tau}{L} d\tau \\ \vdots \\ -\int_0^\infty z_i(\tau,\delta) \cos \dfrac{(N-1)\pi\tau}{2L} d\tau \\ -\int_0^\infty z_i(\tau,\delta) \sin \dfrac{(N-1)\pi\tau}{2L} d\tau \end{pmatrix}, \quad N = 3, 5, \ldots \qquad (3.45)$$

$$h_1 = \begin{pmatrix} \operatorname{sgn}(f_N(t_0,\tilde{x}))/2 \\ \operatorname{sgn}(f_N(t_0,\tilde{x})) \cos \dfrac{\pi t_0}{L} \\ \operatorname{sgn}(f_N(t_0,\tilde{x})) \sin \dfrac{\pi t_0}{L} \\ \vdots \\ \operatorname{sgn}(f_N(t_0,\tilde{x})) \cos \dfrac{(N-1)\pi t_0}{2L} \\ \operatorname{sgn}(f_N(t_0,\tilde{x})) \sin \dfrac{(N-1)\pi t_0}{2L} \end{pmatrix}, \quad N = 3, 5, \ldots \qquad (3.46)$$

$$h_2 = \begin{pmatrix} 0 \\ -\dfrac{\pi}{L} \operatorname{sgn}\left(\dot{f}_N(t_1,\tilde{x})\right) \sin \dfrac{\pi t_1}{L} \\ \dfrac{\pi}{L} \operatorname{sgn}\left(\dot{f}_N(t_1,\tilde{x})\right) \cos \dfrac{\pi t_1}{L} \\ \vdots \\ -\dfrac{(N-1)\pi}{2L} \operatorname{sgn}\left(\dot{f}_N(t_1,\tilde{x})\right) \sin \dfrac{(N-1)\pi t_1}{2L} \\ \dfrac{(N-1)\pi}{2L} \operatorname{sgn}\left(\dot{f}_N(t_1,\tilde{x})\right) \cos \dfrac{(N-1)\pi t_1}{2L} \end{pmatrix}, \quad N = 3, 5, \ldots \qquad (3.47)$$

where $t_0 \in [0, 2L]$ and $t_1 \in [0, 2L]$ denote any time such that $\|f_N(\tilde{x})\|_\infty$ and $\|\dot{f}_N(\tilde{x})\|_\infty$ are achieved, respectively.

3 Matching to Environment Generating Persistent Disturbances   133

Proof. Our first goal is to show that g given in (3.45) is a subgradient for the objective function $\zeta(x)$. g is derived by using the subgradient $\phi^{\text{sg}}(f)$ of the objective functional given in (3.35). Since (3.35) is a linear functional of f, by substituting

$$f_N(t,x) = \frac{1}{2}x_1 + \sum_{k=1}^{(N-1)/2}\left(x_{2k}\cos\frac{k\pi t}{L} + x_{2k+1}\sin\frac{k\pi t}{L}\right) \quad (3.48)$$

for $f(t)$ in (3.35),

$$\phi^{\text{sg}}(f_N(t,x))$$
$$= \phi^{\text{sg}}(1/2)\,x_1 + \phi^{\text{sg}}\left(\cos\frac{\pi t}{L}\right)x_2 + \phi^{\text{sg}}\left(\sin\frac{\pi t}{L}\right)x_3$$
$$+\cdots+\phi^{\text{sg}}\left(\cos\frac{(N-1)\pi t}{2L}\right)x_{N-1} + \phi^{\text{sg}}\left(\sin\frac{(N-1)\pi t}{2L}\right)x_N$$

$$= \begin{pmatrix} -\int_0^\infty z_i(\tau,\delta)/2\,d\tau \\ -\int_0^\infty z_i(\tau,\delta)\cos\frac{\pi\tau}{L}d\tau \\ -\int_0^\infty z_i(\tau,\delta)\sin\frac{\pi\tau}{L}d\tau \\ \vdots \\ -\int_0^\infty z_i(\tau,\delta)\cos\frac{(N-1)\pi\tau}{2L}d\tau \\ -\int_0^\infty z_i(\tau,\delta)\sin\frac{(N-1)\pi\tau}{2L}d\tau \end{pmatrix}^T \times \begin{pmatrix} x_1 \\ x_2 \\ x_3 \\ \vdots \\ x_{N-1} \\ x_N \end{pmatrix}$$

$$= g^T x \quad (3.49)$$

It follows from (3.49) that the vector g in (3.45) is a subgradient for the objective function $\zeta(x)$. Notice that the subgradient g is independent of the point $\tilde{x} \in \mathbb{R}^N$.

Next, h_1 in (3.46) is proved to be a subgradient for the constraint function $\xi_1(x)$. Since $\tilde{f}(t)$ in the infinite-dimensional space corresponds to $f_N(t,\tilde{x})$ in the finite-dimensional subspace, and $\psi_1^{\text{sg}}(f)$ in (3.36) is a linear functional of

f,

$$\psi_1^{\text{sg}}(f_N(t,x))$$
$$= \psi_1^{\text{sg}}(1/2)x_1 + \psi_1^{\text{sg}}\left(\cos\frac{\pi t}{L}\right)x_2 + \psi_1^{\text{sg}}\left(\sin\frac{\pi t}{L}\right)x_3$$
$$+ \cdots + \psi_1^{\text{sg}}\left(\cos\frac{(N-1)\pi t}{2L}\right)x_{N-1} + \psi_1^{\text{sg}}\left(\sin\frac{(N-1)\pi t}{2L}\right)x_N$$

$$= \begin{pmatrix} \operatorname{sgn}(f_N(t_0,\tilde{x}))/2 \\ \operatorname{sgn}(f_N(t_0,\tilde{x}))\cos\frac{\pi t_0}{L} \\ \operatorname{sgn}(f_N(t_0,\tilde{x}))\sin\frac{\pi t_0}{L} \\ \vdots \\ \operatorname{sgn}(f_N(t_0,\tilde{x}))\cos\frac{(N-1)\pi t_0}{2L} \\ \operatorname{sgn}(f_N(t_0,\tilde{x}))\sin\frac{(N-1)\pi t_0}{2L} \end{pmatrix}^T \times \begin{pmatrix} x_1 \\ x_2 \\ x_3 \\ \vdots \\ x_{N-1} \\ x_N \end{pmatrix}$$

$$= h_1^T x \tag{3.50}$$

It follows from (3.50) that h_1 in (3.46) is a subgradient for the constraint function $\xi_1(x)$ at $\tilde{x} \in \mathbb{R}^N$.

Finally, h_2 in (3.47) is proved to be a subgradient for the constraint function $\xi_2(x)$. In the same way as the proof of (3.46),

$$\psi_2^{\text{sg}}(f_N(t,x))$$
$$= \psi_2^{\text{sg}}(1/2)x_1 + \psi_2^{\text{sg}}\left(\cos\frac{\pi t}{L}\right)x_2 + \psi_2^{\text{sg}}\left(\sin\frac{\pi t}{L}\right)x_3$$
$$+ \cdots + \psi_2^{\text{sg}}\left(\cos\frac{(N-1)\pi t}{2L}\right)x_{N-1} + \psi_2^{\text{sg}}\left(\sin\frac{(N-1)\pi t}{2L}\right)x_N$$

$$= \begin{pmatrix} 0 \\ -\frac{\pi}{L}\operatorname{sgn}(\dot{f}_N(t_1,\tilde{x}))\sin\frac{\pi t_1}{L} \\ \frac{\pi}{L}\operatorname{sgn}(\dot{f}_N(t_1,\tilde{x}))\cos\frac{\pi t_1}{L} \\ \vdots \\ -\frac{(N-1)\pi}{2L}\operatorname{sgn}(\dot{f}_N(t_1,\tilde{x}))\sin\frac{(N-1)\pi t_1}{2L} \\ \frac{(N-1)\pi}{2L}\operatorname{sgn}(\dot{f}_N(t_1,\tilde{x}))\cos\frac{(N-1)\pi t_1}{2L} \end{pmatrix}^T \times \begin{pmatrix} x_1 \\ x_2 \\ x_3 \\ \vdots \\ x_{N-1} \\ x_N \end{pmatrix}$$

$$= h_2^T x \tag{3.51}$$

It can be concluded from (3.51) that h_2 in (3.47) is a subgradient for the constraint function $\xi_2(x)$ at $\tilde{x} \in \mathbb{R}^N$. □

3.4 Algorithm for Computing Peak Output

In this section, a numerical algorithm for the computation of the peak output is described.

A number of special algorithms for convex optimisation are available. The ellipsoid algorithm is one of such algorithms and is used here. The original ellipsoid algorithm was derived from a method for solving non-smooth convex programming problems by Shor (1964), and its improved version was developed by Yudin and Nemirovskii (1976). The ellipsoid algorithm came under the spotlight in a paper written by the Soviet mathematician Khachiyan (1979).

Basically, the ellipsoid algorithm generates a sequence of ellipsoids in $\mathbb{R}^n$ such that they are guaranteed to contain a minimiser. Let E_k be an ellipsoid in $\mathbb{R}^n$ centred at x_k at the kth iteration. Given a subgradient g_k of an objective function ϕ at the centre x_k, a half-space that contains the minimiser can be determined since in the half-space defined by

$$E_k \cap \left\{ z : g_k^T(z - x_k) > 0 \right\} \tag{3.52}$$

the values of ϕ always exceed the value of ϕ at x_k. Therefore, the half-space (3.52) can be precluded from a search for a minimiser and another half-space defined by

$$E_k \cap \left\{ z : g_k^T(z - x_k) \leqslant 0 \right\} \tag{3.53}$$

is used for the next iteration; see Figure 3.3. At the $(k+1)$ iteration, the ellipsoid E_{k+1} of minimum volume containing the half ellipsoid (3.53) is generated as shown in Figure 3.4 and E_{k+1} is then sliced to determine a half-space in which the minimiser lies. This procedure is repeated. A brief account of the basic ellipsoid algorithm and extension of the algorithm to constrained convex optimisation problems are given in Boyd and Barratt (1991).

The ellipsoid algorithm to solve the constrained convex optimisation problem (3.12) is given by Algorithm 3.1.

3.5 Numerical Example

In this section, a numerical example that illustrates the design of matched systems subject to persistent inputs is given. The computational method developed in this chapter for the peak output is utilised.

The plant to be controlled here is a hydraulic actuator . The model of a hydraulic actuator used is taken from Niksefat and Sepehri (2001). A schematic diagram of the hydraulic actuator is shown in Figure 3.5 where k_s and d_s are the sensor stiffness and the sensor damping, m_a and d_a are the piston mass and the piston damping, m_e, d_e and k_e are the mass, the damping and the stiffness that represent the environment in contact with

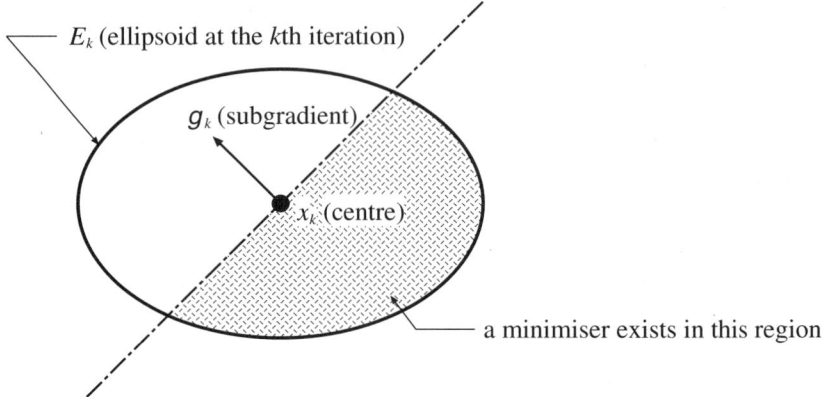

Fig. 3.3. The shaded region is a half-space including a minimiser since the objective function $\phi(x)$ is greater than or equal to $\phi(x_k)$ in the half-space $\{z : g_k^T(z-x) > 0\}$

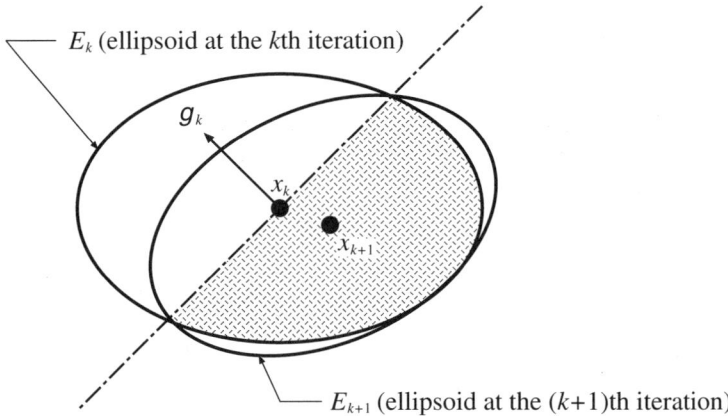

Fig. 3.4. The half-space including a minimiser at the kth iteration is enclosed by ellipsoid E_{k+1} of minimum volume at the $(k+1)$th iteration

the hydraulic actuator, A_i and A_o are the input side and output side piston effective areas, p_i and p_o are the input and output line pressures, q_i and q_o are the fluid flows into and out of the valve, p_e and p_s represent the pump pressure and the return pressure, x, x_e and x_{sp} are the actuator displacement, the environmental displacement and the spool displacement, respectively.

3 Matching to Environment Generating Persistent Disturbances

Algorithm 3.1. Ellipsoid algorithm to solve the convex optimisation problem (3.12)

```
set A > 0 (size and orientation of E) and x (centre of E);
set N > 0 (order of the Fourier series expansion);
set 2L > 0 (period of f_N(t,x); see (3.14));
set ε > 0 (accuracy of the optimisation);
k=0;
while 1
    k=k+1;
    compute the constraint functions ξ₁(x) in (3.40) and ξ₂(x) in (3.41);
    ξ(x) = max {ξ₁(x), ξ₂(x)};
    if ξ(x) > 0
        compute the subgradient h of ξ at x (if ζ₁(x) ⩾ ζ₂(x), h = h₁ in (3.46);
                                                  otherwise h = h₂ in (3.47));
```
$$\tilde{g} = \frac{h}{\sqrt{h^T A h}};$$
```
        if ξ(x) − √(hᵀAh) > 0
            break; (since the feasible set is empty)
        end
    else
        compute the objective function ζ(x) in (3.39);
        compute the subgradient g of ζ at x in (3.45);
```
$$\tilde{g} = \frac{g}{\sqrt{g^T A g}};$$
```
    end
```
$$x = x - \frac{A\tilde{g}}{N+1};$$
$$A = \frac{N^2}{N^2-1}\left(A - \frac{2}{N+1}A\tilde{g}\tilde{g}^T A\right);$$
```
    if (ξ(x) ⩽ 0) & (√(gᵀAg) ⩽ ε)
        break; (since the optimisation problem has been solved)
    end
end
```

The equations of motion are

$$m_a \ddot{x} = f_a - d_s (\dot{x} - \dot{x}_e) - d\dot{x} - k_s (x - x_e) \tag{3.54}$$

$$m_e \ddot{x}_e = d_s (\dot{x} - \dot{x}_e) - d_e \dot{x}_e + k_s (x - x_e) - k_e x_e \tag{3.55}$$

$$f = k_s (x - x_e) \tag{3.56}$$

where f is the sensed force. The force f_a generated by the hydraulic actuator is given by

$$f_a = p_i A_i - p_o A_o \tag{3.57}$$

The linearised fluid flows q_i and q_o are given as

$$q_i = K_s^i x_{sp} - K_p^i p_i \tag{3.58}$$

$$q_o = K_s^o x_{sp} - K_p^o p_o \tag{3.59}$$

where K_s^i and K_s^o are input and output side flow sensitivity gains, and K_p^i and K_p^o are input side and output side pressure sensitivity gains. Depending on the sign of the spool displacement, they are given as the following formulae:

$$K_s^i = c_d w \sqrt{\frac{2}{\rho}(p_s - p_i)} \tag{3.60}$$

$$K_s^o = c_d w \sqrt{\frac{2}{\rho}(p_o - p_e)} \tag{3.61}$$

$$K_p^i = \frac{c_d w x_{sp}}{\sqrt{2\rho(p_s - p_i)}} \tag{3.62}$$

$$K_p^o = \frac{c_d w x_{sp}}{\sqrt{2\rho(p_o - p_e)}} \tag{3.63}$$

for $x_{sp} \geq 0$ (extension) and

$$K_s^i = c_d w \sqrt{\frac{2}{\rho}(p_i - p_e)} \tag{3.64}$$

$$K_s^o = c_d w \sqrt{\frac{2}{\rho}(p_s - p_o)} \tag{3.65}$$

$$K_p^i = \frac{-c_d w x_{sp}}{\sqrt{2\rho(p_i - p_e)}} \tag{3.66}$$

$$K_p^o = \frac{-c_d w x_{sp}}{\sqrt{2\rho(p_s - p_o)}} \tag{3.67}$$

for $x_{sp} < 0$ (retraction) where c_d, w and ρ are the orifice coefficient of discharge, the area gradient that relates the spool displacement and the mass density of the fluid, respectively. On the other hand, continuity equations for oil flow through the cylinder are given as

$$q_i = A_i \frac{dx}{dt} + \frac{1}{\beta} V_i \frac{dp_i}{dt} \tag{3.68}$$

$$q_o = A_o \frac{dx}{dt} - \frac{1}{\beta} V_o \frac{dp_o}{dt} \tag{3.69}$$

where β is the effective bulk modulus of the fluid and V_i and V_o are the volumes of fluid trapped at the input and output sides of the actuator and the volume of fluid, respectively. Although V_i and V_o are generally functions of actuator displacement x, the following approximation is made for the sake of simplicity:

$$\frac{V_i(x)}{\beta} \approx \frac{V_o(x)}{\beta} \approx \frac{1}{\beta}\left(\frac{\bar{V}_i + \bar{V}_o}{2}\right) =: C \tag{3.70}$$

3 Matching to Environment Generating Persistent Disturbances   139

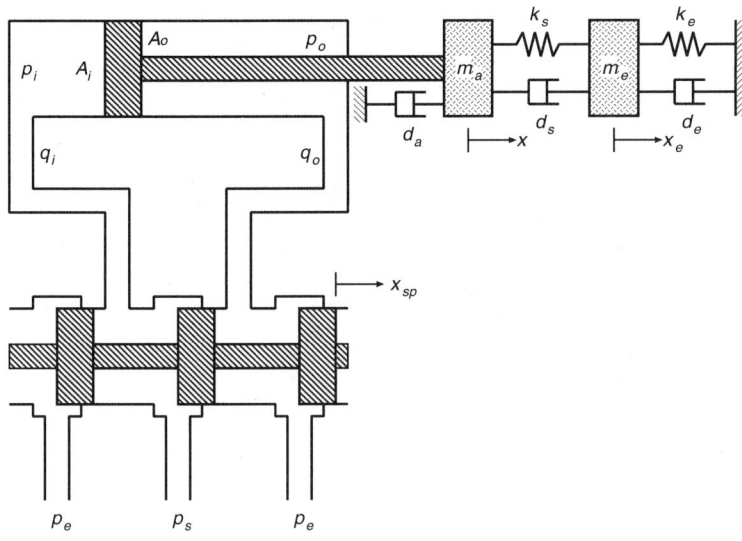

Fig. 3.5. Schematic diagram of a hydraulic actuator (©2001 IEEE)

where $\bar{V}_i$ and $\bar{V}_o$ are the initial volumes trapped in the input and output sides of the actuator. Laplace transforming (3.68) and (3.69) with the use of (3.58), (3.59) and (3.70), and rearranging for the line pressures, the following relationships are obtained:

$$P_i = -\frac{A_i s}{Cs + K_p^i} X + \frac{K_s^i}{Cs + K_p^i} X_{sp} \tag{3.71}$$

$$P_o = -\frac{A_o s}{Cs + K_p^o} X - \frac{K_s^o}{Cs + K_p^o} X_{sp} \tag{3.72}$$

Substituting (3.71) and (3.72) for (3.57), the relation among $F_a(s)$, $X_{sp}(s)$ and $X(s)$ is derived as

$$F_a(s) = \left(\frac{K_s^o A_o}{Cs + K_p^o} + \frac{K_s^i A_i}{Cs + K_p^i} \right) X_{sp}(s) \\ - \left(\frac{A_i^2 s}{Cs + K_p^i} + \frac{A_o^2 s}{Cs + K_p^o} \right) X(s) \tag{3.73}$$

From (3.54), (3.55) and (3.56), the following transfer functions are derived as follows:

$$\frac{F_a(s)}{F(s)} = \frac{(m_a s^2 + ds)(m_e s^2 + (d_e + d_s)s + k_s + k_e)}{k_s(m_e s^2 + d_e s + k_e)} \\ + \frac{(d_s s + k_s)(m_e s^2 + d_e s + k_e)}{k_s(m_e s^2 + d_e s + k_e)} \tag{3.74}$$

Table 3.1. Hydraulic actuator model parameters (©2001 IEEE)

Parameter	Description	Value
k_e	Environmental stiffness	75 kN/m
K_s	Flow sensitivity gain	0.375 m^3/Pa·s
K_p	Pressure sensitivity gain	2.5×10^{-12} m^2/s
C	Volume change of the fluid per unit pressure	1.5×10^{-11} m^3/Pa
d	Sensor damping	700 N/m·s
m_a	Actuator piston mass	20 kg
A_i	Input side piston effective area	0.00203 m^2
A_o	Output side piston effective area	0.00152 m^2
k_{sp}	Valve dynamics gain	0.0012 m/V
τ	Valve dynamics gain	35 ms

$$\frac{X(s)}{F(s)} = \frac{1}{k_s} \frac{m_e s^2 + (d_e + d_s)s + k_s + k_e}{m_e s^2 + d_e s + k_e} \tag{3.75}$$

From (3.73), (3.74) and (3.75), the transfer function from the contact force $F(s)$ to the spool displacement $X_{sp}(s)$ can be obtained. However, to simplify the transfer function, the following three assumptions are made:

- K_s^i and K_s^o are identical and denoted by K_s. Similarly, K_p^i and K_p^o are identical and denoted by K_p.
- The stiffness of the force sensor and the piston rod are high compared to the environmental stiffness and the hydraulic compliance.
- The stiffness k_e dominates the dynamics of the environment.

The resulting transfer function is then written as

$$\frac{F(s)}{X_{sp}(s)} = \frac{K_s k_e (A_i + A_o)}{(K_p + Cs)(m_a s^2 + ds + k_e) + (A_i^2 + A_o^2)s} \tag{3.76}$$

On the other hand, the input voltage u to the proportional valve is written as

$$u = \frac{\tau}{k_{sp}} \frac{dx_{sp}}{dt} + \frac{1}{k_{sp}} x_{sp} \tag{3.77}$$

where τ and k_{sp} are constant gains that describe the valve dynamics.

Hence, from (3.76) and (3.77), the transfer function from the sensed force to the control voltage is obtained as

$$\frac{F(s)}{U(s)} = \frac{k_{sp}}{\tau s + 1} \left[\frac{K_s k_e (A_i + A_o)}{(K_p + Cs)(m_a s^2 + ds + k_e) + (A_i^2 + A_o^2)s} \right] \tag{3.78}$$

Nominal values of the parameters that appear in (3.78) are summarised in Table 3.1. Applying the parameter values in Table 3.1 to (3.78), the transfer

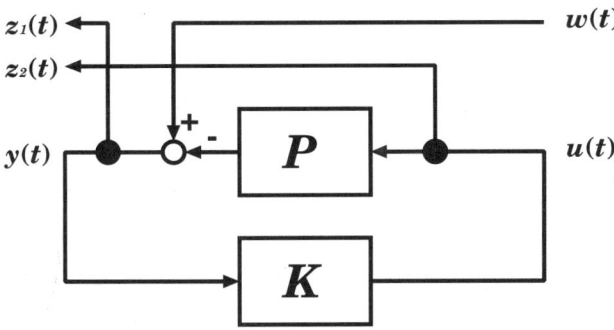

Fig. 3.6. Hydraulic force control system

function from the control voltage to the measured force is written as follows:

$$P(s) := \frac{F(s)}{U(s)} = \frac{1.1411 \times 10^{10}}{(s + 0.0248)(s + 28.57)(s^2 + 35.14s + 2.519 \times 10^4)} \quad (3.79)$$

Suppose that the input w is reference force. Additionally, suppose that the regulated outputs z_1 and z_2 are tracking error and control voltage, respectively. The block diagram for this force control system is depicted in Figure 3.6.

Define the possible input set as

$$\mathcal{F}_\infty(1000, 1000) := \left\{ w : \begin{array}{l} \|w\|_\infty \leqslant 1000 \text{ N}, \\ \|\dot{w}\|_\infty \leqslant 1000 \text{ N/s}, \\ \dot{w} \in C(-\infty, \infty) \end{array} \right\} \quad (3.80)$$

Assume that the purpose of control is to keep the tracking error and the control voltage within 80 N and 0.1 V, respectively. Then the tolerable set is defined as

$$\mathcal{T} := \left\{ w \in C(-\infty, \infty) : \begin{array}{l} \|z_1(w)\|_\infty \leqslant 80 \text{ N}, \\ \|z_2(w)\|_\infty \leqslant 0.1 \text{ V} \end{array} \right\} \quad (3.81)$$

Clearly, a match is achieved if and only if $\hat{z}_1(\mathcal{F}_\infty(1000, 1000)) \leqslant 80$ N and $\hat{z}_2(\mathcal{F}_\infty(1000, 1000)) \leqslant 0.1$ V.

In this example, a 3rd-order controller $K(s, p)$ is designed, having the form:

$$K(s, p) = \frac{p_3 s^2 + p_4 s + p_5}{s(s^2 + p_1 s + p_2)} \quad (3.82)$$

where $p = [p_1\ p_2\ p_3\ p_4\ p_5]^T \in \mathbb{R}^5$ denotes the vector of controller parameters to be determined. Notice that an integrator is included in $K(s, p)$ to remove the steady-state error.

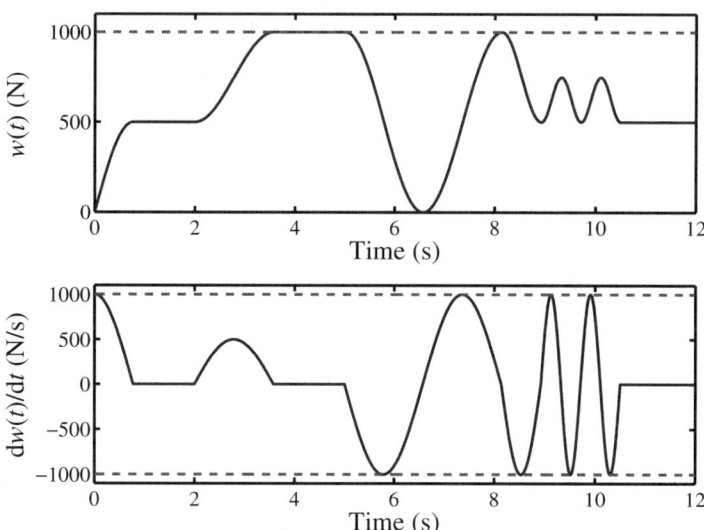

Fig. 3.7. Test reference input $w(t)$

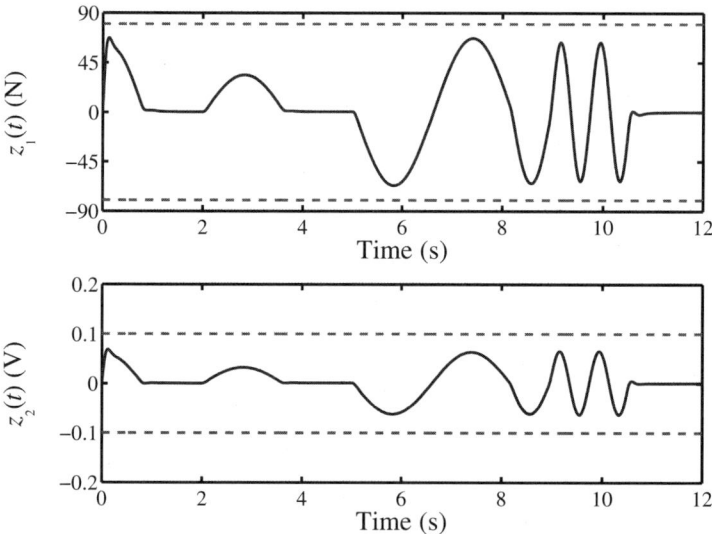

Fig. 3.8. Response $z_1(t)$ and $z_2(t)$ to the test reference input $w(t)$

To evaluate performance via the finite-dimensional convex optimisation problem stated in Section 3.3, the order N of the truncated Fourier series expansion and the time $2L$ must be specified. Here, $N = 501$ and $2L = 3$ are used, and the optimisation problem is solved with Algorithm 3.1.

3 Matching to Environment Generating Persistent Disturbances

The moving boundaries process (Zakian and Al-Naib, 1973) is employed to solve the inequalities in (3.81) and the following controller which achieves the match is obtained:

$$K(s) = \frac{0.35s^2 + 2.05s + 0.0375}{s\left(s^2 + 329s + 2206\right)} \tag{3.83}$$

which gives

$$\hat{z}_1\left(\mathcal{F}_\infty\left(1000, 1000\right)\right) = 75.6\text{N} \tag{3.84}$$
$$\hat{z}_2\left(\mathcal{F}_\infty\left(1000, 1000\right)\right) = 0.08\text{V} \tag{3.85}$$

To verify the match between the system and its environment, a test reference input $w \in \mathcal{F}_\infty\left(1000, 1000\right)$ is imposed, and the closed-loop responses are checked. The test input imposed on the closed-loop system is shown in Figure 3.7. The responses $z_1\left(w\right)$ and $z_2\left(w\right)$ to this test input are shown in Figure 3.8. It can be seen from Figure 3.8 that $\left\|z_1\left(w\right)\right\|_\infty$ and $\left\|z_2\left(w\right)\right\|_\infty$ do not exceed 80 N and 0.1 V respectively. These figures prove that the match $\mathcal{F}_\infty\left(1000, 1000\right) \subseteq \mathcal{T}$ is certainly achieved.

3.6 Conclusions

This chapter has presented a method for computing the peak output for the possible set $\mathcal{F}_\infty\left(M, D\right)$ defined by (3.6).

In Section 3.3, the computation of the peak output $\hat{z}_i\left(\mathcal{F}_\infty\left(M, D\right)\right)$ is initially formulated as an infinite-dimensional convex optimisation problem. Then this infinite-dimensional problem is reduced to a finite-dimensional convex optimisation problem. This reduction is done by considering the input over a sufficiently long period of time (the time taken for the impulse response of the system to become and remain negligibly small) and expressing it as a truncated Fourier series. Hence, the Fourier coefficients are the optimised variables in the resulting finite-dimensional convex optimisation problem.

Section 3.4 provides a numerical algorithm for solving the finite-dimensional convex optimisation problem.

Section 3.5 demonstrates the design of a controller for a hydraulic actuator. A linearised model of a hydraulic actuator is derived and a problem of designing a 3rd-order controller that ensures a match to the environment generating reference inputs is considered. The reference force to the hydraulic actuator is treated as a persistent disturbance. The tolerable set in this case is defined by two closed-loop responses; that is, the tracking error between the force generated by the hydraulic actuator and the reference, and the input voltage. Parameters in the 3rd-order controller (3.82) are searched by the moving boundaries process and a set of parameters are successfully found after a number of iterations. The closed-loop responses to a test reference input indicate that the system and the environment are matched.

References

Birch, B. J. and Jackson, R. (1959). The behaviour of linear systems with inputs satisfying certain bounding conditions, *Journal of Electronics & Control*, 6(4):366–375.

Bongiorno, Jr., J. J. (1967). On the response of linear systems to inputs with limited amplitudes and slopes, *SIAM Review*, 9(3):554–563.

Boyd, S. P. and Barratt, C. H. (1991). *Linear Controller Design: Limits of Performance*, Prentice-Hall, Englewood Cliffs, NJ.

Chang, S. S. L. (1962). Minimal time control with multiple saturation limits, *IRE International Convention Record*, 10(2):143–151.

Horowitz, I. M. (1963). *Synthesis of Feedback Systems*, Academic Press, London.

Khachiyan, L. G. (1979). A polynomial algorithm in linear programming (in Russian), *Doklady Akademii Nauk SSSR*, 244:1093–1096.

Lane, P. G. (1992). *Design of control systems with inputs and outputs satisfying certain bounding conditions*, PhD thesis, University of Manchester Institute of Science and Technology.

Megginson, R. E. (1991). *An Introduction to Banach Space Theory*, Springer-Verlag, New York, NY.

Niksefat, N. and Sepehri, N. (2001). Designing robust force control of hydraulic actuators despite system and environment uncertainties, *IEEE Control Systems Magazine*, 21(2):66–77.

Reinelt, W. (2000). Maximum output amplitude of linear systems for certain input constraints, *Proceedings of the 39th IEEE Conference on Decision and Control*, pp. 1075–1080.

Rockafellar, R. T. (1970). *Convex Analysis*, Princeton University Press, Princeton, NJ.

Shor, N. Z. (1964). *On the structure of algorithms for the numerical solution of optimal planning and design problems (in Russian)*, PhD thesis, Cybernetics Institute, Academy of Sciences of the Ukrainian SSR.

Yudin, D. B. and Nemirovskii, A. S. (1976). Informational complexity and efficient methods for the solution of convex extremal problems (in Russian), *Ekonomika i Matematicheskie Metody*, 12:357–369.

Zakian, V. (1979). New formulation for the method of inequalities, *IEE Proceedings*, 126:579–584.

Zakian, V. (1991). Well matched systems, *IMA Journal of Mathematical Control & Information*, 8:29–38.

Zakian, V. (1996). Perspectives on the principle of matching and the method of inequalities, *International Journal of Control*, 65(1):147–175.

Zakian, V. and U. Al-Naib (1973), Design of dynamical and control systems by the method of inequalities, *IEE Proceedings*, 120(11):1421–1427.

Zeidler, E. (1991). *Applied Functional Analysis: Applications to Mathematical Physics*, Springer-Verlag, New York, NY.

4 LMI-based Design

Takahiko Ono

Abstract. This chapter presents an LMI-based method for design of control systems in accordance with the principle of matching and the principle of inequalities. The inputs are assumed to be persistent and/or transient. From the exponential convergence conditions of the unit impulse and the unit step responses, matrix inequalities are derived as a sufficient condition for ensuring that the system is matched to the environment. By changing the variables, these matrix inequalities are transformed into the set of LMIs and BMIs. Since the BMI can be viewed as an LMI by fixing a single variable, the problem amounts to a convex admissibility problem with a line search over $(0, \infty)$. The advantages and drawbacks of this method are examined with an example of multiobjective critical control system design.

4.1 Introduction

The problem of designing a control system that is matched to its environment (Zakian 1991, 1996) gives rise to a problem, called the admissibility problem, of solving a set of inequalities of the form

$$\phi_i(p) \leq \varepsilon_i \quad (i = 1, 2, \ldots, n) \tag{4.1}$$

The inequalities provide a necessary and sufficient condition for a matched environment-system couple. The admissibility problem is defined as that of determining one or more solutions of the inequalities if one exists, or determining that a solution does not exist. Such a solution characterises a controller for the system. Search methods for solving inequalities (see Chapters in Part III) are usually employed in practice. In this chapter, an alternative to the use of search methods for obtaining a matched environment-system is considered. This is based on linear matrix inequalities (LMIs).

An LMI is an affine functional inequality given in the form of

$$F(x) := F_0 + \sum_{i=1}^{m} x_i F_i > 0 \tag{4.2}$$

where $x := [x_1, \ldots, x_m]^t$ is a variable and F_i is a given symmetric matrix. The important feature of an LMI is that the set of its admissible solutions forms a convex set. Taking advantage of this, an LMI admissibility problem can be reduced to a convex programming problem. From the viewpoint of the matched environment-system design, this means that if the design specifications, stated as a conjunction of inequalities (4.1), can be expressed as an LMI, the admissible solutions can be found by convex programming. This is in contrast to the approach based on search methods, in which the problem leads to a nonconvex programming problem since $\phi_i(p)$ is not a convex function and the set of admissible solutions to (4.1) is not a convex set. In general, the convex programming problem can be solved efficiently. In this sense, the LMI-based approach has the potential of realising a *computationally efficient* method for matched environment-system design.

The aim of this chapter is to develop an LMI-based approach to matched environment-system design. To do this, first the inequalities that represent the design specifications (4.1) are translated into to a set of matrix inequalities. Then a change of variables is performed to transform them into the LMIs (precisely, the set comprising the LMIs and the bilinear matrix inequalities (BMIs)). Owing to many researchers' work, LMI solvers and their user-friendly software packages are now easily available. Bearing in mind the use of these software packages, the problem is formulated with LMIs involving matrix variables, not scalar variables as shown in (4.2). However, since these LMIs provide only a sufficient condition for the inequalities (4.1), the method developed in this chapter does not overcome an inherent limitation with search methods for solving inequalities, which is that, if a solution does not exist, an exhaustive search has to be carried out to confirm this. Accordingly, the LMI approach considered here is not, at this stage of its development, an effective general method in practice. In a numerical example, an application to design of a multiobjective critical control system is presented. With this example, the advantages and drawbacks of this method are examined. Particularly, this chapter focuses on how to formulate the matched environment-system design as an LMI problem and does not mention the algorithms to solve it.

4.2 Preliminary

4.2.1 Notations and Definitions

Let $\mathbb{R}$ and $\mathbb{R}_+$ denote the set of all real numbers and the set of nonnegative real numbers, respectively. Let $\|x\|_p$ represent the p-norm of a function $x : \mathbb{R}_+ \mapsto \mathbb{R}$.

$$\|x\|_p := \begin{cases} \left(\int_0^\infty |x(t)|^p dt\right)^{1/p} & \text{for } p = 1, 2 \\ \sup\{|x(t)| : t \in \mathbb{R}_+\} & \text{for } p = \infty \end{cases}$$

For a linear time-invariant system with an input $w \in \mathcal{F}$ and an output z, let $z(t, w)$ denote the time response to the input w at time t and let $\hat{z}(\mathcal{F})$ denote the peak norm of z over $t \in \mathbb{R}_+$ and $w \in \mathcal{F}$:

$$\hat{z}(\mathcal{F}) := \sup\{|z(t,w)| : t \in \mathbb{R}_+,\ w \in \mathcal{F}\}$$

The ith row vector and the jth column vector of a matrix M are denoted by row(M,i) and col(M,j), respectively. The symbol $\Psi(X,Y)$ means the operation defined by $\Psi(X,Y) := XY + (XY)^t$.

4.2.2 LMI

Before developing the LMI approach, the basic concepts of LMI are summarised below.

- If a symmetric matrix $P \in \mathbb{R}^{m \times m}$ satisfies $x^t P x > 0$ for all nonzero $x \in \mathbb{R}^m$, P is said to be positive definite and it is expressed as $P > 0$. The positiveness of P is also equivalent to that all of the eigenvalues of P are positive. If all of the eigenvalues of P are negative, $-P$ is positive definite and it is expressed simply as $P < 0$. If $x^t P x \geq 0$ holds for all nonzero $x \in \mathbb{R}^m$, P is said to be semi-positive definite, and this is equivalent to that all of the eigenvalues of P are non-negative. The semi-positiveness of P is expressed simply as $P \geq 0$.
- The set of all solutions to the LMI in (4.2) forms a convex set: suppose that $F(x_1) > 0$ and $F(x_2) > 0$. Then, for any $a \in [0,1]$,

$$F(ax_1 + (1-a)x_2) = aF(x_1) + (1-a)F(x_2) > 0 \tag{4.3}$$

- Several LMIs can be integrated into a single LMI by stacking them as a block diagonal functional inequality:

$$F_1(x) > 0, \ldots, F_n(x) > 0 \iff \text{block diag}(F_1(x), \ldots, F_n(x)) > 0 \tag{4.4}$$

This feature is useful especially for multi-objective design.
- Two matrices X and Y are said to be congruent if there exists a nonsingular matrix T such that $Y = TXT^t$. If X and Y are congruent, $Y > 0$ if and only if $X > 0$:

$$X > 0 \iff Y = TXT^t > 0 \tag{4.5}$$

This transformation, which converts X to Y, is referred to as congruence transformation.
- The matrix inequality involving an affine function of matrix variables can be converted into the form of (4.2). For instance, consider the inequality

$$F(X) = AX + XA^t + BB^t < 0 \tag{4.6}$$

where $A \in \mathbb{R}^{2 \times 2}$ and $B \in \mathbb{R}^{2 \times 1}$ are given matrices and $X = X^t \in \mathbb{R}^{2 \times 2}$ is a variable. The basis of the space of all symmetric matrices in $\mathbb{R}^{2 \times 2}$ is given as

$$\{E_1, E_2, E_3\} := \left\{ \begin{bmatrix} 1 & 0 \\ 0 & 0 \end{bmatrix}, \begin{bmatrix} 0 & 1 \\ 1 & 0 \end{bmatrix}, \begin{bmatrix} 0 & 0 \\ 0 & 1 \end{bmatrix} \right\} \tag{4.7}$$

Then X is parameterised by the variable $x = [x_1\ x_2\ x_3]^t \in \mathbb{R}^3$ as follows:

$$X = x_1 E_1 + x_2 E_2 + x_3 E_3 \tag{4.8}$$

Substituting (4.8) into (4.6), the inequality $F(X) < 0$ can be expressed in the form of (4.2), where $F_0 = -BB^t$ and $F_i = -AE_i - E_i A^t$. Essentially, the matrix inequality depending affinely on matrix variables is viewed as an LMI.

- Supposed that P is a real-valued symmetric matrix. Then the following three statements are equivalent:

  1. $\begin{bmatrix} P_{11} & P_{12} \\ P_{12}^t & P_{22} \end{bmatrix} > 0$

  2. $P_{11} > 0$ and $P_{22} - P_{12}^t P_{11}^{-1} P_{12} > 0$

  3. $P_{22} > 0$ and $P_{11} - P_{12} P_{22}^{-1} P_{12}^t > 0$

This is called Schur complement. Using this relationship, some nonlinear matrix inequality can be converted to an LMI. Schur complement can be derived from the following congruence transformation:

$$\begin{bmatrix} I & -P_{12} P_{22}^{-1} \\ 0 & I \end{bmatrix} \begin{bmatrix} P_{11} & P_{12} \\ P_{12}^t & P_{22} \end{bmatrix} \begin{bmatrix} I & 0 \\ -P_{22}^{-1} P_{12}^t & I \end{bmatrix} = \begin{bmatrix} P_{11} - P_{12} P_{22}^{-1} P_{12}^t & 0 \\ 0 & P_{22} \end{bmatrix}$$

- The problems arising in LMI-based design are mainly grouped into the following three types:

1. Convex admissibility problem:
    find x
    subject to $F(x) > 0$
    where $F(x)$ is a symmetric affine matrix function of $x \in \mathbb{R}^m$.

2. Convex optimisation problem:
    minimise $c^t x$
    subject to $F(x) > 0$
    where $c \in \mathbb{R}^m$ is a given vector and $F(x)$ is a symmetric affine matrix function of $x \in \mathbb{R}^m$.

3. Quasi-convex optimisation problem:
    minimise λ
    subject to $\lambda F_1(x) - F_2(x) > 0$, $F_1(x) > 0$, $F_3(x) > 0$ and $\lambda > 0$
    where $F_i(x)$ ($i = 1, 2, 3$) is a symmetric and affine function of $x \in \mathbb{R}^m$.

The problem encountered in this chapter is a convex admissibility problem. Generally, it can be handled as a convex optimisation problem:
    minimising $\tilde{c}^t \tilde{x}$
    subject to $F(\tilde{x}) + \lambda I > 0$
where $\tilde{c} = [0, \ldots, 0, 1]^t \in \mathbb{R}^{m+1}$ and $\tilde{x} := [x^t\ \lambda]^t \in \mathbb{R}^{m+1}$. In this problem, if $c^t \tilde{x}\, (= \lambda)$ is negative, x is the admissible solution to $F(x) > 0$.

4.3 Problem Formulation

Consider a linear time-invariant continuous-time system, which consists of a plant G with the state space realisation

$$\begin{bmatrix} \dot{x}_c(t) \\ z(t) \\ y(t) \end{bmatrix} = \begin{bmatrix} A_c & B_w & B_u \\ C_z & D_w & D_u \\ C_y & D_y & 0 \end{bmatrix} \begin{bmatrix} x_c(t) \\ w(t) \\ u(t) \end{bmatrix} \quad (4.9)$$

and a controller K with the state space realisation

$$\begin{bmatrix} \dot{\xi}(t) \\ u(t) \end{bmatrix} = \begin{bmatrix} A_K & B_K \\ C_K & D_K \end{bmatrix} \begin{bmatrix} \xi(t) \\ y(t) \end{bmatrix} \quad (4.10)$$

where $z(t) := [z_1(t) \ldots z_{n_z}(t)]^t$ is a vector of outputs to be controlled and $w(t) := [w_1(t) \ldots w_{n_w}(t)]^t$ is a vector of exogenous inputs. Assume that (A_c, B_u, C_y) is stabilisable and detectable and that the initial states of G and K are zero: $x(0) = 0$ and $\xi(0) = 0$. Assume in addition that the exogenous input at each channel forms a linear combination of four elements:

$$w_j(t) = w_{j1}(t) + w_{j2}(t) + w_{j3}(t) + w_{j4}(t) \quad (4.11)$$

where w_{jq} is known only to the extent that it belongs to the space $\mathcal{F}_q(M_{jq})$:

$$w_{jq} \in \mathcal{F}_q(M_{jq}) \quad (q = 1, \ldots, 4) \quad (4.12)$$

The space $\mathcal{F}_1(M_{j1})$ is the set of unknown but bounded piecewise continuous inputs with bound M_{j1} on the magnitude:

$$\mathcal{F}_1(M_{j1}) := \left\{ f : \mathbb{R}_+ \mapsto \mathbb{R} : \begin{array}{l} f \text{ is piecewise continuous,} \\ \|f\|_\infty \leq M_{j1} \end{array} \right\} \quad (4.13)$$

and $\mathcal{F}_2(M_{j2})$ is the set of all piecewise continuous transient inputs whose energy is less than and equal to M_{j2}:

$$\mathcal{F}_2(M_{j2}) := \left\{ f : \mathbb{R}_+ \mapsto \mathbb{R} : \begin{array}{l} f \text{ is piecewise continuous,} \\ \|f\|_2 \leq M_{j2} \end{array} \right\} \quad (4.14)$$

The space $\mathcal{F}_3(M_{j3})$ is the set of all piecewise smooth inputs with bound M_{j3} on the rate of change and zero initial condition:

$$\mathcal{F}_3(M_{j3}) := \left\{ f : \mathbb{R}_+ \mapsto \mathbb{R} : \begin{array}{l} f \text{ is piecewise smooth,} \\ \|\dot{f}\|_\infty \leq M_{j3} \text{ and } f(0) = 0 \end{array} \right\} \quad (4.15)$$

and $\mathcal{F}_4(M_{j4})$ is the set of all piecewise smooth inputs with bound M_{j4} on the 2-norm of the rate of change and zero initial condition:

$$\mathcal{F}_4(M_{j4}) := \left\{ f : \mathbb{R}_+ \mapsto \mathbb{R} : \begin{array}{l} f \text{ is piecewise smooth,} \\ \|\dot{f}\|_2 \leq M_{j4} \text{ and } f(0) = 0 \end{array} \right\} \quad (4.16)$$

The spaces $\mathcal{F}_1(M_{j1})$ and $\mathcal{F}_2(M_{j2})$ are often used as models of persistent and transient elements of disturbances, respectively (Zakian 1996). The spaces $\mathcal{F}_3(M_{j3})$ and $\mathcal{F}_4(M_{j4})$ can be used as models of slow-changing elements of inputs encountered in the tracking problems or slow-changing disturbance rejection problems. Throughout the chapter, the set of all inputs having the characteristics above is denoted by $\mathcal{F}(M)$, where

$$M := \begin{bmatrix} M_{11} & \cdots & M_{14} \\ \vdots & \ddots & \vdots \\ M_{n_w 1} & \cdots & M_{n_w 4} \end{bmatrix} \tag{4.17}$$

But note that, if the element w_{jq} does not exist, $M_{jq} = 0$.

The purpose is to design the internally stabilising linear controller K which matches the control system to its environment characterised by $\mathcal{F}(M)$; the problem is formulated into the admissibility problem involving the inequality constraints

$$\hat{z}_i(\mathcal{F}(M)) \leq \varepsilon_i \quad (i = 1, 2, \ldots, n_z) \tag{4.18}$$

If the exogenous input including a nonzero element w_{j1} or w_{j2} is applied to the system with a nonzero feedthrough term D_w, the problem, described by (4.18), may not be solvable. For instance, consider the situation such that the $(1, 1)$ entry of D_w is 1.0, $w_{j1} \in \mathcal{F}_1(1.0)$ and $\varepsilon_1 = 0.1$. In this case, the peak norm of z_1 is greater than 1.0 and the inequality, $\hat{z}_1(\mathcal{F}(M)) \leq 0.1$, is not met for any controller. The same situation occurs if the nonzero element w_{j2} is applied to the system since the space $\mathcal{F}_2(M_{j2})$ includes the input with large magnitude like an impulse. Therefore, the solvability of the problem depends on the problem setting itself. To avoid such a trivial situation, a further assumption is made: if $M_{j1} \neq 0$ or $M_{j2} \neq 0$, the plant G meets $D_{ij} = 0$ and $F_j = 0$, where D_{ij} is the (i, j) element of D_w and $F_j := \text{col}(D_y, j)$.

4.4 Controller Design via LMI

Let T_{ij} denote the subsystem from the jth input channel to the ith output channel of the closed-loop system and let $\zeta_{ij}(t, w_{jq})$ denote the response of T_{ij} to the input w_{jq} at time t. Since the feedback system, which consists of G and K, is linear, the ith output can be expressed as

$$z_i(t, w) = \sum_{j=1}^{n_w} \sum_{q=1}^{4} \zeta_{ij}(t, w_{jq}) \tag{4.19}$$

Then the peak norm of z_i can be given by

$$\hat{z}_i(\mathcal{F}(M)) = \sum_{j=1}^{n_w} \sum_{q=1}^{4} \hat{\zeta}_{ij}(\mathcal{F}_q(M_{jq})) \tag{4.20}$$

From this equality, it follows that the condition (4.18) is equivalent to the conjunction of the inequalities

$$\hat{\zeta}_{ij}(\mathcal{F}_q(M_{jq})) \leq \varepsilon_{ijq} \tag{4.21}$$

$$\sum_{j=1}^{n_w}\sum_{q=1}^{4} \varepsilon_{ijq} \leq \varepsilon_i \tag{4.22}$$

where $i \in N_z := \{1, \ldots, n_z\}$, $j \in N_w := \{1, \ldots, n_w\}$ and $q \in N := \{1, 2, 3, 4\}$. First, let us consider the matrix inequality condition which ensures (4.21). The state space representation of T_{ij} is given as follows:

$$\begin{bmatrix} \dot{x}(t) \\ \zeta_{ij}(t, w_{jq}) \end{bmatrix} = \begin{bmatrix} \hat{A} & \hat{B}_j \\ \hat{C}_i & \hat{D}_{ij} \end{bmatrix} \begin{bmatrix} x(t) \\ w_{jq}(t) \end{bmatrix} \tag{4.23}$$

where $\hat{A}$, $\hat{B}_j$, $\hat{C}_i$ and $\hat{D}_{ij}$ are defined by

$$\hat{A} := \begin{bmatrix} A_c + B_u D_K C_y & B_u C_K \\ B_K C_y & A_K \end{bmatrix}$$

$$\hat{B}_j := [\,(B_j + B_u D_K F_j)^t \quad (B_K F_j)^t\,]^t \tag{4.24}$$

$$\hat{C}_i := [\,C_i + E_i D_K C_y \quad E_i C_K\,]$$

$$\hat{D}_{ij} := D_{ij} + E_i D_K F_j$$

in which $B_j := \mathrm{col}(B_w, j)$, $C_i := \mathrm{row}(C_z, i)$ and $E_i := \mathrm{row}(D_u, i)$. The controller is chosen so that the closed-loop system is internally stable, namely, all subsystems T_{ij}'s are stable. The internal stability of T_{ij} is equivalent to the existence of P_{ij} such that

$$\hat{A}^t P_{ij} + P_{ij} \hat{A} < 0, \quad P_{ij} = P_{ij}^t > 0 \tag{4.25}$$

Basically, the controller must meet (4.25) for a certain P_{ij}. The following four propositions give the matrix inequality conditions for both of (4.21) and (4.25).

Proposition 4.1. *The inequalities (4.21) and (4.25) hold for $w_{j1} \in \mathcal{F}_1(M_{j1})$, if there exist a positive definite symmetric matrix P_{ij} and a positive scalar λ_{ij} such that*

$$\hat{A}^t P_{ij} + P_{ij} \hat{A} + 2\lambda_{ij} P_{ij} < 0 \tag{4.26a}$$

$$\begin{bmatrix} P_{ij} & P_{ij}\hat{B}_j \\ \hat{B}_j^t P_{ij} & \lambda_{ij}\xi_{ij1} \end{bmatrix} > 0 \tag{4.26b}$$

$$\begin{bmatrix} \hat{P}_{ij} & \hat{C}_i^t \\ \hat{C}_i & \lambda_{ij}\xi_{ij1} \end{bmatrix} > 0 \tag{4.26c}$$

where $\xi_{ij1} := \varepsilon_{ij1}/M_{j1}$.

Proof. From the assumption, $D_{ij} = 0$. That is, the subsystem T_{ij} is strictly proper. The necessary and sufficient condition for (4.21) is then given by

$$\|\zeta_{ij}(\delta)\|_1 \leq \xi_{ij1} \tag{4.27}$$

where $\zeta_{ij}(\delta)$ means the unit impulse response of T_{ij}. It is easily seen that the time trajectory of $\zeta_{ij}(\delta)$, that is, $\zeta_{ij}(t,\delta)$ coincides with $\zeta_\delta(t)$, which is the initial response of the unforced system

$$\begin{cases} \dot{x}_\delta(t) = \hat{A} x_\delta(t), \quad x_\delta(0) = \hat{B}_j \\ \zeta_\delta(t) = \hat{C}_i x_\delta(t) \end{cases} \tag{4.28}$$

Thus (4.27) is equivalent to

$$\|\zeta_\delta\|_1 \leq \xi_{ij1} \tag{4.29}$$

Define the positive function as $V(t) := x_\delta^t(t) P_{ij} x_\delta(t)$. The inequalities in (4.26) indicate that

$$\dot{V}(t) + 2\lambda_{ij} V(t) < 0, \quad V(0) < \lambda_{ij} \xi_{ij1}, \quad \zeta_\delta^2(t) < \lambda_{ij} \xi_{ij1} V(t) \tag{4.30}$$

These inequalities yield

$$|\zeta_\delta(t)| \leq \lambda_{ij} \xi_{ij1} e^{-\lambda_{ij} t} \tag{4.31}$$

and then (4.29) is ensured. Accordingly, if there exist $P_{ij} = P_{ij}^t > 0$ and $\lambda_{ij} > 0$ satisfying (4.26), the inequality (4.21) is ensured. It is obvious that (4.25) holds if (4.26a) is met. □

Proposition 4.2. *The inequalities (4.21) and (4.25) hold for $w_{j2} \in \mathcal{F}_2(M_{j2})$, if there exists a positive definite symmetric matrix P_{ij} such that*

$$\begin{bmatrix} \hat{A}^t P_{ij} + P_{ij} \hat{A} & P_{ij} \hat{B}_j \\ \hat{B}_j^t P_{ij} & -\xi_{ij2} \end{bmatrix} < 0 \tag{4.32a}$$

$$\begin{bmatrix} P_{ij} & \hat{C}_i^t \\ \hat{C}_i & \xi_{ij2} \end{bmatrix} > 0 \tag{4.32b}$$

where $\xi_{ij2} := \varepsilon_{ij2}/M_{j2}$.

Proof. From the assumption, $D_{ij} = 0$, namely, T_{ij} is strictly proper. Then the necessary and sufficient condition for (4.21) is given as

$$\|\zeta_{ij}(\delta)\|_2 \leq \xi_{ij2} \tag{4.33}$$

First let us show the sufficiency (*i.e.*, $\exists P_{ij} = P_{ij}^t > 0$ satisfying (4.32) ⇒ (4.33)). The inequalities in (4.32) mean

$$\hat{A}' P_{ij} + P_{ij} \hat{A} + \frac{1}{\xi_{ij2}} P_{ij} \hat{B}_j \hat{B}_j' P_{ij} < 0, \quad \xi_{ij2} - \hat{C}_i P_{ij}^{-1} \hat{C}_i' > 0 \tag{4.34}$$

Multiplying these two inequalities by P_{ij}^{-1} from both sides and replacing P_{ij}^{-1} with Q, they are changed to

$$\hat{A}Q + Q\hat{A}' + \frac{1}{\xi_{ij2}}\hat{B}_j\hat{B}_j' < 0, \quad \xi_{ij2} - \hat{C}_i Q \hat{C}_i' > 0 \tag{4.35}$$

Meanwhile, the impulse response of T_{ij} is computed as $\zeta_{ij}(t,\delta) = \hat{B}_j^t e^{\hat{A}^t t} \hat{C}_i^t$, which is equivalent to the output of the unforced system

$$\begin{cases} \dot{x}_\delta(t) = \hat{A}^t x_\delta(t), \quad x_\delta(0) = \hat{C}_i^t \\ \zeta_\delta(t) = \hat{B}_j^t x_\delta(t) \end{cases} \tag{4.36}$$

Introduce the Lyapunov function as $V(t) := x_\delta^t(t) Q x_\delta(t)$. Then the inequalities in (4.35) are rewritten as

$$\zeta_\delta^2(t) \leq -\xi_{ij2}\dot{V}(t), \quad V(0) \leq \xi_{ij2} \tag{4.37}$$

This leads to

$$\int_0^t \zeta_\delta^2(\tau) d\tau \leq \xi_{ij2}(\xi_{ij2} - V(t)) \tag{4.38}$$

It is found out from (4.35) that $\hat{A}Q + Q\hat{A}' < 0$ since $\hat{B}_j \hat{B}_j' \geq 0$. This means that $\dot{V}(t)$ is always negative and $V(t)$ converges on 0 as t increases. Therefore, it follows from (4.38) that the inequality (4.33) holds. Next show the necessity (*i.e.*, (4.33) $\Rightarrow \exists P_{ij} = P_{ij}^t > 0$ satisfying (4.32)). The 2-norm of the impulse response of T_{ij} can be computed as

$$\begin{aligned}\|\zeta_{ij}(\delta)\|_2 &= \left(\hat{C}_i \int_0^\infty e^{\hat{A}t} \hat{B}_j \hat{B}_j^t e^{\hat{A}^t t} dt\, \hat{C}_i^t\right)^{\frac{1}{2}} \\ &= \left(\hat{C}_i \tilde{Q} \hat{C}_i^t\right)^{\frac{1}{2}}\end{aligned} \tag{4.39}$$

where $\tilde{Q}$ is the solution of the Lyapunov equation

$$\hat{A}\tilde{Q} + \tilde{Q}\hat{A}^t + \hat{B}_j \hat{B}_j^t = 0 \tag{4.40}$$

Since $\|\zeta_{ij}(\delta)\|_2$ is assumed to be finite, $\hat{A}$ is stable. Then $\tilde{Q} \geq 0$ since $\hat{B}\hat{B}^t \geq 0$. When the Lyapunov equation above has the solution $\tilde{Q}(= \tilde{Q}^t \geq 0)$, there exists a matrix $Q(= Q^t > 0)$ such that

$$\hat{A}Q + Q\hat{A}^t + \hat{B}_j \hat{B}_j^t < 0 \tag{4.41}$$

Multiplying (4.41) by Q^{-1} from both sides and replacing Q^{-1} with P_{ij}/ξ_{ij2}, it is changed to

$$\hat{A}^t P_{ij} + P_{ij}\hat{A} + \frac{1}{\xi_{ij2}} P_{ij} \hat{B}_j \hat{B}_j^t P_{ij} < 0 \tag{4.42}$$

Meanwhile, from the assumption that $\|\zeta_{ij}(\delta)\|_2 \le \xi_{ij2}$,

$$\xi_{ij2}^2 - \hat{C}_i Q \hat{C}_i^t > 0 \tag{4.43}$$

Replacing Q with $P_{ij}^{-1}\xi_{ij2}$, (4.43) is expressed as

$$\xi_{ij2} - \hat{C}_i P_{ij}^{-1} \hat{C}_i^t > 0 \tag{4.44}$$

Applying Schur complement to (4.42) and (4.44), (4.32) is obtained. □

Proposition 4.3. *Suppose that T_{ij} has integral action, that is, the output $z_i(t)$ converges to 0 when the unit step input is applied to T_{ij}. Then the inequalities in (4.21) and (4.25) are satisfied for $w_{j3} \in \mathcal{F}_3(M_{j3})$, if there exist a positive definite symmetric matrix P_{ij} and a positive scalar λ_{ij} such that*

$$\hat{A}^t P_{ij} + P_{ij} \hat{A} + 2\lambda_{ij} P_{ij} < 0 \tag{4.45a}$$

$$\begin{bmatrix} P_{ij} & P_{ij}\hat{B}_j \\ \hat{B}_j^t P_{ij} & \lambda_{ij}\xi_{ij3} \end{bmatrix} > 0 \tag{4.45b}$$

$$\frac{1}{\lambda_{ij}\xi_{ij3}} \hat{C}_i^t \hat{C}_i - \hat{A}^t P_{ij} \hat{A} < 0 \tag{4.45c}$$

where $\xi_{ij3} := \varepsilon_{ij3}/M_{j3}$.

Proof. The necessary and sufficient condition for (4.21) is given by

$$\|\zeta_{ij}(h)\|_1 \le \xi_{ij3} \tag{4.46}$$

where $\zeta_{ij}(h)$ means the unit step response of T_{ij}. It can be verified that the time trajectory of $\zeta_{ij}(h)$, that is, $\zeta_{ij}(t,h)$ equals the output of the system

$$\begin{cases} \dot{x}_h(t) = \begin{bmatrix} \hat{A} & \hat{B}_j \\ 0 & 0 \end{bmatrix} x_h(t) + \begin{bmatrix} 0 \\ 1 \end{bmatrix} \delta(t) \\ \zeta_h(t) = [\hat{C}_i \ \hat{D}_{ij}] x_h(t) \end{cases} \tag{4.47}$$

where $\delta(t)$ is the unit impulse. Thus (4.46) can also be written as $\|\zeta_h\|_1 \le \xi_{ij3}$. Meanwhile, from the assumption on the integral behavior of T_{ij} and the final value theorem,

$$T_{ij}(0) = \hat{D}_{ij} - \hat{C}_i \hat{A}^{-1} \hat{B}_j = 0 \tag{4.48}$$

where $T_{ij}(s)$ is the transfer function of T_{ij}. Substituting $\hat{C}_i \hat{A}^{-1} \hat{B}_j$ for D_{ij} in (4.47), the state space representation in (4.47) can be expressed as

$$\begin{bmatrix} \dot{x}(t) \\ \zeta_h(t) \end{bmatrix} = \begin{bmatrix} \hat{A} & \hat{B}_j \\ \hat{C}_i \hat{A}^{-1} & 0 \end{bmatrix} \begin{bmatrix} x(t) \\ \delta(t) \end{bmatrix} \tag{4.49}$$

which is equivalent to the following system description.

$$\begin{cases} \dot{x}(t) = \hat{A}x(t), \quad x(0) = \hat{B}_j \\ \zeta_h(t) = \hat{C}_i \hat{A}^{-1} x(t) \end{cases} \tag{4.50}$$

This is similar to the proof of Proposition 4.1. Accordingly, the condition for (4.46) can be given as the existential condition of $P_{ij} = P_{ij}^t > 0$ and $\lambda_{ij} > 0$ that meet (4.45). □

Proposition 4.4. *Suppose that T_{ij} has integral action. Then the inequalities (4.21) and (4.25) hold for $w_{j4} \in \mathcal{F}_4(M_{j4})$, if there exist a positive definite symmetric matrix P_{ij} such that*

$$\begin{bmatrix} \hat{A}^t P_{ij} + P_{ij}\hat{A} & P_{ij}\hat{B}_j \\ \hat{B}_j^t P_{ij} & -\xi_{ij4} \end{bmatrix} < 0 \tag{4.51a}$$

$$\frac{1}{\xi_{ij4}} \hat{C}_i^t \hat{C}_i - \hat{A}^t P_{ij} \hat{A}^t < 0 \tag{4.51b}$$

where $\xi_{ij4} := \varepsilon_{ij4}/M_{j4}$.

Proof. The necessary and sufficient condition for (4.21) is given as

$$\|\zeta_{ij}(h)\|_2 \leq \xi_{ij4} \tag{4.52}$$

The time trajectory of $\zeta_{ij}(h)$ coincides with the initial response of the unforced system

$$\begin{cases} \dot{x}_h(t) = \hat{A}^t x_h(t), \quad x_h(0) = (\hat{C}_i \hat{A}^{-1})^t \\ \zeta_h(t) = \hat{B}_j^t x_h(t) \end{cases} \tag{4.53}$$

So the condition (4.52) can be rewritten as $\|\zeta_h\|_2 \leq \xi_{ij4}$. This situation is the same as in the proof in Proposition 4.2, except that $\hat{C}_i$ in (4.36) is replaced with $\hat{C}_i \hat{A}^{-1}$ in (4.53). By following the same process of the proof of Proposition 4.2, (4.51) is proved. □

Notice that the inequalities in Propositions 4.1, 4.3 and 4.4 are the sufficient conditions for (4.21). It is also noticed that the inequalities (4.26a) and (4.45a) can be written as

$$(\hat{A} + \lambda_{ij}I)^t P_{ij} + P_{ij}(\hat{A} + \lambda_{ij}I) < 0 \tag{4.54}$$

This indicates that $\hat{A} + \lambda_{ij}I$ is stable, that is, all eigenvalues of $\hat{A}$ are placed on the left side of $s = -\lambda_{ij}$. Thus λ_{ij} can be utilised as an adjustable variable for pole placement of the system. Notice also that the control system needs integral action if the exogenous input has the element w_{j3} or w_{j4}.

Next move on to the controller design based on the previous four propositions. The aim is to determine the parameters A_K, B_K, C_K and D_K so that the inequalities in (4.26), (4.32), (4.45), (4.51) and (4.22) are simultaneously satisfied. However, they can still not be obtained in the framework of an LMI admissibility problem since those inequalities are not affine to them. One solution is to take a change of variables. Here, a change of variables that is applicable to a wide variety of output feedback control problems (Chilali and Gahinet, 1996; Gahinet, 1996; Scherer, Gahinet and Chilali, 1997), is performed.

Generally, all P_{ij}'s are allowed to be different. In this change of variables, however, they are treated as identical to preserve the convexity:

$$P_{ij} = P \tag{4.55}$$

In this case, the inequalities (4.26a) and (4.45a) are identical to each other. For simplicity, λ_{ij} is also treated as identical:

$$\lambda_{ij} = \lambda \tag{4.56}$$

Under these additive conditions, partition P and P^{-1} as

$$P = \begin{bmatrix} \tilde{Y} & \tilde{N} \\ \tilde{N}^t & \star \end{bmatrix}, \quad P^{-1} = \begin{bmatrix} \tilde{X} & \tilde{M} \\ \tilde{M}^t & \star \end{bmatrix} \tag{4.57}$$

where $\tilde{X}$ and $\tilde{Y}$ are nonsingular and symmetric matrices with the same dimension as A_c and the symbol $\star$ means any symmetric matrix. From $P^{-1}P = I$, it follows that $\tilde{M}$ and $\tilde{N}$ are chosen so that

$$\tilde{M}\tilde{N}^t = I - \tilde{X}\tilde{Y} \tag{4.58}$$

Now consider the following change of variables:

$$\begin{cases} \tilde{A} := \tilde{N}A_K\tilde{M}^t + \tilde{N}B_KC_y\tilde{X} + \tilde{Y}B_uC_K\tilde{M}^t + \tilde{Y}(A_c + B_uD_KC_y)\tilde{X} \\ \tilde{B} := \tilde{N}B_K + \tilde{Y}B_uD_K \\ \tilde{C} := C_K\tilde{M}^t + D_KC_y\tilde{X} \\ \tilde{D} := D_K \end{cases} \tag{4.59}$$

With this transformation, the set of the original variables (A_K, B_K, C_K, D_K, P, λ, ε_{ijq}'s) is mapped to the set of the new variables ($\tilde{A}$, $\tilde{B}$, $\tilde{C}$, $\tilde{D}$, $\tilde{X}$, $\tilde{Y}$, $\tilde{M}$, $\tilde{N}$, λ, ε_{ijq}'s). If $\tilde{M}$ and $\tilde{N}$ are the nonsingular matrices that satisfy (4.58), the controller parameters can be determined uniquely by $\tilde{A}$, $\tilde{B}$, $\tilde{C}$, $\tilde{D}$, $\tilde{X}$, $\tilde{Y}$, $\tilde{M}$ and $\tilde{N}$ as follows:

$$\begin{cases} D_K = \tilde{D} \\ C_K = (\tilde{C} - D_KC_y\tilde{X})\tilde{M}^{-t} \\ B_K = \tilde{N}^{-1}(\tilde{B} - \tilde{Y}B_uD_K) \\ A_K = \tilde{N}^{-1}(\tilde{A} - \tilde{N}B_KC_y\tilde{X} - \tilde{Y}B_uC_K\tilde{M}^t \\ \qquad\qquad - \tilde{Y}(A_c + B_uD_KC_y)\tilde{X})\tilde{M}^{-t} \end{cases} \tag{4.60}$$

Accordingly, the inequalities of Propositions 4.1 to 4.4 can be rewritten with the new variables without loss of generality. To rewrite them with the new variables, the congruence transformation is performed. Let us define Π_1 and Π_2 as follows:

$$\Pi_1 := \begin{bmatrix} \tilde{X} & I \\ \tilde{M}^t & 0 \end{bmatrix}, \quad \Pi_2 := \begin{bmatrix} I & \tilde{Y} \\ 0 & \tilde{N}^t \end{bmatrix} \tag{4.61}$$

Then the following relationships are obtained.

$$\Pi_1^t P \Pi_1 = \Pi_1^t \Pi_2 = \begin{bmatrix} \tilde{X} & I \\ I & \tilde{Y} \end{bmatrix}$$

$$\Pi_1^t P \hat{A} \Pi_1 = \Pi_2^t \hat{A} \Pi_1 = \begin{bmatrix} A_c \tilde{X} + B_u \tilde{C} & A_c + B_u \tilde{D} C_y \\ \tilde{A} & \tilde{Y} A_c + \tilde{B} C_y \end{bmatrix} \tag{4.62}$$

$$\Pi_1^t P \hat{B}_j = \Pi_2^t \hat{B}_j = \begin{bmatrix} B_j + B_u \tilde{D} F_j \\ \tilde{Y} B_j + \tilde{B} F_j \end{bmatrix}$$

$$\hat{C}_i \Pi_1 = [\, C_i \tilde{X} + E_i \tilde{C} \quad C_i + E_i \tilde{D} C_y \,]$$

Perform the congruence transformation to (4.26a) with Π_1, and (4.26b) and (4.26c) with block diag(Π_1, I). By using (4.62), they can be expressed as

$$\begin{bmatrix} \Psi(A_c, \tilde{X}) + \Psi(B_u, \tilde{C}) + 2\lambda \tilde{X} & * \\ \tilde{A} + (A_c + B_u \tilde{D} C_y)^t + 2\lambda I & \Psi(A_c^t, \tilde{Y}) + \Psi(\tilde{B}, C_y) + 2\lambda \tilde{Y} \end{bmatrix} < 0 \tag{4.63a}$$

$$\begin{bmatrix} \tilde{X} & I & B_j + B_u \tilde{D} F_j \\ * & \tilde{Y} & \tilde{Y} B_j + \tilde{B} F_j \\ * & * & \dfrac{\lambda \varepsilon_{ij1}}{M_{j1}} \end{bmatrix} > 0 \tag{4.63b}$$

$$\begin{bmatrix} \tilde{X} & * & * \\ I & \tilde{Y} & * \\ C_i \tilde{X} + E_i \tilde{C} & C_i + E_i \tilde{D} C_y & \dfrac{\lambda \varepsilon_{ij1}}{M_{j1}} \end{bmatrix} > 0 \tag{4.63c}$$

where $*$ means the symmetric block. These inequalities depend affinely on $\tilde{X}$, $\tilde{Y}$, $\tilde{A}$, $\tilde{B}$, $\tilde{C}$, $\tilde{D}$ and ε_{ij1}, but have the cross terms $\lambda \tilde{X}$, $\lambda \tilde{Y}$ and $\lambda \varepsilon_{ij1}$. However, they can be handled as LMIs if λ is fixed. In a similar way, applying the congruence transformation to the inequalities in (4.32) with block diag(Π_1, I), they are changed to

$$\begin{bmatrix} \Psi(A_c, \tilde{X}) + \Psi(B_u, \tilde{C}) & \tilde{A}^t + A_c + B_u \tilde{D} C_y & B_j + B_u \tilde{D} F_j \\ * & \Psi(A'_c, \tilde{Y}) + \Psi(\tilde{B}, C_y) & \tilde{Y} B_j + \tilde{B} F_j \\ * & * & -\dfrac{\varepsilon_{ij2}}{M_{j2}} \end{bmatrix} < 0 \quad (4.64a)$$

$$\begin{bmatrix} \tilde{X} & * & * \\ I & \tilde{Y} & * \\ C_i \tilde{X} + E_i \tilde{C} & C_i + E_i \tilde{D} C_y & \dfrac{\varepsilon_{ij2}}{M_{j2}} \end{bmatrix} > 0 \quad (4.64b)$$

In this case, these two inequalities are LMIs to the new variables. Next consider applying the same transformation to the inequalities in Proposition 4.3. However, (4.45c) cannot be changed to the inequality which is affine in new variables even if λ is fixed. To circumvent this problem, the congruence transform is applied not to (4.45) but to its sufficient condition given in the following proposition.

Proposition 4.5. *Set $P_{ij} = P$ and $\lambda_{ij} = \lambda$. If there exist a positive definite symmetric matrix P and a positive real number λ such that*

$$\begin{bmatrix} \hat{A}^t P + P \hat{A} + 2\lambda P & * \\ \hat{C}_i & -2\lambda^2 \xi_{ij2} \end{bmatrix} < 0 \quad (4.65a)$$

$$\begin{bmatrix} P & P \hat{B}_j \\ * & \lambda \xi_{ij2} \end{bmatrix} > 0 \quad (4.65b)$$

then the inequalities in (4.45) hold for such P and λ.

Proof. The inequality (4.65a) is equivalent to

$$\frac{1}{\lambda \xi_{ij2}} \hat{C}_i^t \hat{C}_i + 2\lambda (\hat{A}^t P + P \hat{A} + 2\lambda P) < 0 \quad (4.66)$$

Generally the following inequality holds for any $P = P^t > 0$ and any $\lambda > 0$.

$$(\hat{A} + 2\lambda I)^t P (\hat{A} + 2\lambda I) = \hat{A}^t P \hat{A} + 2\lambda (\hat{A}^t P + P \hat{A} + 2\lambda P) > 0 \quad (4.67)$$

From (4.66) and (4.67), it can be readily seen that (4.45c) is ensured if (4.66) holds. The inequality (4.45a) is also ensured since $\hat{C}_i^t \hat{C}_i \geq 0$. □

Performing the congruence transformation with block diag(Π_1, I) for both of (4.65b) and (4.65b), they are changed to

$$\begin{bmatrix} \Psi(A_c, \tilde{X}) + \Psi(B_u, \tilde{C}) + 2\lambda\tilde{X} & * & * \\ \tilde{A} + (A_c + B_u \tilde{D} C_y)^t + 2\lambda I & \Psi(A_c^t, \tilde{Y}) + \Psi(\tilde{B}, C_y) + 2\lambda\tilde{Y} & * \\ C_i \tilde{X} + E_i \tilde{C} & C_i + E_i \tilde{D} C_y & -\dfrac{2\lambda^2 \varepsilon_{ij2}}{M_{j2}} \end{bmatrix} < 0 \quad (4.68a)$$

$$\begin{bmatrix} \tilde{X} & I & B_j + B_u \tilde{D} F_j \\ * & \tilde{Y} & \tilde{Y} B_j + \tilde{B} F_j \\ * & * & \dfrac{\lambda \varepsilon_{ij2}}{M_{j2}} \end{bmatrix} > 0 \quad (4.68b)$$

In this case, if λ is fixed, these inequalities can be treated as LMIs. Next consider applying the congruence transform to the inequalities in Proposition 4.4. However, (4.51b) is not transformed in the affine form to new variables. Like the case of Proposition 4.3, the congruence transform is performed for the following sufficient condition.

Proposition 4.6. Set $P_{ij} = P$ and $\lambda_{ij} = \lambda$. If there exist a positive-definite symmetric matrix P and a positive scalar λ such that

$$\begin{bmatrix} \hat{A}^t P + P\hat{A} + 2\lambda P & * \\ \hat{C}_i & -2\lambda \xi_{ij4} \end{bmatrix} < 0 \quad (4.69a)$$

$$\begin{bmatrix} P & P\hat{B}_j \\ * & 2\lambda \xi_{ij4} \end{bmatrix} > 0 \quad (4.69b)$$

then the inequalities in (4.51) hold for such P and λ.

Proof. It is clear that (4.51b) holds for P and λ satisfying (4.69a). Furthermore,

$$\hat{A}^t P + P\hat{A} + 2\lambda P < 0 \quad (4.70)$$

also holds. Meanwhile, (4.69b) is equivalent to

$$P - \dfrac{1}{2\lambda \xi_{ij4}} P\hat{B}_j \hat{B}_j^t P > 0 \quad (4.71)$$

The inequalities (4.70) and (4.71) yield

$$\hat{A}' P + P\hat{A} + \dfrac{1}{\xi_{ij2}} P\hat{B}_j \hat{B}'_j P < 0 \quad (4.72)$$

which is equivalent to (4.51a). □

Performing the congruence transform with block diag(Π_1, I) for both of (4.69a) and (4.69b), they are changed to

$$\begin{bmatrix} \Psi(A_c, \tilde{X}) + \Psi(B_u, \tilde{C}) + 2\lambda \tilde{X} & * & * \\ \tilde{A} + (A_c + B_u \tilde{D} C_y)' + 2\lambda I & \Psi(A'_c, \tilde{Y}) + \Psi(\tilde{B}, C_y) + 2\lambda \tilde{Y} & * \\ C_i \tilde{X} + E_i \tilde{C} & C_i + E_i \tilde{D} C_y & -\dfrac{2\lambda \varepsilon_{ij4}}{M_{j4}} \end{bmatrix} < 0 \quad (4.73a)$$

$$\begin{bmatrix} \tilde{X} & I & B_j + B_u \tilde{D} F_j \\ * & \tilde{Y} & \tilde{Y} B_j + \tilde{B} F_j \\ * & * & \dfrac{2\lambda \varepsilon_{ij4}}{M_{j4}} \end{bmatrix} > 0 \quad (4.73b)$$

Finally, perform the congruence transformation to the inequality $P > 0$. Then it is changed to the LMI

$$\begin{bmatrix} \tilde{X} & I \\ I & \tilde{Y} \end{bmatrix} > 0 \quad (4.74)$$

After the congruence transformation, all inequalities in (4.63), (4.64), (4.68), (4.73), (4.74) and (4.22) can be viewed as LMIs if λ is fixed. Accordingly, the problem of finding a solution set ($\tilde{X}$, $\tilde{Y}$, $\tilde{A}$, $\tilde{B}$, $\tilde{C}$, $\tilde{D}$, $\tilde{M}$, $\tilde{N}$, ε_{ijq}'s) can be solved in the framework of the LMI admissibility problem involving the line search of λ over $(0, \infty)$. For convenience, let Λ_{ijq} be defined as follows:

Λ_{ij1} : the set of inequalities (4.63) for T_{ij}
Λ_{ij2} : the set of inequalities (4.64) for T_{ij}
Λ_{ij3} : the set of inequalities (4.68) for T_{ij}
Λ_{ij4} : the set of inequalities (4.73) for T_{ij}
Λ_P : the inequality (4.74)
Λ_{ε_i} : the inequalities (4.22) for ε_i

The procedure of controller design is summarised below.

Step 1: Give the tolerable bound ε_i of the controlled output z_i.
Step 2: Estimate each bound M_{jq} of inputs and model the set of exogenous inputs as $\mathcal{F}(M)$.
Step 3: If $M_{j3} \neq 0$ or $M_{j4} \neq 0$, configure the system so that it has integral action.
Step 4: Construct all the inequalities set Λ_{ijq}, Λ_P and Λ_{ε_i} for $i \in N_z$, $j \in N_w$ and $q \in N$ such that $M_{jq} \neq 0$.
Step 5: Set the lower and upper bounds of λ to λ_{lb} and λ_{ub}, respectively.
Step 6: Fix $\lambda \in [\lambda_{lb}, \lambda_{ub}]$.

Step 7: Search for a solution set ($\tilde{A}$, $\tilde{B}$, $\tilde{C}$, $\tilde{D}$, $\tilde{X}$, $\tilde{Y}$, ε_{ijq}'s) subject to the LMIs constructed in Step 4.

Step 8: If an admissible solution set is found, determine $\tilde{M}$ and $\tilde{N}$ so that (4.58) is met, and calculate the controller parameters from (4.60). If a solution set is not found, go to Step 9.

Step 9: If $\lambda < \lambda_{ub}$, increase the value of λ and go back to Step 7. Otherwise stop searching. In this case, there is not an admissible solution to the LMIs constructed in Step 4.

Bear in mind that this method requires the system to have integral action if the exogenous input includes the element w_{j3} or w_{j4}. Therefore, before searching for admissible solutions, it is necessary to configure the system so that the subsystem T_{ij} has integral behaviour: the closed-loop transfer function satisfies $T_{ij}(0) = 0$.

4.5 Numerical Example

An example of matched environment-system design using LMI is presented. With this example, the advantages and drawbacks of the method are examined.

The system to be designed is the unity feedback control system shown in Fig. 4.1. The plant P is the third-order linear time-invariant system consisting of the actual plant and the actuator, which is described by the following transfer function (Whidborne and Liu, 1993)

$$P(s) = \frac{27697}{s(s^2 + 1429s + 42653)} \quad (4.75)$$

The exogenous inputs are the reference command signal $r(t)$ which belongs to $\mathcal{F}_3(1.0)$ and the disturbance $v(t)$ which belongs to $\mathcal{F}_1(0.5)$. Defining the vector of exogenous inputs as $w := [\, r \; v \,]^t$, the set of inputs can be modelled as $\mathcal{F}(M)$, where

$$M = \begin{bmatrix} 0 & 0 & 1.0 & 0 \\ 0.5 & 0 & 0 & 0 \end{bmatrix} \quad (4.76)$$

The objective is to design the internally stabilising linear controller $K(s)$ which maintains the tracking error $e(t) (:= r(t) - y_p(t))$ and the control input $u(t)$ within the ranges ± 0.02094 and ± 10.0 for all inputs in $\mathcal{F}(M)$. Since the maximum amplitude of the disturbance is 0.5, the constraint on the control input is equivalent to $|u_K(t)| \leq 9.5$, where $u_K(t)$ is the output of the controller. Defining the vector of outputs as $z := [\, e \; u_K \,]^t$, the condition for the system to be matched to the environment characterised by $\mathcal{F}(M)$ is given as

$$\hat{z}_1(\mathcal{F}(M)) \leq 0.02094, \quad \hat{z}_2(\mathcal{F}(M)) \leq 9.5 \quad (4.77)$$

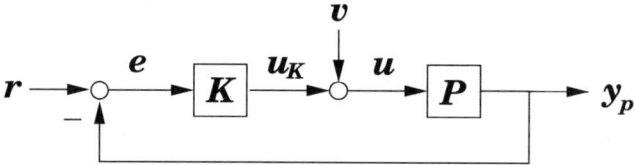

Fig. 4.1. Unity feedback control system

Since $P(s)$ has a pole at the origin, the subsystems T_{11} and T_{21} have the integral actions if $K(s)$ is a stabilising controller. Therefore the LMI-based method can be applicable to this problem without prior configuration of the system. The controller is designed so that the set of inequalities (Λ_{113}, Λ_{121}, Λ_{213}, Λ_{221}, Λ_P, Λ_{ε_1}, Λ_{ε_2}) are satisfied in the framework of the convex admissibility problem with a line search of λ. The design follows the procedure shown in Step 1 to Step 9.

As increasing λ from 1.0, the search was tried by using the function `feasp` in the MATLAB LMI CONTROL TOOLBOX, which is the LMI solver implementing the projective algorithm (Nesterov and Nemirovsky, 1994). However, the admissible solution was not found. This may be because the LMI constraint is the sufficient condition for (4.77). Next, the tolerable bounds ε_i's were relaxed and the search was performed again. As a result of this, some controllers have been found. For instance, the controller, which was obtained for $(\lambda, \varepsilon_1, \varepsilon_2) = (78, 209.44, 9500)$, is given by

$$K(s) = 793.5 \frac{(s+1398.5)(s+104.8)(s+47.5)}{(s+1385.3)(s+220.2)(s+107.4)} \quad (4.78)$$

Generally, the controllers obtained for the relaxed bounds do not necessarily ensure the specification. It is thus necessary to check whether (4.77) is satisfied. The controller in (4.78) achieves

$$\hat{z}_1(\mathcal{F}(M)) \simeq 0.01928, \quad \hat{z}_2(\mathcal{F}(M)) \simeq 6.28 \quad (4.79)$$

In this case, the system regulated by (4.78) is matched to the environment characterised by $\mathcal{F}(M)$.

The simulation was done for the designed control system. The signals $r(t)$ and $v(t)$ are shown in Figures 4.2 and 4.3, respectively. The maximum bound of the slope of $r(t)$ is 1.0 and the maximum bound of $v(t)$ is 0.5. The responses $e(t)$ and $u_K(t)$ to these inputs are illustrated in Figures 4.4 and 4.5, respectively. The dotted lines in these figures show the tolerable bounds $\varepsilon_1(=0.02094)$ and $\varepsilon_2(=9.5)$. As the figures indicate, the system is matched to the environment.

In this example, the order of the controller is fixed to three, which is the same order as the plant transfer function. In practice, however, a lower-order controller is preferable. If the lower-order controller might be required, further treatment is necessary. Sometimes the model-order reduction is effective. For

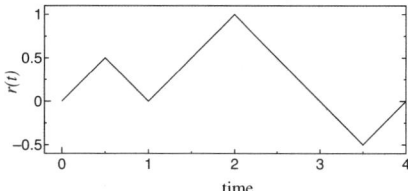

Fig. 4.2. Reference command signal

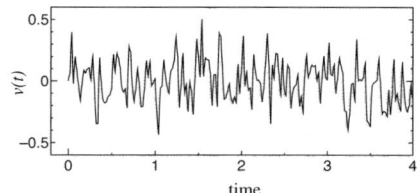

Fig. 4.3. Disturbance at the input side

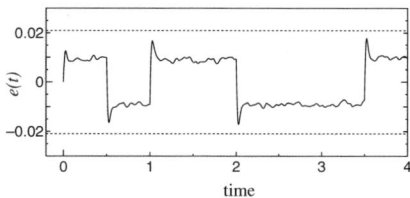

Fig. 4.4. Tracking error $e(t)$

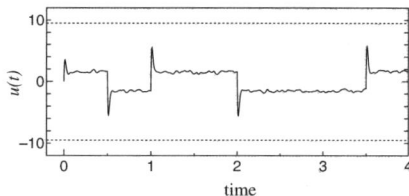

Fig. 4.5. Control input $u_K(t)$

instance, by using the function schmr in the MATLAB ROBUST CONTROL TOOLBOX, the controller in (4.78) is reduced to

$$K(s) = 793.5 \frac{s + 44.0}{s + 214.3} \tag{4.80}$$

which attains

$$\hat{z}_1(\mathcal{F}(M)) \simeq 0.01953, \quad \hat{z}_2(\mathcal{F}(M)) \simeq 6.68 \tag{4.81}$$

But note that the general order reduction methods do not always preserve the environment-system match. In this sense, this method has a limitation in practice. Furthermore, admissible controllers cannot necessarily be found with this method even if they exist. This is because the controller is designed under the sufficient condition. Accordingly, like the example above, even when the admissible solutions are not found for the original bounds ε_i's, by relaxing them, it may be possible to find them. However, there is numerical advantage: if the LMIs are solvable, the admissible solutions can be obtained efficiently.

4.6 Conclusion

Design using LMI is considered in this chapter. Compared to the approaches that use search methods for solving inequalities, the LMI approach is still in a development stage and is not considered as an effective general method in practice. One reason is that it is not flexible in choosing the structure and the order of a controller; The method allows only a controller of the same order as a plant to be designed. The other reason is that it is conservative since the LMI is a sufficient condition for (4.18). Accordingly, even if admissible solutions are not found with this method, it cannot be concluded that a solution to the inequalities (4.18) does not exist. From the numerical aspect, however, it is attractive: if the admissible set of the LMI is not empty, it is possible to obtain the solutions efficiently.

References

Chilali, M.,P. Gahinet, (1996) H_∞ design with pole placement constraints: An LMI approach. *IEEE Trans. Automat. Contr.*, 41:358–367.

Gahinet, P., (1996) Explicit controller formulas for LMI-base H^∞ synthesis. *Automatica*, 32:1007–1014.

Nesterov, Y., A. Nemirovsky, (1994) *Interior Point Polynomial Methods in Convex Programming: Theory and Applications*, SIAM, Philadelphia.

Scherer, C., P. Gahinet, M. Chilali, (1997) Multiobjective output-feedback control via LMI optimization. *IEEE Trans. Automat. Contr.*, 42:896–911.

Whidborne, J. F., G. P. Liu, (1993) *Critical Control Systems: Theory, Design and Applications*, Research Studies Press, Taunton, U.K.

Zakian, V., (1991) Well matched systems. *IMA Journal of Mathematical Control & Information*, 8:29–38. See also Corrigendum. (1992), 9:101.

Zakian, V., (1996) Perspectives on the principle of matching and the method of inequalities. *Int. J. Contr.*, 65:147–175.

5 Design of a Sampled-data Control System

Takahiko Ono

Abstract. This chapter deals with the design of sampled-data control systems. The aim is to ensure that the system is matched to its environment. By applying the lifting technique, the system is treated as a linear time-invariant discrete-time system that completely retains the intersample behaviour. A numerically tractable condition for matching the system to its environment is derived in the form of inequalities involving a unit step response. Design parameters that satisfy the matching condition are found by an inequality solver. However, instead of using the exact matching condition, the corresponding inequalities obtained from the fast-discretised system are used in computation. This involves an approximation, but provides an efficient numerical procedure for computer-aided design. The design procedure is illustrated with a numerical example.

5.1 Introduction

This chapter is concerned with the design of sampled-data control systems in accordance with the principle of matching and the principle of inequalities (Zakian 1991, 1996). To this end, an approximate but effective design procedure is given to ensure that the system is matched to the environment.

At the end of the 1950s, the first industrial computer-controlled systems appeared in the field of process control. Since then, computer control has increasingly been used in industries and in university laboratories. A computer-controlled system is a hybrid system, which chiefly consists of continuous-time plants and actuators, digital controllers, analogue-to-digital (A/D) converters and digital-to-analogue (D/A) converters. In control engineering, such a hybrid system is traditionally called a sampled-data system. The feature of a sampled-data system is that the control action can be updated only at the moment of the sampling instants. That is, no new control action can be given immediately over the sampling interval even if the system is perturbed by uncertainties like disturbances or parameter changes. Accordingly the designer is required to design the control system so that the system behaviour is acceptable not only at the sampling instants but also over the sampling intervals. This requirement is especially essential in design for critical systems (*e.g.* Zakian 1989), in which the response of the system must be maintained within tolerable ranges.

The aim of design is to find the controller that satisfies a set of inequalities

$$\varphi_i(p) \leq \varepsilon_i, \quad i = 1, 2, \ldots, n$$

which provide a practical, necessary and sufficient condition for the environment and the system to be matched. Generally, there are three approaches to design of controllers for sampled-data systems and each can be used in the context of the principle of matching.

- **Digital redesign:** First, design the continuous-time controller so that the matching condition for continuous-time plants is ensured. Next, discretise it and evaluate the sampled-data performance. If it is not acceptable, repeat the continuous-time design until the desired performance is achieved.
- **Discrete design:** First, design the discrete-time controller so that the matching condition for discrete-time plants is satisfied. Then evaluate the sampled-data performance. If it does not ensure a match between the system and the environment, repeat the discrete-time design until the sampled-data performance is satisfactory.
- **Direct design:** Design the discrete-time controller directly subject to the matching condition for sampled-data systems. This method has no approximation process.

The first two approaches above involve the approximation process. It is thus necessary to confirm that the control system is matched to its environment. This confirmation is done by evaluating the matching condition required in the direct design. For this reason, this chapter first considers the exact matching condition for sampled-data systems, especially for the persistent and/or transient inputs restricted in rate of change. Whidborne and Liu (1993) treats the sampled-data system as a time-varying one in design of critical control systems. In this chapter, however, the system is treated as a time-invariant discrete-time system that completely retains intersample behaviour by using lifting (Bamieh *et al*, 1991; Bamieh and Pearson, 1992). The matching condition is derived from such a time-invariant system as an inequality condition on the unit step response. Next an approximate but simple method for evaluating the exact matching condition is developed. In this method, the fast-discretisation technique (Keller and Anderson, 1992; Chen and Francis, 1995) plays a central role. Some practical techniques for computer-aided design for matched systems are also given. In a numerical example, the matched environment-system design based on this method is demonstrated.

5.1.1 Notations and Definitions

Let h be the sampling period, $\mathbb{R}$ be the set of real numbers, $\mathbb{R}_+$ be the set of non-negative real numbers and $\mathbb{Z}_+$ be the set of non-negative integers. A real-valued Euclidean space is represented simply by $\mathcal{R}$. $\mathcal{L}(\mathbb{X}, \mathbb{R})$ denotes the vector space which consists of all piecewise smooth and continuous functions from a domain $\mathbb{X} \subseteq \mathbb{R}_+$ to $\mathbb{R}$. Particularly, for $p \in \{2, \infty\}$, $\mathcal{L}_p(\mathbb{X}, \mathbb{R})$ means the subspace of $\mathcal{L}(\mathbb{X}, \mathbb{R})$ such that the following norm is finite.

$$\|f\|_{\mathcal{L}_p(\mathbb{X}, \mathbb{R})} := \begin{cases} \left(\int_{\mathbb{X}} |f(t)|^2 dt \right)^{\frac{1}{2}} & \text{for } p = 2 \\ \sup\{|f(t)| : t \in \mathbb{X}\} & \text{for } p = \infty \end{cases}$$

$\ell(\mathbb{Z}_+, \mathcal{K})$ denotes the vector space which consists of all function-space valued sequences whose elements belong to $\mathcal{K} := \mathcal{L}([0, h), \mathbb{R})$. For $p \in \{2, \infty\}$, $\ell_p(\mathbb{Z}_+, \mathcal{K}_p)$ means the subspace of $\ell(\mathbb{Z}_+, \mathcal{K})$ such that each element belongs to $\mathcal{K}_p := \mathcal{L}_p([0, h), \mathbb{R})$ and the following norm is finite.

$$\|\hat{f}\|_{\ell_p(\mathbb{Z}_+, \mathcal{K}_p)} := \begin{cases} \left(\sum_{k=0}^{\infty} \|\hat{f}(k)\|_{\mathcal{K}_2}^2 \right)^{\frac{1}{2}} & \text{for } p = 2 \\ \sup\{\|\hat{f}(k)\|_{\mathcal{K}_\infty} : k \in \mathbb{Z}_+\} & \text{for } p = \infty \end{cases}$$

$\ell_p(\mathbb{Z}_+, \mathbb{R}^n)$ denotes the vector space comprising all n-dimensional real number sequences such that the following norm is finite.

$$\|\hat{f}\|_{\ell_p(\mathbb{Z}_+, \mathbb{R}^n)} := \begin{cases} \left(\sum_{k=0}^{\infty} \|\hat{f}[k]\|^2 \right)^{\frac{1}{2}} & \text{for } p = 2 \\ \sup\{\|\hat{f}[k]\|_\infty : k \in \mathbb{Z}_+\} & \text{for } p = \infty \end{cases}$$

where $\|\cdot\|$ is the Euclidean norm and $\|\hat{f}[k]\|_\infty := \max\{|f_i[k]| : i = 1, \ldots, n\}$. Throughout this chapter, the kth elements of a function-valued sequence and a real number sequence are expressed as $\hat{f}(k)$ and $\hat{f}[k]$, respectively. The Dirac delta function and the unit step function are represented by $\delta(t)$ and $1(t)$, respectively.

5.2 Design for SISO Systems

Begin with design for SISO systems. The result in this section is the basis of design for MIMO systems, which is presented in the next section.

5.2.1 Problem Formulation

Consider the sampled-data system shown in Figure 5.1. The plant, denoted by G, is a linear time-invariant continuous-time system with the realisation

$$G : \begin{cases} \dot{x}(t) = A_c\, x(t) + B_w\, w(t) + B_u\, u(t) \\ z(t) = C_z\, x(t) + D_w\, w(t) + D_u\, u(t) \\ y(t) = C_y\, x(t) + D_y\, w(t) \end{cases} \qquad (5.1)$$

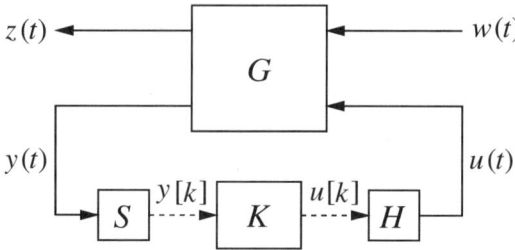

Fig. 5.1. Sampled-data control system

where $z(t) \in \mathbb{R}$ is an output to be regulated and $w(t) \in \mathbb{R}$ is an exogenous input like a disturbance, a reference command signal or a sensor noise. The plant is controlled by $u(t)$ and the output is measured as $y(t)$ through the sensors. Generally, G is the integrated system consisting of an actual plant, actuators, sensors, anti-aliasing filters and so on. In mathematical treatment, A/D and D/A converters are often called a sampler and a holder, respectively. The sampler, denoted by S, samples $y(t)$ with the constant period h to yield the discrete-time signal $y[k]$:

$$y[k] := y(kh) \tag{5.2}$$

while the holder, denoted by H, yields the control input $u(t)$ by holding the discrete-time signal $u[k]$ constant over the sampling interval:

$$u(t) := u[k] \quad \text{for} \quad kh \leq t < (k+1)h \tag{5.3}$$

The signal $u[k]$ is updated by the discrete-time controller K, which is described by the difference equation

$$K : \begin{cases} \xi[k+1] = A_K\,\xi[k] + B_K\,y[k] \\ u[k] = C_K\,\xi[k] + D_K\,y[k] \end{cases} \tag{5.4}$$

Assume that both S and H synchronise each other at the period h and that the initial states of G and K are zero: $x(0) = 0$ and $\xi[0] = 0$. Assume in addition that the environment of the system is characterised by the exogenous inputs having the form of

$$w(t) = w_1(t) + w_2(t) \tag{5.5}$$

The first element of w is known only to the extent that it belongs to the space $\mathcal{F}_1(M_1)$, comprising all piecewise smooth and continuous signals with bound M_1 on the rate of change and zero initial condition:

$$w_1 \in \mathcal{F}_1(M_1) := \left\{ f \in \mathcal{L}(\mathbb{R}_+, \mathbb{R}) : \|\dot{f}\|_{\mathcal{L}_\infty(\mathbb{R}_+, \mathbb{R})} \leq M_1,\ f(0) = 0 \right\}$$

The second element is known only to the extent that it belongs to the space $\mathcal{F}_2(M_2)$, comprising all piecewise smooth and continuous signals with bound M_2 on $\mathcal{L}_2$-norm of the rate of change and zero initial condition:

$$w_2 \in \mathcal{F}_2(M_2) := \left\{ f \in \mathcal{L}(\mathbb{R}_+, \mathbb{R}) : \|\dot{f}\|_{\mathcal{L}_2(\mathbb{R}_+, \mathbb{R})} \leq M_2,\ f(0) = 0 \right\}$$

The space of all exogenous inputs satisfying the assumption above is symbolised as $\mathcal{F}(M)$, where $M := [M_1\ M_2]$. This function space can widely be utilised as a model of persistent inputs, transient inputs or both. For instance, it is useful for tracking problems or slow-changing disturbance rejection problems.

This section considers the design for a sampled-data control system to be matched to the environment characterised by $\mathcal{F}(M)$, that is, to ensure

$$\|z(\mathcal{F}(M))\|_{\text{peak}} := \sup\{|z(t, w)| : t \in \mathbb{R}_+,\ w \in \mathcal{F}(M)\} \leq \varepsilon \tag{5.6}$$

Design parameters that satisfy (5.6) are determined by inequality solvers described in Part III. For this reason, a numerically tractable expression of (5.6), called a matching condition, is required. First, by using the lifting framework, the matching condition for $\mathcal{F}(M)$ is given. Next, some useful techniques for computer-aided design are presented.

5.2.2 Matching Condition for $\mathcal{F}(M)$

For the block diagram of Figure 5.1, the closed-loop system from w to z is characterised as a time-varying system. In general, however, time-invariant systems are easier to design and analyse than time-varying ones. To treat it as a time-invariant system, lifting is introduced. Roughly speaking, lifting is the cutting operation that converts a continuous-time signal to a function-space valued sequence. This section considers the lifting, L, that maps $f \in \mathcal{L}(\mathbb{R}_+, \mathbb{R})$ to $\hat{f} \in \ell(\mathbb{Z}_+, \mathcal{K})$ in the following manner:

$$L: f \mapsto \hat{f} := \{\hat{f}(k, \theta)\}_{k=0}^{\infty},\ \hat{f}(k, \theta) := f(kh + \theta),\ \theta \in [0, h) \tag{5.7}$$

Especially, the lifting mapping $\mathcal{L}_p(\mathbb{R}_+, \mathbb{R})$ to $\ell_p(\mathbb{Z}_+, \mathcal{K})$ is isomorphism, so the p-norm of a signal is unchanged after lifting:

$$\|f\|_{\mathcal{L}_p(\mathbb{R}_+, \mathbb{R})} = \|\hat{f}\|_{\ell_p(\mathbb{Z}_+, \mathcal{K}_p)} \tag{5.8}$$

This implies that maximum amplitude or energy of a signal is invariant after lifting. This enables the matched environment-system design, in which the exogenous inputs are characterised by p-norm of the rate of change, in the lifting framework.

Preliminary to the matching condition, consider the input-output relationship of a sampled-data system in the lifting framework. Let $\hat{z}$ and $\hat{w}$ be the lifted output and the lifted input, respectively:

$$\begin{aligned}&\hat{z} := \{\hat{z}(k, \theta)\}_{k=0}^{\infty},\ \hat{z}(k, \theta) = z(kh + \theta) \\ &\hat{w} := \{\hat{w}(k, \theta)\}_{k=0}^{\infty},\ \hat{w}(k, \theta) = w(kh + \theta)\end{aligned} \tag{5.9}$$

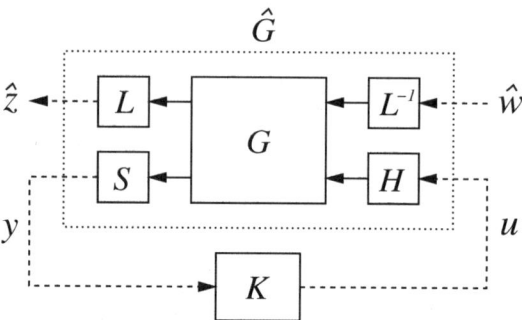

Fig. 5.2. Lifted system (L: lifting operator, L^{-1}: inverse operator of L)

To clarify the causality between $\hat{z}$ and $\hat{w}$, let the output for $\hat{w}$ be expressed explicitly as $\hat{z}(k, \theta, \hat{w})$. Then the realisation of G can be rewritten as $\hat{G}$, which is invariant to k:

$$\hat{G} : \begin{cases} x[k+1] = A_d\, x[k] + \hat{B}_w\, \hat{w}(k,\theta) + B_d\, u[k] \\ \hat{z}(k,\theta,\hat{w}) = \hat{C}_z\, x[k] + \hat{D}_w\, \hat{w}(k,\theta) + \hat{D}_u\, u[k] \\ y[k] = C_y\, x[k] + \hat{D}_y\, \hat{w}(k,\theta) \end{cases} \qquad (5.10)$$

where

$$A_d : \mathcal{R} \mapsto \mathcal{R}, \quad A_d\, v := e^{A_c h} v$$

$$\hat{B}_w : \mathcal{K} \mapsto \mathcal{R}, \quad \hat{B}_w\, v(\cdot) := \int_0^h e^{A_c(h-\tau)} B_w v(\tau) d\tau$$

$$B_d : \mathcal{R} \mapsto \mathcal{R}, \quad B_d\, v := \int_0^h e^{A_c \tau} d\tau\, B_u v$$

$$\hat{C}_z : \mathcal{R} \mapsto \mathcal{K}, \quad (\hat{C}_z\, v)(\theta) := C_z e^{A_c \theta} v$$

$$\hat{D}_w : \mathcal{K} \mapsto \mathcal{K}, \quad (\hat{D}_w\, v(\cdot))(\theta) := \int_0^\theta C_z e^{A_c(\theta-\tau)} B_w v(\tau) d\tau + D_w v(\theta)$$

$$\hat{D}_u : \mathcal{R} \mapsto \mathcal{K}, \quad (\hat{D}_u\, v)(\theta) := \int_0^\theta C_z e^{A_c \tau} d\tau\, B_u v + D_u v$$

$$\hat{D}_y : \mathcal{K} \mapsto \mathcal{R}, \quad \hat{D}_y\, v(\cdot) := \int_0^h D_y \delta(\tau) v(\tau) d\tau$$

The block diagram of the sampled-data system after lifting is shown in Figure 5.2. In this figure, the underlying base periods of $\hat{z}$, $\hat{w}$, y and u are identical and the dynamics of G is completely preserved in $\hat{G}$, so the closed-loop system from $\hat{w}$ to $\hat{z}$ can be viewed as a time-invariant discrete-time system that completely retains the intersample behaviour. Thus $\hat{z}$ can be related with $\hat{w}$ in convolution form. However, since w is characterised by rate of change, it is more natural to consider the relationship between z and $\dot{w}$. For the lifted system, it corresponds to the relationship between $\hat{z}$ and $\hat{w}^{(1)}$, where

$$\hat{w}^{(1)} := \{\hat{w}^{(1)}(k,\theta)\}_{k=0}^{\infty}, \quad \hat{w}^{(1)}(k,\theta) := \frac{d\hat{w}(k,\theta)}{d\theta} \tag{5.11}$$

Such a relationship is given as follows.

Lemma 5.1. *Consider the sampled-data system shown in Figure 5.2. Then $\hat{z}$ and $\hat{w}^{(1)}$ are related by*

$$\hat{z}(k,\theta,\hat{w}) = \sum_{l=0}^{k} \int_{0}^{h} \hat{z}(k-l,\theta,\hat{1}(\tau))\hat{w}^{(1)}(l,\tau)d\tau \tag{5.12}$$

where $\hat{1}(\tau)$ is the lifted signal of a unit step input with delay τ:

$$\hat{1}(\tau) := \{1(\theta - \tau),\ 1,\ 1,\ 1,\ldots\}, \quad \theta \in [0,h]$$

Proof. The equation (5.12) can be proven directly by calculating the unit step response. However, it would be easier to obtain (5.12) from the following the input-output relationship (Ono et al, 2000):

$$\hat{z}(k,\theta,\hat{w}) = (\hat{D}\hat{w})(k,\theta) + \sum_{l=0}^{k-1} \hat{C}(\theta)\hat{A}^{k-l-1}\hat{B}\hat{w}(l,\cdot) \tag{5.13}$$

where

$$\hat{A} := \begin{bmatrix} e^{Ah} + B_u(h)D_K C_y & B_u(h)C_K \\ B_K C_y & A_K \end{bmatrix} \quad B_u(\theta) := \int_0^\theta e^{A\tau} d\tau B_u$$

$$\hat{B}\hat{w}(l,\cdot) := \begin{bmatrix} \int_0^h \left\{ e^{A(h-\tau)} B_w + B_u(h) D_K D_y \delta(\tau) \right\} \hat{w}(l,\tau) d\tau \\ \int_0^h B_K D_y \delta(\tau) \hat{w}(l,\tau) d\tau \end{bmatrix}$$

$$\hat{C}(\theta) := [\, C_z e^{A\theta} + (D_u + C_z B_u(\theta))D_K C_y \quad (D_u + C_z B_u(\theta))C_K \,]$$

$$(\hat{D}\hat{w})(k,\theta) := C_z \int_0^\theta e^{A(\theta-\tau)} B_w \hat{w}(k,\tau) d\tau + D_w \hat{w}(k,\theta)$$

$$+ (D_u + C_z B_u(\theta))D_K D_y \hat{w}(k,0)$$

It can be verified from (5.13) that

$$\hat{z}(k,\theta,\hat{w}) = -\sum_{l=0}^{k} \int_0^h \frac{d}{d\tau}\hat{z}(k-l,\theta,\hat{1}(\tau))\hat{w}(l,\tau)d\tau \tag{5.14}$$

Integrating each term of the right hand by parts under the condition that $\hat{w}(0,0) = 0$, $\hat{w}(k,h) = \hat{w}(k+1,0)$ and $\hat{z}(k,\theta,\hat{1}(h)) = \hat{z}(k-1,\theta,\hat{1}(0))$, the equation (5.12) is obtained. □

Next consider the matching condition for $\mathcal{F}(M)$ in the lifting framework. Let $\hat{\mathcal{F}}(M)$ denote the space of all lifted inputs given in the form of

$$\hat{w} = \hat{w}_1 + \hat{w}_2 \tag{5.15}$$

where $\hat{w}_1$ and $\hat{w}_2$ are the lifted inputs of w_1 and w_2 respectively, that is,

$$\hat{w}_1 \in \hat{\mathcal{F}}_1(M_1) := \left\{ \hat{f} \in \ell(\mathbb{Z}_+,\mathcal{K}) : \|\hat{f}^{(1)}\|_{\ell_\infty(\mathbb{Z}_+,\mathcal{K}_\infty)} \leq M_1,\ \hat{f}(0,0) = 0 \right\}$$

and

$$\hat{w}_2 \in \hat{\mathcal{F}}_2(M_2) := \left\{ \hat{f} \in \ell(\mathbb{Z}_+,\mathcal{K}) : \|\hat{f}^{(1)}\|_{\ell_2(\mathbb{Z}_+,\mathcal{K}_2)} \leq M_2,\ \hat{f}(0,0) = 0 \right\}$$

Define the peak norm of $\hat{z}$ as

$$\|\hat{z}(\hat{\mathcal{F}}(M))\|_{\text{peak}} := \sup\left\{ |\hat{z}(k,\theta,\hat{w})| : k \in \mathbb{Z}_+,\ \theta \in [0,h],\ \hat{w} \in \hat{\mathcal{F}}(M) \right\} \tag{5.16}$$

Since $\mathcal{F}_i(M_i)$ and $\hat{\mathcal{F}}_i(M_i)$ are isomorphic,

$$\|z(\mathcal{F}(M))\|_{\text{peak}} = \|\hat{z}(\hat{\mathcal{F}}(M))\|_{\text{peak}} \tag{5.17}$$

Therefore the inequality (5.6) is identical to

$$\|\hat{z}(\hat{\mathcal{F}}(M))\|_{\text{peak}} \leq \varepsilon \tag{5.18}$$

The numerically tractable expression of (5.18) is derived from (5.12) as follows.

Theorem 5.1. *The matching condition for $\mathcal{F}(M)$, that is, the necessary and sufficient condition for (5.6), is given as*

$$\sup\{M_1\phi_1(\theta) + M_2\phi_2(\theta) : \theta \in [0,h]\} \leq \varepsilon \tag{5.19}$$

where $\phi_i(\theta)$ is of the following form.

$$\phi_i(\theta) = \begin{cases} \sum_{k=0}^{\infty} \int_0^h |\hat{z}(k,\theta,\hat{1}(\tau))|d\tau & \text{for}\ \ i=1 \\ \left(\sum_{k=0}^{\infty} \int_0^h |\hat{z}(k,\theta,\hat{1}(\tau))|^2 d\tau\right)^{\frac{1}{2}} & \text{for}\ \ i=2 \end{cases} \tag{5.20}$$

Proof. From (5.16),

$$\|\hat{z}(\hat{\mathcal{F}}(M))\|_{\text{peak}} = \sup_{\theta \in [0,h)} \sup \left\{ |\hat{z}(k,\theta,\hat{w})| : k \in \mathbb{Z}_+, \hat{w} \in \hat{\mathcal{F}}(M) \right\} \quad (5.21)$$

Comparing (5.21) with (5.19), it is sufficient to prove

$$\sup \left\{ |\hat{z}(k,\theta,\hat{w})| : k \in \mathbb{Z}_+, \hat{w} \in \hat{\mathcal{F}}(M) \right\} = M_1 \phi_1(\theta) + M_2 \phi_2(\theta) \quad (5.22)$$

It turns out from (5.12) that, for any $k \in \mathbb{Z}_+$ and $\hat{w} \in \hat{\mathcal{F}}(M)$,

$$|\hat{z}(k,\theta,\hat{w})| \leq \sum_{l=0}^{k} \int_0^h |\hat{z}(k-l,\theta,\hat{1}(\tau))| |\hat{w}_1^{(1)}(l,\tau)| d\tau$$

$$+ \left(\sum_{l=0}^{k} \int_0^h |\hat{z}(k-l,\theta,\hat{1}(\tau))|^2 d\tau \right)^{\frac{1}{2}} \left(\sum_{l=0}^{k} \int_0^h |\hat{w}_2^{(1)}(l,\tau)|^2 d\tau \right)^{\frac{1}{2}}$$

$$\leq M_1 \phi_1(\theta) + M_2 \phi_2(\theta)$$

Now introduce the input $\hat{v}_{k,\theta} := \hat{v}_{1k,\theta} + \hat{v}_{2k,\theta} \in \hat{\mathcal{F}}(M)$, where $\hat{v}_{ik,\theta}$ belongs to $\hat{\mathcal{F}}_i(M_i)$ and satisfies

$$\hat{v}_{1k,\theta}^{(1)}(l,\tau) := \begin{cases} M_1 \, \text{sgn}\{\hat{z}(k-l,\theta,\hat{1}(\tau))\} & \text{for } 0 \leq l \leq k \\ 0 & \text{for } l > k \end{cases}$$

$$\hat{v}_{2k,\theta}^{(1)}(l,\tau) := \begin{cases} \dfrac{M_2 \hat{z}(k-l,\theta,\hat{1}(\tau))}{N(k,\theta)} & \text{for } 0 \leq l \leq k \\ 0 & \text{for } l > k \end{cases}$$

in which

$$N(k,\theta) := \left(\sum_{l=0}^{k} \int_0^h |\hat{z}(l,\theta,\hat{1}(\tau))|^2 d\tau \right)^{\frac{1}{2}}$$

For any positive real number δ, $|\hat{z}(\infty, \theta, \hat{v}_{\infty,\theta})| > M_1 \phi_1(\theta) + M_2 \phi_2(\theta) - \delta$, so (5.22) is proved. $\square$

For (5.19) to be satisfied, $\phi_i(\theta)$ must be at least finite. This requires a control system to have integral action, that is, $z(t) \to 0 \, (t \to \infty)$ when applying the unit step input to a system. This requirement is essential to the matched environment-system design for $\mathcal{F}(M)$.

Next consider the special case, the unity feedback control system depicted in Figure 5.3. Supposed that w is the reference command signal. If z is the

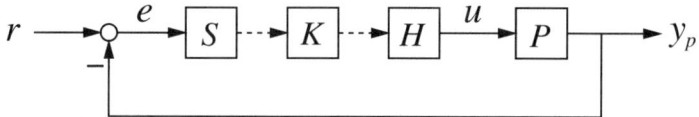

Fig. 5.3. Unit feedback control system

tracking error e, $\phi_i(\theta)$ can be rewritten simply as

$$\phi_i(\theta) = \begin{cases} h \sum_{k=0}^{\infty} |\hat{z}(k,\theta,\hat{1}(0))| + \theta & \text{for } i = 1 \\ \left(h \sum_{k=0}^{\infty} |\hat{z}(k,\theta,\hat{1}(0))|^2 + \theta \right)^{\frac{1}{2}} & \text{for } i = 2 \end{cases} \quad (5.23)$$

If z is the control input u,

$$\phi_i(\theta) = \begin{cases} h \sum_{k=0}^{\infty} |\hat{z}(k,0,\hat{1}(0))| & \text{for } i = 1 \\ \left(h \sum_{k=0}^{\infty} |\hat{z}(k,0,\hat{1}(0))|^2 \right)^{\frac{1}{2}} & \text{for } i = 2 \end{cases} \quad (5.24)$$

The function $\phi_i(\theta)$ can be calculated only from the response to the unit step with no delay. Compared to the general case, calculation cost is saved. The equations (5.23) and (5.24) can be derived from (5.20). For more details of their proofs, see Ono et al (2002). The matching condition for a unity feedback system is also seen in Whidborne and Liu (1993).

5.2.3 Evaluation of the Matching Condition using the Fast-discretisation

To evaluate the matching condition (5.19), it is necessary to calculate $\phi_i(\theta)$. Numerical integration is one way to do it. However, this section shows an alternative method. It is based on the fast-discretisation, a technique for discretisation of a system.

Zero-order hold approximation is often used when obtaining a discrete-time model of a system. It discretises a system by inserting the sample and zero-order hold operators fictitiously to a system. Figure 5.4 explains the discretisation of G based on zero-order hold approximation. Fictitious sample and zero-order hold operators are inserted at the channels of z and w, respectively. The system G_d, which is enclosed by a dotted line, is referred to as a discrete-time model of G. The fast-discretisation is similar to this approximation. The difference is that the fictitious sample and hold operators with

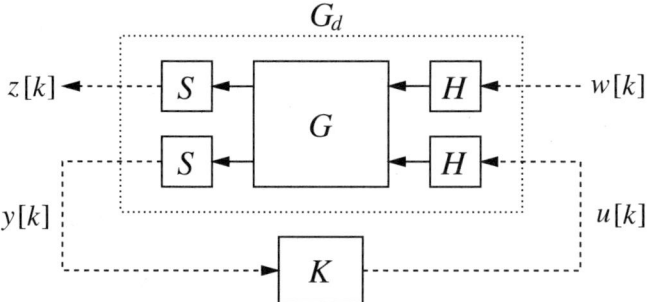

Fig. 5.4. Conventional discretisation of G

period h_f ($:=h/n$, n: positive integer), which work n-times faster than the actual sampler and holder, are inserted. In this case, the closed-loop system is time-varying since the base period of fictitiously sampled input and output differs from that of $\{y[k]\}$ and $\{u[k]\}$. To convert such a time-varying system to a time-invariant one, lifting for discrete-time signals is introduced. In this section, let us simply call it discrete lifting.

The discrete lifting, L_n, is the stacking operator that maps $\ell(\mathbb{Z}_+, \mathbb{R})$ to $\ell(\mathbb{Z}_+, \mathbb{R}^n)$ in the following manner:

$$L_n : \{f[0], f[1], f[2], \cdots\} \mapsto \left\{ \begin{bmatrix} f[0] \\ f[1] \\ \vdots \\ f[n-1] \end{bmatrix}, \begin{bmatrix} f[n] \\ f[n+1] \\ \vdots \\ f[2n-1] \end{bmatrix}, \cdots \right\}$$

Applying the discrete lifting to the fictitiously sampled input and output, they are changed into n-dimensional sequences with the same period as $\{y[k]\}$ and $\{u[k]\}$. Therefore the closed-loop system with such lifted input and output can be viewed as a time-invariant system. Now let $w[k, m]$ and $z[k, m]$ be the input and the output fictitiously sampled at time $t = kh + mh_f$, where $m \in M_n := \{0, 1, \ldots, n-1\}$. Let $\hat{w}$ and $\hat{z}$ be the lifted input and the lifted output, respectively:

$$\hat{w} := \{\hat{w}[0], \hat{w}[1], \hat{w}[2], \cdots\}, \quad \hat{z} := \{\hat{z}[0], \hat{z}[1], \hat{z}[2], \cdots\}$$

where

$$\hat{w}[k] := \begin{bmatrix} w[k, 0] \\ w[k, 1] \\ \vdots \\ w[k, n-1] \end{bmatrix}, \quad \hat{z}[k] := \begin{bmatrix} z[k, 0] \\ z[k, 1] \\ \vdots \\ z[k, n-1] \end{bmatrix}$$

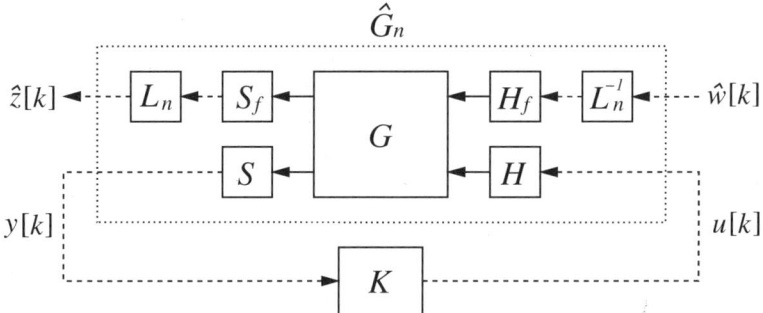

Fig. 5.5. Fast-discretised system (S_f: Fictitious sampler with period h_f, H_f: Fictitious holder with period h_f, L_n: Lifting operator, L_n^{-1}: inverse operator of L_n)

The block diagram explaining the fast-discretisation of G is shown in Figure 5.5. The system $\hat{G}_n$, enclosed by a dotted line, is referred to as a fast-discretised model of G. The realisation of $\hat{G}_n$ is given by

$$\hat{G}_n : \begin{cases} x[k+1] = A_d\, x[k] + \hat{B}_w\, \hat{w}[k] + B_d\, u[k] \\ \hat{z}[k] = \hat{C}_z\, x[k] + \hat{D}_w\, \hat{w}[k] + \hat{D}_u u[k] \\ y[k] = C_y\, x[k] + \hat{D}_y\, \hat{w}[k] \end{cases} \quad (5.25)$$

where

$$A_d := e^{A_c h}, \quad B_d := \int_0^h e^{A_c \tau} d\tau\, B_u$$

$$\hat{B}_w := [\, A_f^{n-1} B_{wf} \quad A_f^{n-2} B_{wf} \quad \cdots \quad B_{wf}\,]$$

$$\hat{C}_z := \begin{bmatrix} C_z \\ C_z A_f \\ \vdots \\ C_z A_f^{n-1} \end{bmatrix}, \quad \hat{D}_w := \begin{bmatrix} D_w & 0 & \cdots & 0 \\ C_z B_{wf} & D_w & \cdots & 0 \\ \vdots & \vdots & \ddots & 0 \\ C_z A_f^{n-2} B_{wf} & C_z A_f^{n-3} B_{wf} & \cdots & D_w \end{bmatrix}$$

$$\hat{D}_u := \begin{bmatrix} D_u \\ C_z B_{uf} + D_u \\ \vdots \\ \sum_{k=0}^{n-2} C_z A_f^k B_{uf} + D_u \end{bmatrix}, \quad \hat{D}_y := [\, D_y \quad 0 \quad \cdots \quad 0\,]$$

in which

$$A_f := e^{A_c h_f}, \quad B_{wf} := \int_0^{h_f} e^{A_c \tau} d\tau\, B_w, \quad B_{uf} := \int_0^{h_f} e^{A_c \tau} d\tau\, B_u$$

The system $\hat{G}_n$ partially includes the dynamics of G between the sampling instants if $n > 1$. If $n = 1$, the intersample response of G is completely ignored and $\hat{G}_n$ is identical to G_d. Accordingly, the fast-discretisation technique covers the conventional zero-order hold approximation. A technique similar to the fast-discretisation is utilised in linear simulation of sampled-data systems. The same idea is also seen in modified z-transform.

The relationship between $\hat{z}$ and $\hat{w}$ can be given in the form of a convolution equation since the closed-loop system comprising $\hat{G}_n$ and K is time-invariant. However, like the equation (5.12), it is natural to relate $\hat{z}$ with the rate of change of inputs. To do this, some definitions are given. The rate of change of $w[k, m]$ is defined as

$$\Delta w[k, m] := \frac{w[k, m] - w[k, m - 1]}{h_f} \tag{5.26}$$

and $\Delta \hat{w}[k]$ as the n-dimensional vector as follows.

$$\Delta \hat{w}[k] := \begin{bmatrix} \Delta w[k, 0] \\ \Delta w[k, 1] \\ \vdots \\ \Delta w[k, n-1] \end{bmatrix} \tag{5.27}$$

The lifted signal of a discrete unit step with j steps delay is defined as

$$\hat{1}_j := \left\{ \underset{(j+1)}{\begin{bmatrix} 0 \\ \vdots \\ 0 \\ 1 \\ \vdots \\ 1 \end{bmatrix}}, \begin{bmatrix} 1 \\ \vdots \\ \vdots \\ \vdots \\ 1 \end{bmatrix}, \begin{bmatrix} 1 \\ \vdots \\ \vdots \\ \vdots \\ 1 \end{bmatrix}, \cdots \right\} \tag{5.28}$$

Furthermore, let the response to $\hat{w}$ be expressed as $z[k, m, \hat{w}]$. Then $\hat{z}$ and $\Delta \hat{w}$ are related as follows.

Lemma 5.2. *Consider the discrete-time closed-loop system comprising $\hat{G}_n$ and K. Then the input-output relationship is given by*

$$z[k, m, \hat{w}] = h_f \sum_{l=0}^{k} \sum_{j=0}^{n-1} z[k - l, m, \hat{1}_j] \Delta w[l, j] \tag{5.29}$$

Proof. The input-output relationship can be given in the form of

$$\hat{z}[k, \hat{w}] = \sum_{l=0}^{k} \Phi[k - l] \hat{w}[l] \tag{5.30}$$

where $\Phi[k]$ is an n-by-n real matrix and

$$\hat{z}[k,\hat{w}] := \begin{bmatrix} z[k,0,\hat{w}] \\ z[k,1,\hat{w}] \\ \vdots \\ z[k,n-1,\hat{w}] \end{bmatrix} \tag{5.31}$$

Defining $\Phi_{(i,j)}[k]$ as the (i,j) entry of $\Phi[k]$, (5.30) can be expanded as

$$z[k,m,\hat{w}] = \sum_{l=0}^{k}\sum_{j=0}^{n-1} \Phi_{(m+1,j+1)}[k-l]w[l,j] \tag{5.32}$$

Consider the responses to the inputs $\hat{1}_j$ and $\hat{1}_{j+1}$. It turns out from (5.32) that

$$z[k,m,\hat{1}_j] - z[k,m,\hat{1}_{j+1}] = \Phi_{(m+1,j+1)}[k]. \tag{5.33}$$

Substituting (5.33) into (5.32),

$$z[k,m,\hat{w}] = \sum_{l=0}^{k}\sum_{j=0}^{n-1} \{z[k-l,m,\hat{1}_j] - z[k-l,m,\hat{1}_{j+1}]\} w[l,j] \tag{5.34}$$

Rearranging (5.34) under the condition that $w[0,0] = 0$, $z[k,m,\hat{1}_n] = z[k-1,m,\hat{1}_0]$ and $z[0,m,\hat{1}_{m+1}] = z[0,m,\hat{1}_{m+2}] = \cdots = z[0,m,\hat{1}_{n-1}] = 0$, (5.29) is obtained. $\square$

As can be seen from Figure 5.5, the exogenous input of the actual plant G is $H_f L_n^{-1} \hat{w}$, that is, a step-like input which is generated by letting the input sequence $L_n^{-1}\hat{w} = \{w[0,0], w[0,1], w[0,2], \ldots\}$ through the holder H_f. The equation (5.29) gives the output of G at time $t = kh + mh_f$ for such a step-like input.

Next consider the maximum of an output of the system (5.29). Let $\hat{\mathcal{F}}(M,n)$ be the space of all inputs having the form of

$$\hat{w}[k] = \hat{w}_1[k] + \hat{w}_2[k] \tag{5.35}$$

where $\hat{w}_1$ belongs to the vector space

$$\hat{\mathcal{F}}_1(M_1, n) := \left\{ \hat{f} \in \ell(\mathbb{Z}_+, \mathbb{R}^n) : \|\Delta \hat{f}\|_{\ell_\infty(\mathbb{Z}_+, \mathbb{R}^n)} \leq M_1,\ f[0,0] = 0 \right\}$$

and $\hat{w}_2$ belongs to the vector space

$$\hat{\mathcal{F}}_2(M_2, n) := \left\{ \hat{f} \in \ell(\mathbb{Z}_+, \mathbb{R}^n) : \|\Delta \hat{f}\|_{\ell_2(\mathbb{Z}_+, \mathbb{R}^n)} \leq \frac{M_2}{\sqrt{h_f}},\ f[0,0] = 0 \right\}$$

The space $\hat{\mathcal{F}}(M,n)$ is the counterpart of $\mathcal{F}(M)$ in the sense that $w \in \mathcal{F}(M)$ can be approximated by $H_f L_n^{-1} \hat{w}$ for $\hat{w} \in \hat{\mathcal{F}}(M,n)$. Therefore, $\|z(\mathcal{F}(M))\|_{\text{peak}}$ can be approximated by the peak norm of (5.29), which is defined by

$$\|\hat{z}(\hat{\mathcal{F}}(M,n))\|_{\text{peak}} := \sup\left\{|z[k,m,\hat{w}]| : k \in \mathbb{Z}_+,\ m \in M_n,\ \hat{w} \in \hat{\mathcal{F}}(M,n)\right\} \tag{5.36}$$

Especially, if n is sufficiently large, it can be well approximated by (5.36). In fact, it is known that (Ono et al, 2002)

$$\|\hat{z}(\hat{\mathcal{F}}(M,n))\|_{\text{peak}} \to \|z(\mathcal{F}(M))\|_{\text{peak}} \quad (n \to \infty) \tag{5.37}$$

The computable expression of (5.36) is given as follows.

Theorem 5.2. *Consider the discrete-time system whose input-output relationship is given by (5.29). Then (5.36) is calculated by*

$$\|\hat{z}(\hat{\mathcal{F}}(M,n))\|_{peak} = \max\left\{M_1\phi_1(m,n) + M_2\phi_2(m,n) : m \in M_n\right\} \tag{5.38}$$

where $\phi_i(m,n)$ is of the following form.

$$\phi_i(m,n) = \begin{cases} h_f \sum_{j=0}^{n-1} \sum_{k=0}^{\infty} |z[k,m,\hat{1}_j]| & \text{for}\ \ i=1 \\ \sqrt{h_f} \left(\sum_{j=0}^{n-1} \sum_{k=0}^{\infty} |z[k,m,\hat{1}_j]|^2 \right)^{\frac{1}{2}} & \text{for}\ \ i=2 \end{cases} \tag{5.39}$$

Proof. The process of proving (5.38) is similar to that of Theorem 5.1 except that (5.38) is for a fast-discretised system, so let us prove only

$$\sup\left\{|z[k,m,\hat{w}]| : k \in \mathbb{Z}_+,\ \hat{w} \in \hat{\mathcal{F}}(M,n)\right\} = M_1\phi_1(m,n) + M_2\phi_2(m,n)$$

It follows from (5.29) that, for any $k \in \mathbb{Z}_+$ and $\hat{w} \in \hat{\mathcal{F}}(M,n)$,

$$|z[k,m,\hat{w}]| \le h_f \sum_{l=0}^{k} \sum_{j=0}^{n-1} |z[k-l,m,\hat{1}_j]||\Delta w_1[l,j]|$$

$$+ h_f \left(\sum_{l=0}^{k}\sum_{j=0}^{n-1} |z[k-l,m,\hat{1}_j]|^2\right)^{\frac{1}{2}} \left(\sum_{l=0}^{k}\sum_{j=0}^{n-1} |\Delta w_2[l,j]|^2\right)^{\frac{1}{2}}$$

$$\le M_1\phi_1(m,n) + M_2\phi_2(m,n)$$

Introduce the input $\hat{v}_{k,m} := \hat{v}_{1k,m} + \hat{v}_{2k,m} \in \hat{\mathcal{F}}(M,n)$, where $\hat{v}_{ik,m}$ belongs to $\hat{\mathcal{F}}_i(M_i,n)$ and satisfies

$$\Delta v_{1k,m}[l,j] := \begin{cases} M_1 \operatorname{sgn}\{z[k-l,m,\hat{\imath}_j]\} & \text{for } 0 \le l \le k \\ 0 & \text{for } l > k \end{cases}$$

$$\Delta v_{2k,m}[l,j] := \begin{cases} \dfrac{M_2 z[k-l,m,\hat{\imath}_j]}{\sqrt{h_f} N[k,m]} & \text{for } 0 \le l \le k \\ 0 & \text{for } l > k \end{cases}$$

in which

$$N[k,m] := \left(\sum_{l=0}^{k} \sum_{j=0}^{n-1} |z[l,m,\hat{\imath}_j]|^2 \right)^{\frac{1}{2}}$$

For any positive real number δ, $|z[\infty,m,\hat{v}_{\infty,m}]| > M_1 \phi_1(m,n) + M_2 \phi_2(m,n) - \delta$. Thus the equation (5.38) holds. $\square$

For a unity feedback system shown in Figure 5.3, $\phi_i(m,n)$ can be calculated more easily. Suppose that w is the reference command signal r. If z is the tracking error e,

$$\phi_i(m,n) = \begin{cases} h \left(\displaystyle\sum_{k=0}^{\infty} |z[k,m,\hat{\imath}_0]| + \dfrac{m}{n} \right) & \text{for } i = 1 \\ \sqrt{h} \left(\displaystyle\sum_{k=0}^{\infty} |z[k,m,\hat{\imath}_0]|^2 + \dfrac{m}{n} \right)^{\frac{1}{2}} & \text{for } i = 2 \end{cases} \tag{5.40}$$

If z is the control input u,

$$\phi_i(m,n) = \begin{cases} h \displaystyle\sum_{k=0}^{\infty} |z[k,0,\hat{\imath}_0]| & \text{for } i = 1 \\ \sqrt{h} \left(\displaystyle\sum_{k=0}^{\infty} |z[k,0,\hat{\imath}_0]|^2 \right)^{\frac{1}{2}} & \text{for } i = 2 \end{cases} \tag{5.41}$$

The equations (5.40) and (5.41) can be obtained from (5.39). The details of their proofs can be found in Ono et al (2002).

Consequently the matching condition (5.19) can be checked nearly by

$$\max \{M_1 \phi_1(m,n) + M_2 \phi_2(m,n) : m \in M_n\} \le \varepsilon \tag{5.42}$$

Especially, if n is sufficiently large, the matching condition can be evaluated reliably by (5.42). Also, since the fast-discretisation includes the conventional zero-order hold approximation, (5.42) can be interpreted as a matching condition for discrete-time systems incorporating the intersample behaviour.

5.2.4 Matched Environment-system Design

There are three approaches to matched environment-system design: digital redesign, discrete design and direct design. In each approach, the design parameters are determined so that the matching condition (5.19) is met. Particularly, in the direct design, they are searched by an inequality solver subject to (5.19). In the discrete design, they are searched subject to the matching condition for a discrete-time system (5.42) until (5.19) is met. For $n = 1$, (5.42) is identical to the matching condition for the discrete-time system consisting of G_d and K. If n is sufficiently large, it is almost equivalent to (5.19). Thus the discrete design involving (5.42) covers the direct design in practice. For this reason, some practical and useful techniques to evaluate (5.42), which can be used in computer-aided design, are presented.

Sampling period. Consider the unity feedback system. Suppose that w is a reference command signal and z is a tracking error. Since $z[0, 0, \hat{1}_0] = 1$,

$$M_1 \phi_1(m, n) + M_2 \phi_2(m, n) \geq M_1 h + M_2 \sqrt{h} \tag{5.43}$$

Thus the sampling period h must be chosen so that the inequality

$$M_1 h + M_2 \sqrt{h} \leq \varepsilon \tag{5.44}$$

is at least satisfied. The condition (5.44) can also be derived from (5.23). However, it is just the necessary condition. When determining the sampling period, it is necessary to consider other aspects such as the natural frequencies of a plant or the bandwidth of an anti-arising filter if it is integrated in P.

Calculation of the unit step response. When evaluating (5.42), it is necessary to compute the unit step response $\hat{z}[k, \hat{1}_j]$. Supposing that the closed-loop system has integral action, a method for calculating $\hat{z}[k, \hat{1}_j]$ is presented. Let F_j denote the n-dimensional vector where up to jth element is 0 and the rest of entries are 1:

$$F_j := [\underbrace{0, \ldots, 0}_{j}, \underbrace{1, \ldots, 1}_{n-j}]^t \tag{5.45}$$

and let A_{cl}, B_{cl}, C_{cl} and D_{cl} denote the coefficient matrices of the state-space model of the closed-loop system comprising $\hat{G}_n$ and K:

$$A_{cl} := \begin{bmatrix} A_d + B_d D_K C_y & B_d C_K \\ B_K C_y & A_K \end{bmatrix}$$

$$B_{cl} := \begin{bmatrix} \hat{B}_w + B_d D_K \hat{D}_y \\ B_K \hat{D}_y \end{bmatrix}$$

$$C_{cl} := \begin{bmatrix} \hat{C}_z + \hat{D}_u D_K C_y & \hat{D}_u C_K \end{bmatrix}$$

$$D_{cl} := \hat{D}_w + \hat{D}_u D_K \hat{D}_y.$$

The initial states of G and K are zero, so $\hat{z}[0, \hat{1}_j] = D_{cl}F_j$. For $k \geq 1$, $\hat{z}[k, \hat{1}_j]$ is equal to $\hat{\zeta}[k-1]$, which is the output of the system

$$\begin{cases} \varphi[i+1] = A_{cl}\,\varphi[i] + B_{cl}F_0\,v[i] \\ \hat{\zeta}[i] = C_{cl}\,\varphi[i] + D_{cl}F_0\,v[i] \end{cases} \tag{5.46}$$

where $v[i] = 1$ and $\varphi[0] = B_{cl}F_j$. Let $T(z)$ denote the transfer function matrix from v to $\hat{\zeta}$, that is, $T(z) := C_{cl}(zI - A_{cl})^{-1}B_{cl}F_0 + D_{cl}F_0$. Since the closed-loop system has integral action, A_{cl} is stable and $\hat{\zeta}[\infty] = 0$. Therefore, from the final value theorem,

$$\hat{\zeta}[\infty] = T(1) = C_{cl}(I - A_{cl})^{-1}B_{cl}F_0 + D_{cl}F_0 = 0 \tag{5.47}$$

Combining (5.46) with (5.47),

$$\hat{\zeta}[i] = C_{cl}A_{cl}^i\{\varphi[0] - (I - A_{cl})^{-1}B_{cl}F_0\} \tag{5.48}$$

Recalling $\hat{z}[k, \hat{1}_j] = \hat{\zeta}[k-1]$, $\hat{z}[k, \hat{1}_j]$ is computed by

$$\begin{cases} \eta[k+1] = A_{cl}\,\eta[k], \quad \eta[0] = B_{cl}F_j - (I - A_{cl})^{-1}B_{cl}F_0 \\ \hat{z}[k+1, \hat{1}_j] = C_{cl}\,\eta[k], \quad \hat{z}[0, \hat{1}_j] = D_{cl}F_j \end{cases} \tag{5.49}$$

For a unity feedback control system, $\phi_i(m, n)$ is computed from $\{\hat{z}[k, \hat{1}_0]\}_{k=0}^{\infty}$. In this case, $\hat{z}[k, \hat{1}_0]$ can be calculated as $\hat{\zeta}[k]$ by setting $v[i] = 1$ and $\varphi[0] = 0$ in (5.46). Accordingly it is obtained from the difference equation

$$\begin{cases} \eta[k+1] = A_{cl}\,\eta[k], \quad \eta[0] = -(I - A_{cl})^{-1}B_{cl}F_0 \\ \hat{z}[k, \hat{1}_0] = C_{cl}\,\eta[k] \end{cases} \tag{5.50}$$

Evaluation of the matching condition based on finite repeated calculation. Evaluation of (5.42) requires the calculation of the unit step response $\hat{z}[k, \hat{1}_j]$ for all $k \in \mathbb{Z}_+$. However, if $\hat{z}[\infty, \hat{1}_j] = 0$, it is possible to avoid such an infinite-time computation. Here is discussed a finite repeated calculation for evaluation of (5.42). Let γ be the maximum singular value of A_{cl} and

$$\alpha := \max\{\|c_i\| \|B_{cl}F_j - (I - A_{cl})^{-1}B_{cl}F_0\| : i \in \{1, 2, \ldots, n\}\} \tag{5.51}$$

where c_i is the ith row vector of C_{cl}. From (5.49),

$$|z[k+1, m, \hat{1}_j]| \leq \alpha\gamma^k \tag{5.52}$$

Introduce the function $E(m, n)$, which is defined by

$$E(m, n) := \sum_{i=1}^{2} M_i \phi_i(m, n) - \sum_{i=1}^{2} M_i \bar{\phi}_{i,N}(m, n) \tag{5.53}$$

where N is a positive integer and

$$\bar{\phi}_{i,N}(m,n) := \begin{cases} h_f \sum_{j=0}^{n-1} \sum_{k=0}^{N} |z[k,m,\hat{1}_j]| & \text{for } i=1 \\ \sqrt{h_f} \left(\sum_{j=0}^{n-1} \sum_{k=0}^{N} |z[k,m,\hat{1}_j]|^2 \right)^{\frac{1}{2}} & \text{for } i=2 \end{cases} \quad (5.54)$$

The function $E(m,n)$ means the loss when $M_1\phi_1(m,n) + M_2\phi_2(m,n)$ is replaced by $M_1\bar{\phi}_{1,N}(m,n) + M_2\bar{\phi}_{2,N}(m,n)$. The closed-loop system has integral behaviour, so γ is less than 1. This yields

$$E(m,n) < \left(\frac{M_1 h}{1-\gamma} + M_2 \sqrt{\frac{h}{1-\gamma^2}} \right) \alpha \gamma^N \quad (5.55)$$

Now consider the positive integer N such that the right hand of (5.55) is smaller than a real number μ ($0 < \mu < \varepsilon$). It lies in the range

$$N > \frac{\log \mu - \log \alpha \left(\frac{M_1 h}{1-\gamma} + M_2 \sqrt{\frac{h}{1-\gamma^2}} \right)}{\log \gamma} \quad (5.56)$$

For such N, $E(m,n)$ is smaller than μ. This means that, if

$$\max \{ M_1 \bar{\phi}_{1,N}(m,n) + M_2 \bar{\phi}_{2,N}(m,n) : m \in M_n \} < \varepsilon - \mu \quad (5.57)$$

is met for the integer N satisfying (5.56), (5.42) is also met. That is, if (5.57) is checked, it is not necessary to perform an infinite-time-calculation of the unit step response. But note that the condition (5.57) as well as (5.42) is neither a necessary and sufficient condition nor a sufficient condition for (5.6). To confirm the environment-system match exactly, it is necessary to evaluate (5.19).

Determination of n. If n is sufficiently large, $\|\hat{z}(\hat{\mathcal{F}}(M,n))\|_{\text{peak}}$ gives a good approximation of the actual peak norm but much calculation time. To save the calculation time, how should n be chosen? Although limited only to the unity feedback system design, an answer to this question is given.

Consider the unity feedback system shown in Figure 5.3. Suppose that w is a reference command signal and z is a tracking error. It follows from (5.40) that

$$\phi_i(m, 2^i) = \phi_i(2m, 2^{i+1}) \quad (5.58)$$

Define the sequence $\Phi(n)$ as

$$\Phi(n) := \{ M_1 \phi_1(m,n) + M_2 \phi_2(m,n) \}_{m=0}^{n-1} \quad (5.59)$$

Then, $\Phi(2^i) \subset \Phi(2^{i+1})$. This means that

$$\|\hat{z}(\hat{\mathcal{F}}(M, 2^i))\|_{\text{peak}} \leq \|\hat{z}(\hat{\mathcal{F}}(M, 2^{i+1}))\|_{\text{peak}} \tag{5.60}$$

From (5.37) and (5.60), it turns out that $\{\|\hat{z}(\hat{\mathcal{F}}(M, 2^i))\|_{\text{peak}}\}_{i=0}^{\infty}$ is the monotone increasing sequence converging on $\|z(\mathcal{F}(M))\|_{\text{peak}}$. Denoting the set of all design parameters satisfying (5.19) and (5.42) by $\mathcal{S}$ and $\mathcal{S}(n)$ respectively, this indicates that

$$\mathcal{S}(2^i) \supseteq \mathcal{S}(2^{i+1}) \tag{5.61}$$

and

$$\mathcal{S}(2^i) \to \mathcal{S} \quad (i \to \infty) \tag{5.62}$$

In short, $\mathcal{S}(n)$ converges on $\mathcal{S}$ asymptotically from outside as n increases twice. This leads us to a method for choosing n in design of a unity feedback system. Now set $n = 2^i$, where i is a non-negative integer. In the admissible problem of finding the design parameter p that satisfies (5.42), the inequality solver stops the search when p moves into the inside of $\mathcal{S}(2^i)$. However, it is not necessarily the solution to (5.19). If it is not so, the inequality solver will change the initial search point of p and start the search again. But, instead, by updating n to $n = 2^{i+1}$, it is possible to carry on the search from the last stopped point since the boundary of $\mathcal{S}(2^{i+1})$ moves into the inside of $\mathcal{S}(2^i)$. In short, a further search can be continued by updating n twice. Compared to the case where n is fixed to be large or n is fixed to be small but the initial starting point is frequently changed, the calculation cost would be saved. On the other hand, if z is a control input, $\mathcal{S}(n) = \mathcal{S}$ for any n. But note that, in more general case, (5.61) does not hold.

5.3 Design for MIMO Systems

This section extends the results for SISO systems to the case of MIMO systems. As in the preceding section, mathematical symbols with *hat* represent lifted signals.

Consider the sampled-data system shown in Figure 5.1, where $z(t)$ is an output of dimension n_z and $w(t)$ is an input of dimension n_w:

$$z(t) = [z_1(t), \ldots, z_{n_w}(t)]^t, \quad w(t) = [w_1(t), \ldots, w_{n_w}(t)]^t$$

Assume that each component of $w(t)$ is given in the form of a linear combination of two inputs:

$$w_j(t) = w_{j1}(t) + w_{j2}(t) \tag{5.63}$$

where $w_{j1} \in \mathcal{F}_1(M_{j1})$ and $w_{j2} \in \mathcal{F}_2(M_{j2})$. The space of all n_w-dimensional inputs is expressed as $\mathcal{F}(M)$, where

$$M := \begin{bmatrix} M_{11} & M_{12} \\ \vdots & \vdots \\ M_{n_w 2} & M_{n_w 2} \end{bmatrix}$$

The purpose of matched environment-system design for MIMO systems is to design the controller which ensures the set of inequalities

$$\|z_i(\mathcal{F}(M))\|_{\text{peak}} \leq \varepsilon_i \quad (i = 1, 2, \ldots, n_z) \tag{5.64}$$

where the left hand of (5.64) is defined just the same as (5.6). Let us consider the matching condition for $\mathcal{F}(M)$, that is, the numerically tractable expression of (5.64) in the lifting framework.

Define T_{ij} as a subsystem from the jth input channel to the ith output channel, and define $\zeta_{ij}(t, w_j)$ as a response of T_{ij} at time t to w_j. The system is linear, so the lifted output of the closed-loop system can be given as sum of the lifted outputs of all subsystems:

$$\hat{z}_i(k, \theta, \hat{w}) = \sum_{j=1}^{n_w} \sum_{q=1}^{2} \hat{\zeta}_{ij}(k, \theta, \hat{w}_{jq}) \tag{5.65}$$

From the results for SISO systems given in the preceding section, it is immediately found that the supremum of the absolute value of $\hat{\zeta}_{ij}(k, \theta, \hat{w}_{jq})$ over $k \in \mathbb{Z}_+$ and $\hat{w}_{jq} \in \hat{\mathcal{F}}_q(M_{jq})$ is equal to $M_{jq}\phi_{ijq}(\theta)$, where

$$\phi_{ijq}(\theta) = \begin{cases} \sum_{k=0}^{\infty} \int_0^h |\hat{\zeta}_{ij}(k, \theta, \hat{1}(\tau))| d\tau & \text{for} \quad q = 1 \\ \left(\sum_{k=0}^{\infty} \int_0^h |\hat{\zeta}_{ij}(k, \theta, \hat{1}(\tau))|^2 d\tau \right)^{\frac{1}{2}} & \text{for} \quad q = 2 \end{cases} \tag{5.66}$$

Therefore,

$$\|z_i(\mathcal{F}(M))\|_{\text{peak}} = \sup \left\{ \sum_{j=1}^{n_w} \sum_{q=1}^{2} M_{jq}\phi_{ijq}(\theta) : \theta \in [0, h] \right\} \tag{5.67}$$

In (5.67), $M_{jq}\phi_{ijq}(\theta)$ can be well approximated by the maximum output of the fast-discretised model of T_{ij} if the sampling period of S_f and H_f is sufficiently small. Let $\zeta_{ij}[k, m, \hat{w}_{jq}]$ denote the output of the fast-discretised model of T_{ij} for an input $\hat{w}_{jq}$. Then the supremum of the absolute value of $\zeta_{ij}[k, m, \hat{w}_{jq}]$ over $k \in \mathbb{Z}_+$ and $\hat{w}_{jq} \in \hat{\mathcal{F}}_q(M_{jq}, n)$ is equal to $M_{jq}\phi_{ijq}(m, n)$,

where

$$\phi_{ijq}(m,n) = \begin{cases} h_f \sum_{r=0}^{n-1} \sum_{k=0}^{\infty} |\zeta_{ij}[k,m,\hat{1}_r]| & \text{for} \quad q=1 \\ \sqrt{h_f} \left(\sum_{r=0}^{n-1} \sum_{k=0}^{\infty} |\zeta_{ij}[k,m,\hat{1}_r]|^2 \right)^{\frac{1}{2}} & \text{for} \quad q=2 \end{cases} \quad (5.68)$$

Thus, (5.67) can be approximated by

$$\|z_i(\mathcal{F}(M))\|_{\text{peak}} \simeq \max \left\{ \sum_{j=1}^{n_w} \sum_{q=1}^{2} M_{jq} \phi_{ijq}(m,n) : m \in M_n \right\} \quad (5.69)$$

This approximation requires calculation of $\zeta_{ij}[k,m,\hat{1}_j]$. It can be calculated by the difference equation (5.49) just by replacing $z[k,m,\hat{1}_j]$ with $\zeta_{ij}[k,m,\hat{1}_j]$.

Consequently, the matching condition of MIMO systems for $\mathcal{F}(M)$ is given as a conjunction of inequalities

$$\sup \left\{ \sum_{j=1}^{n_w} \sum_{q=1}^{2} M_{jq} \phi_{ijq}(\theta) : \theta \in [0,h] \right\} \leq \varepsilon_i, \quad i=1,2,\ldots,n_z \quad (5.70)$$

If n is sufficiently large, the following condition is reliable to evaluate (5.70).

$$\max \left\{ \sum_{j=1}^{n_w} \sum_{q=1}^{2} M_{jq} \phi_{ijq}(m,n) : m \in M_n \right\} \leq \varepsilon_i, \quad i=1,2,\ldots,n_z \quad (5.71)$$

In design, the design parameters are determined so that (5.70) is met. Particularly, in the direct design, they are searched by an inequality solver subject to (5.70). In the discrete design, they are searched subject to (5.71) until (5.70) is met.

5.4 Design Example

The matched environment-system design for a sampled-data system is demonstrated. The control system is designed by the discrete design based on the fast-discretisation developed in this chapter.

The control system is designed as a unity feedback system shown in Figure 5.3. The system P is the nominal plant comprising the actual plant to be controlled and the actuator. Its transfer function is given by (Whidborne and Liu, 1993)

$$P(s) = \frac{27697}{s(s^2 + 1429s + 42653)} \quad (5.72)$$

The exogenous input is the reference command signal r, which belongs to $\mathcal{F}(M)$, where $M = [1.0\ 0.01]$:

$$r \in \mathcal{F}(M) = \{f_1 + f_2 : f_1 \in \mathcal{F}_1(1.0),\ f_2 \in \mathcal{F}_2(0.01)\}$$

The regulated output is the tracking error $e := r - y_p$. The sampling period is 0.01, which satisfies the necessary condition (5.44). The goal of design for environment-system match is to design the linear time-invariant discrete-time controller K which maintains the tracking error within ± 0.0175 for all references in $\mathcal{F}(M)$ under the restriction $|u(t)| \le 10.0$. Defining $z := [e\ u]^t$, the specification is given as a set of inequalities

$$\|z_1(\mathcal{F}(M))\|_{\text{peak}} \le 0.0175, \quad \|z_2(\mathcal{F}(M))\|_{\text{peak}} \le 10.0 \tag{5.73}$$

Accordingly this is an example of design for a single-input multi-output sampled-data control system.

The environment-system match, which is stated by (5.73), requires the system to have integral action. This requirement is fulfilled if K stabilises the system since $P(s)$ includes an integrator. For this reason, K is parametrised by a real-valued vector $p := [p_1\ p_2\ p_3] \in \mathbb{R}^3$ as follows.

$$K(z) = \frac{p_1(z + p_2)}{z + p_3} \tag{5.74}$$

The procedure of matched environment-system design based on the fast-discretisation is described as follows:

Step 1: Set $n = 2^i$ and initialise p.

Step 2: Find the vector p that satisfies the inequalities

$$\|\hat{z}_1(\hat{\mathcal{F}}(M, n))\|_{\text{peak}} \le 0.0175, \quad \|\hat{z}_2(\hat{\mathcal{F}}(M, n))\|_{\text{peak}} \le 10.0 \quad (5.75)$$

using an inequality solver. To avoid the infinite iteration in calculating $\phi_i(m, n)$, not (5.42) but (5.57) is evaluated for $\mu = 1.0 \times 10^{-8}$, where N is determined by (5.56) at each trial point and the unit step response is computed by (5.50). If the solution to (5.75) is found, go to Step 3. Otherwise, return to Step 1.

Step 3: Check the matching condition (5.19) for the vector p obtained in Step 2. If it is met, the controller design is completed. If it is not, set $n = 2^{i+1}$ and go back to Step 2.

As a result of the procedure above, some solutions could be found for $n = 2^4$. For example, one of them gives

$$K(z) = \frac{541.1(z - 0.737)}{z + 0.492} \tag{5.76}$$

According to (5.23) and (5.24),

$$\|\hat{z}_1(\mathcal{F}(M))\|_{\text{peak}} \approx 0.0174, \quad \|\hat{z}_2(\mathcal{F}(M))\|_{\text{peak}} \approx 9.96 \tag{5.77}$$

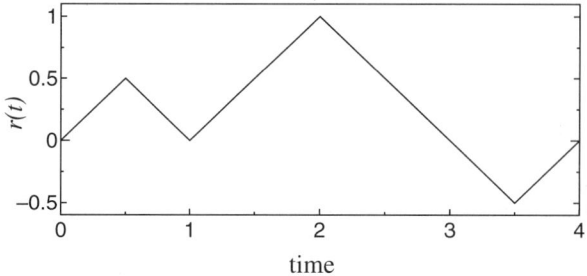

Fig. 5.6. Reference command signal

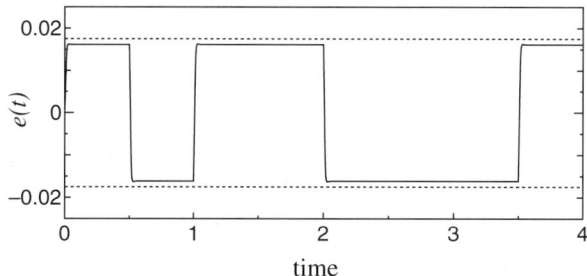

Fig. 5.7. Tracking error

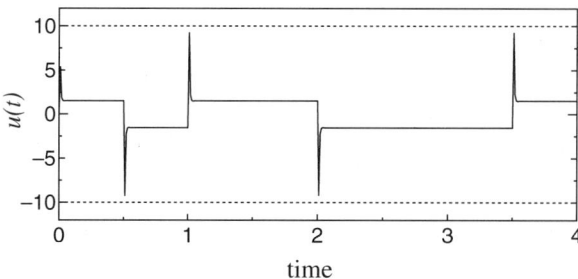

Fig. 5.8. Control input

so the system is matched to the environment by the controller (5.76).

The simulation results for the designed system are given in Figures 5.6-5.8. The reference command signal belongs to $\mathcal{F}([1.0\ 0.01])$. As these figures indicate, the tracking error is maintained within the tolerable bounds under the restriction $|u(t)| \leq 10.0$.

5.5 Conclusion

Design for a sampled-data control system matched to the environment was considered. The matching condition for $\mathcal{F}(M)$ is given as an inequality constraint on the unit step response. Essentially, integral action is necessary to achieve a match. Fast-discretisation provides the approximate but useful procedure to evaluate the matching condition. The validity and applicability of the procedure was demonstrated in the example.

References

Bamieh, B., J.B. Pearson, B.A. Francis, A. Tannenbaum, (1991) A lifting technique for linear periodic systems with applications to sampled-data control. *Systems & Control Letters*, 17:79–88.

Bamieh, B.A., J.B. Pearson, (1992) A general framework for linear periodic systems with applications to H^∞ sampled-data control. *IEEE Transactions Automatic Control*, 37:418–435.

Chen, T., B.A. Francis, (1995) *Optimal sampled-data control systems*. Springer-Verlag, London.

Keller, J.P., B.D.O. Anderson, (1992) A new approach to the discretization of continuous-time controllers. *IEEE Transactions on Automatic Control*, 37:214–223.

Ono, T., T. Ishihara, H. Inooka, (2000) Design of sampled-data critical control systems with persistent/transient inputs using lifting technique. *IEEE Transactions on Automatic Control*, 45:2090–2094.

Ono, T., T. Ishihara, H. Inooka, (2002) Design of sampled-data critical control systems based on the fast-discretization technique. *International Journal of Control*, 75:572–581.

Whidborne, J.F., G.P. Liu, (1993) *Critical Control Systems: Theory, Design and Applications*, Research Studies Press, Taunton, U.K.

Zakian, V., (1979) New formulation for the method of inequalities. *Proceedings of IEE*, 126:579–584.

Zakian, V., (1989) Critical systems and tolerable inputs. *International Journal of Control*, 49:1285–1289.

Zakian, V., (1991) Well matched systems. *IMA Journal of Mathematical Control & Information*, 8:29–38. See also Corrigendum (1992), 9:101.

Zakian, V., (1996) Perspectives on the principle of matching and the method of inequalities. *International Journal of Control*, 65:147–175.

Part III

Search Methods (with Numerical Tests)

6 A Numerical Evaluation of the Node Array Method

Toshiyuki Satoh

Abstract. A new global search method, proposed by Zakian and called the node array method, is evaluated numerically. The global search capability of the method is tested on fourteen known or specially devised problems containing various degrees of search difficulty. The test results show that the method is globally convergent and efficient.

6.1 Introduction

A description of the node array (NA) method is given by Zakian (Section 1.7). The main purpose of this chapter is to provide a numerical evaluation of the efficiency of the method, by testing it on fourteen challenging problems, some of which were specially devised for this purpose, in collaboration with Zakian. In devising some of the test problems, special care was taken to ensure that the global search capacity of the method would be tested. The same test problems are used elsewhere (Whidborne, Chapter 7; Liu and Ishihara, Chapter 8) to test two other methods. The results of those tests provide a basis upon which the efficiency of the node array method can be assessed.

The particular node array method considered in this chapter employs the moving boundaries process (MBP) (see Appendix 6.A) to make the local searches. More specifically, the MBP considered here employs the Rosenbrock (R) trial generator and this generator is described in detail in Appendix 6.A. Consequently, the entire method considered here is denoted by NA & MBP(R). By referring to the detailed operation of the Rosenbrock trial generator, two simple conditions were derived, in collaboration with Zakian, for deciding when a local search by the MBP(R) becomes stuck. Accordingly, a search is deemed to be stuck when both conditions hold simultaneously.

The node array method is known from theoretical considerations to be globally convergent (Zakian, Section 1.7). The test results confirm this and also indicate that the method is relatively efficient, when compared with the results obtained by the two other methods (Whidborne, Chapter 7; Liu and Ishihara, Chapter 8).

6.2 Detection of Stuck Local Search

For full details of the node array method, see Zakian, Section 1.7.

Now suppose that every component of the design vector c_i is a real parameter that lies in the corresponding segment of the real axis $\left[c_i^L, c_i^U\right]$. In order to express the search space in its normalised form, the following transformation of variables is applied to c_i:

$$p_i = \frac{c_i - c_i^L}{c_i^U - c_i^L}, \quad i = 1, 2, \ldots, N \tag{6.1}$$

It follows from (6.1) that the search space is transformed into the normalised search space, which is the hypercube S_N defined by

$$S_N := \left\{ p \in \mathbb{R}^N : p_i \in [0, 1], \quad i = 1, 2, \ldots, N \right\} \tag{6.2}$$

It is necessary to establish criteria for detecting when a local search becomes stuck, so that the local search is abandoned and a new local search is started from another node. To this end, care has to be taken to ensure that the criteria are not too stringent, for then the local search would stop prematurely. Similarly, the criteria must not be too lax, for then the search would continue in a relatively ineffectual manner.

The criteria for the local search, using the MBP(R) method, are obtained by reference to the detailed operation of the Rosenbrock trial generator (see Appendix 6.A)

Recall that this trial generation has the following two features:

- a trial point $\hat{c}^k$ is generated by $\hat{c}^k = c^k + e_j V_j^r$ (see (6.62)).
- if $\hat{c}^k$ is not a successful trial, the real number e_j is replaced by $-0.5 e_j$; otherwise e_j is replaced by $3 e_j$.

It follows from this that a trial point $\hat{c}^k$ converges to the current point c^k after a large number of failed trials. Hence, the size of $e_j V_j^r$, which is called the *step length*, can be used to detect when a search is getting stuck. Additionally, it can be seen that a search direction is changed after a failed trial, since e_j is replaced by $-0.5 e_j$. Therefore, if two consecutive failed trials occur in each orthogonal direction, it indicates that the local search is stuck around the current point c^k.

Hence the following two criteria, used in conjunction, are used to decide when a local search is stuck:

- For a given small positive number δ, all the step lengths in the hypercube S_N have become less than δ.

- $2N$ consecutive failed trials have occurred.

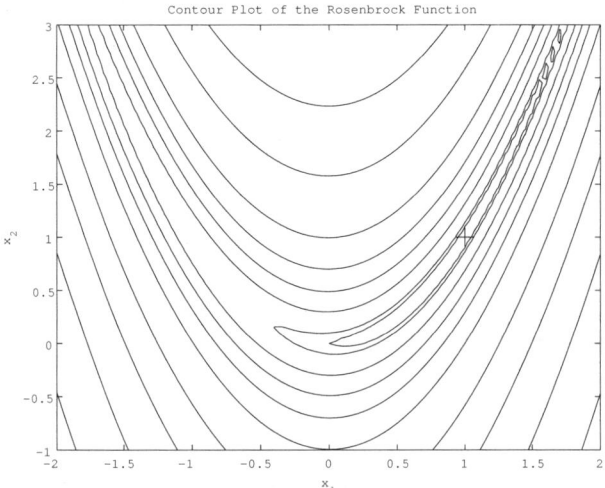

Fig. 6.1. The contour plot of the Rosenbrock function where the global minimum is indicated by the '+' mark

For the test results shown in this chapter, $\delta = 10^{-5}$.

The step length is computed in the following way. Suppose that $e_j V_j^r$ in (6.62) is written as

$$e_j V_j^r = (d_1, d_2, \ldots, d_N)^T \tag{6.3}$$

By applying the transformation in (6.1), each component d_i of the vector $e_j V_j^r$ locates at $(d_i - c_i^L)/(c_i^U - c_i^L)$ in the normalised search space S_N. Similarly, each component of the origin locates at $-c_i^L/(c_i^U - c_i^L)$ in S_N. Hence, according to the definition of the Euclidean norm, the step length of the vector transformed into S_N, which is here denoted by Δc, can be computed as

$$\Delta c = \sqrt{\sum_{i=1}^{N} \left(\frac{d_i - c_i^L}{c_i^U - c_i^L} - \frac{-c_i^L}{c_i^U - c_i^L} \right)^2}$$

$$= \sqrt{\sum_{i=1}^{N} \left(\frac{d_i}{c_i^U - c_i^L} \right)^2} \tag{6.4}$$

6.3 Special Test Problems

In this section, fourteen special test problems are described. Each problem is obtained by defining a scalar valued objective function, which maps a rectangle in the real plane into the real line. This gives rise to the problem

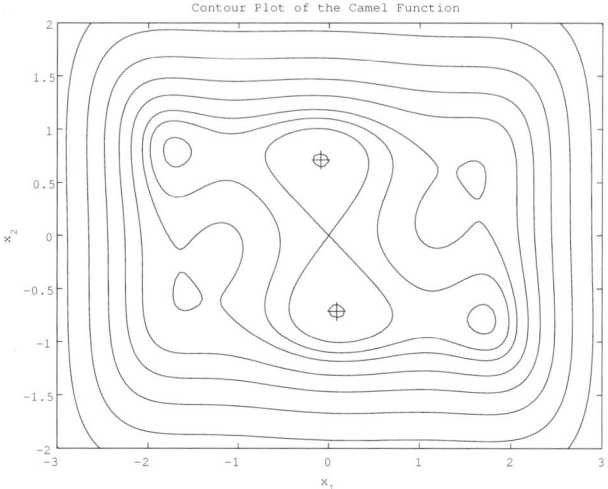

Fig. 6.2. The contour plot of the camel function where the two global minima are indicated by the '+' marks

of finding a solution of the inequality, formed by requiring that the value of the objective function does not exceed a specified number.

The two basic objective functions, from which the other functions are derived, are as follows. First is the well-known Rosenbrock test function (Rosenbrock, 1960) (originally devised for testing methods for minimising a scalar valued function). The shape of this function can be seen in Figure 6.1 and is, for obvious reasons, sometimes referred to as the banana function. The function has a single global minimum. Second is another known test function termed the six-hump camel back function (Dixon and Szego, 1978) , which is here called the camel function and its shape can be seen in Figure 6.2. The camel function has two global minima and four other local minima (such a function is said to be multi-modal, in this case with a total of six minima). These two functions, respectively, give rise to Test Problems 1 and 2.

Each of these two test functions is then used to generate further test functions, as follows. The function and four shifted, but otherwise identical copies of the function are superimposed to make a single function with five global minima or ten global minima plus twenty other local minima, respectively. For each function, all the global minima have identical depth. This gives rise to Test Problems 3 and 4, respectively.

Then, each of the two multi-modal functions obtained in this way is adjusted in five different ways, by altering the height of one, or one pair, respectively, of the global minima. This results in ten distinct functions. Five functions have only one global minimum and four other local minima. Each of the other five functions has one pair of global minima and twenty-eight local minima, respectively. This gives rise to Test Problems 5.1 – 5.5 and 6.1 – 6.5, respectively.

Test Problem 1

The test function 1 is the Rosenbrock function defined by

$$f_1(x_1, x_2) := 100\left(x_2 - x_1^2\right)^2 + (1 - x_1)^2$$
$$-2 \leqslant x_1 \leqslant 2 \tag{6.5}$$
$$-1 \leqslant x_2 \leqslant 3$$

The function f_1 has a deep valley along the parabola $x_2 = x_1^2$. $f_1(x_1, x_2)$ has one global minimum The global minimum value of $f_1(x_1, x_2)$ is 0 at

$$(x_1, x_2) = (1, 1)$$

The test problem 1 using $f_1(x_1, x_2)$ is defined as follows:

$$\begin{array}{ll} \text{Find} & (x_1, x_2) \\ \text{such that} & f_1(x_1, x_2) \leqslant 10^{-5} \end{array} \tag{6.6}$$

Test Problem 2

The test function 2 is the camel function defined by

$$f_2(x_1, x_2) := \left(4 - 2.1x_1^2 + x_1^4/3\right)x_1^2 + x_1x_2 + \left(-4 + 4x_2^2\right)x_2^2$$
$$-3 \leqslant x_1 \leqslant 3 \tag{6.7}$$
$$-2 \leqslant x_2 \leqslant 2$$

The function f_2 has four local minima and two global minima. The global minimum value of $f_2(x_1, x_2)$ is -1.031628 at

$$(x_1, x_2) = (-0.0898, 0.7126)$$
$$(0.0898, -0.7126)$$

The test problem 2 using $f_2(x_1, x_2)$ is now defined as follows:

$$\begin{array}{ll} \text{Find} & (x_1, x_2) \\ \text{such that} & f_2(x_1, x_2) \leqslant -1.03162 \end{array} \tag{6.8}$$

Test Problem 3

The test function 3 comprises five shifted Rosenbrock functions, resulting in five equal global minima defined as

$$f_3(x_1, x_2) := 100\left(x_2 - (x_1 - 4(k-1) + 8)^2\right)^2$$
$$+ (1 - (x_1 - 4(k-1) + 8))^2 \qquad (6.9)$$
$$-10 \leqslant x_1 \leqslant 10$$
$$-1 \leqslant x_2 \leqslant 3$$

where k is defined in accordance with the value of x_1 in the following way:

$$\begin{cases} k = 1 & \text{if } -10 \leqslant x_1 < -5.998751561 \\ k = 2 & \text{if } -5.998751561 \leqslant x_1 < 1.998751561 \\ k = 3 & \text{if } -1.998751561 \leqslant x_1 < 2.001248439 \\ k = 4 & \text{if } 2.001248439 \leqslant x_1 < 6.001248439 \\ k = 5 & \text{if } 6.001248439 \leqslant x_1 \leqslant 10 \end{cases} \qquad (6.10)$$

The function f_3 consists of five segments, and each segment is a shifted Rosenbrock function. It follows that there are five global minima in total and that each segment has one global minimum. The global minimum value of $f_3(x_1, x_2)$ is 0 at

$$(x_1, x_2) = \begin{cases} (-7, 1) \\ (-3, 1) \\ (1, 1) \\ (5, 1) \\ (9, 1) \end{cases}$$

Now the test problem 3 using $f_3(x_1, x_2)$ is stated as follows:

$$\begin{array}{ll} \text{Find} & (x_1, x_2) \\ \text{such that} & f_3(x_1, x_2) \leqslant 10^{-5} \end{array} \qquad (6.11)$$

Test Problem 4

The test function 4 comprises five shifted camel functions with five pairs of global minima and twenty other local minima defined as

6 A Numerical Evaluation of the Node Array Method

$$f_4(x_1, x_2) := \left(4 - 2.1\left(x_1 - 6(k-1) + 12\right)^2 + \left(x_1 - 6(k-1) + 12\right)^4/3\right)$$
$$\times \left(x_1 - 6(k-1) + 12\right)^2 + \left(x_1 - 6(k-1) + 12\right)x_2$$
$$+ \left(4x_2^2 - 4\right)x_2^2$$
$$-15 \leqslant x_1 \leqslant 15$$
$$-2 \leqslant x_2 \leqslant 2$$

(6.12)

where k is defined in accordance with the value of x_1 as follows:

$$\begin{cases} k = 1 & \text{if } -15 \leqslant x_1 < -9 \\ k = 2 & \text{if } -9 \leqslant x_1 < -3 \\ k = 3 & \text{if } -3 \leqslant x_1 < 3 \\ k = 4 & \text{if } 3 \leqslant x_1 < 9 \\ k = 5 & \text{if } 9 \leqslant x_1 \leqslant 15 \end{cases} \qquad (6.13)$$

The function f_4 consists of five segments, and each segment is the shifted camel function. It follows that there are ten global minima in total and that each segment has two global minima. The global minimum value of $f_4(x_1, x_2)$ is -1.031628 at

$$(x_1, x_2) = \begin{cases} (-12.0898, 0.7126) \\ (-11.9102, -0.7126) \\ (-6.0898, 0.7126) \\ (-5.9102, -0.7126) \\ (-0.0898, 0.7126) \\ (0.0898, -0.7126) \\ (5.9102, 0.7126) \\ (6.0898, -0.7126) \\ (11.9102, 0.7126) \\ (12.0898, -0.7126) \end{cases}$$

The test problem using $f_4(x_1, x_2)$ is stated as follows:

Find (x_1, x_2)
such that $f_4(x_1, x_2) \leqslant -1.03162$

(6.14)

Test Problem 5.1

The test function 5.1 comprises five shifted Rosenbrock functions with one global minimum defined as

$$f_{5.1}(x_1, x_2) := 100\left(x_2 - (x_1 - 4(k-1) + 8)^2\right)^2 \\ + \left(1 - (x_1 - 4(k-1) + 8)\right)^2 + c(k) \\ -10 \leqslant x_1 \leqslant 10 \\ -1 \leqslant x_2 \leqslant 3 \quad (6.15)$$

where k is decided in accordance with the value of x_1 as follows:

$$\begin{cases} k = 1 & \text{if } -10 \leqslant x_1 < -5.983147263 \\ k = 2 & \text{if } -5.983147263 \leqslant x_1 < -1.998751561 \\ k = 3 & \text{if } -1.998751561 \leqslant x_1 < 2.001248439 \\ k = 4 & \text{if } 2.001248439 \leqslant x_1 < 6.001248439 \\ k = 5 & \text{if } 6.001248439 \leqslant x_1 \leqslant 10 \end{cases} \quad (6.16)$$

and $c(k)$ is a function of k and taken as

$$c(k) = \begin{cases} -100 & \text{if } k = 1 \\ 0 & \text{otherwise} \end{cases} \quad (6.17)$$

The function $f_{5.1}$ consists of five segments, and each segment is a shifted Rosenbrock function. However, the depth of the segment 1 is adjusted by subtracting the number 100 from the value of the function. Hence, the global minimum exists only in the segment 1. The global minimum value of $f_{5.1}(x_1, x_2)$ is -100 at

$$(x_1, x_2) = (-7, 1)$$

The test problem using the test function 5.1 is stated as follows:

$$\begin{array}{ll} \text{Find} & (x_1, x_2) \\ \text{such that} & f_{5.1}(x_1, x_2) \leqslant -99.99999 \end{array} \quad (6.18)$$

Test Problem 5.2

The test function 5.2 is also the five shifted Rosenbrock functions with one global minimum defined as

$$f_{5.2}(x_1, x_2) := 100\left(x_2 - (x_1 - 4(k-1) + 8)^2\right)^2 \\ + \left(1 - (x_1 - 4(k-1) + 8)\right)^2 + c(k) \\ -10 \leqslant x_1 \leqslant 10 \\ -1 \leqslant x_2 \leqslant 3 \quad (6.19)$$

where k is decided in accordance with the value of x_1 as follows:

$$\begin{cases} k = 1 & \text{if } -10 \leqslant x_1 < -6.014356315 \\ k = 2 & \text{if } -6.014356315 \leqslant x_1 < -1.983147263 \\ k = 3 & \text{if } -1.983147263 \leqslant x_1 < 2.001248439 \\ k = 4 & \text{if } 2.001248439 \leqslant x_1 < 6.001248439 \\ k = 5 & \text{if } 6.001248439 \leqslant x_1 \leqslant 10 \end{cases} \quad (6.20)$$

and $c(k)$ is a function of k and taken as

$$c(k) = \begin{cases} -100 & \text{if } k = 2 \\ 0 & \text{otherwise} \end{cases} \quad (6.21)$$

The function $f_{5.2}$ consists of five segments, and each segment is a shifted Rosenbrock function. However, the depth of the segment 2 is adjusted by subtracting the number 100 from the value of the function. It follows that the global minimum exists only in the segment 2. The global minimum value of $f_{5.2}(x_1, x_2)$ is -100 at

$$(x_1, x_2) = (-3, 1)$$

Now the test problem using the test function 5.2 is stated as follows:

$$\begin{array}{ll} \text{Find} & (x_1, x_2) \\ \text{such that} & f_{5.2}(x_1, x_2) \leqslant -99.99999 \end{array} \quad (6.22)$$

Test Problem 5.3

The test function 5.3 is also the five shifted Rosenbrock functions with one global minimum defined as

$$\begin{aligned} f_{5.3}(x_1, x_2) := & 100 \left(x_2 - (x_1 - 4(k-1) + 8))^2\right)^2 \\ & + (1 - (x_1 - 4(k-1) + 8))^2 + c(k) \\ & -10 \leqslant x_1 \leqslant 10 \\ & -1 \leqslant x_2 \leqslant 3 \end{aligned} \quad (6.23)$$

where k is decided in accordance with the value of x_1 as follows:

$$\begin{cases} k = 1 & \text{if } -10 \leqslant x_1 < -5.998751561 \\ k = 2 & \text{if } -5.998751561 \leqslant x_1 < -2.014356315 \\ k = 3 & \text{if } -2.014356315 \leqslant x_1 < 2.016852737 \\ k = 4 & \text{if } 2.016852737 \leqslant x_1 < 6.001248439 \\ k = 5 & \text{if } 6.001248439 \leqslant x_1 \leqslant 10 \end{cases} \quad (6.24)$$

and $c(k)$ is a function of k and taken as

$$c(k) = \begin{cases} -100 & \text{if } k = 3 \\ 0 & \text{otherwise} \end{cases} \quad (6.25)$$

The function $f_{5.3}$ consists of five segments, and each segment is a shifted Rosenbrock function. However, the depth of the segment 3 is adjusted by subtracting the number 100 from the value of the function. It follows that the global minimum exists only in the segment 3. The global minimum value of $f_{5.3}(x_1, x_2)$ is -100 at

$$(x_1, x_2) = (1, 1)$$

The test problem using the test function 5.3 is stated as follows:

$$\begin{aligned} &\text{Find} & &(x_1, x_2) \\ &\text{such that} & &f_{5.3}(x_1, x_2) \leqslant -99.99999 \end{aligned} \quad (6.26)$$

Test Problem 5.4

The test function 5.4 is also the five shifted Rosenbrock functions with one global minimum defined as

$$\begin{aligned} f_{5.4}(x_1, x_2) := &100 \left(x_2 - (x_1 - 4(k-1) + 8))^2\right)^2 \\ &+ (1 - (x_1 - 4(k-1) + 8))^2 + c(k) \\ &- 10 \leqslant x_1 \leqslant 10 \\ &- 1 \leqslant x_2 \leqslant 3 \end{aligned} \quad (6.27)$$

where k is decided in accordance with the value of x_1 as follows:

$$\begin{cases} k = 1 & \text{if } -10 \leqslant x_1 < -5.998751561 \\ k = 2 & \text{if } -5.998751561 \leqslant x_1 < -1.998751561 \\ k = 3 & \text{if } -1.998751561 \leqslant x_1 < 1.985643685 \\ k = 4 & \text{if } 1.985643685 \leqslant x_1 < 6.016852737 \\ k = 5 & \text{if } 6.016852737 \leqslant x_1 \leqslant 10 \end{cases} \quad (6.28)$$

and $c(k)$ is a function of k and taken as

$$c(k) = \begin{cases} -100 & \text{if } k = 4 \\ 0 & \text{otherwise} \end{cases} \quad (6.29)$$

The function $f_{5.4}$ consists of five segments, and each segment is a shifted Rosenbrock function. However, the depth of the segment 4 is adjusted by subtracting the number of 100 from the value of the function. Hence, the global minimum exists only in the segment 4. The global minimum value of $f_{5.4}(x_1, x_2)$ is -100 at

$$(x_1, x_2) = (5, 1)$$

The test problem using the test function 5.4 is now stated as follows:

$$\begin{array}{ll} \text{Find} & (x_1, x_2) \\ \text{such that} & f_{5.4}(x_1, x_2) \leqslant -99.99999 \end{array} \tag{6.30}$$

Test Problem 5.5

The test function 5.5 is the five shifted Rosenbrock functions with one global minimum defined as

$$\begin{aligned} f_{5.5}(x_1, x_2) := & 100\left(x_2 - (x_1 - 4(k-1) + 8))^2\right)^2 \\ & + (1 - (x_1 - 4(k-1) + 8))^2 + c(k) \\ & -10 \leqslant x_1 \leqslant 10 \\ & -1 \leqslant x_2 \leqslant 3 \end{aligned} \tag{6.31}$$

where k is decided in accordance with the value of x_1 as follows:

$$\begin{cases} k = 1 & \text{if } -10 \leqslant x_1 < -5.998751561 \\ k = 2 & \text{if } -5.998751561 \leqslant x_1 < -1.998751561 \\ k = 3 & \text{if } -1.998751561 \leqslant x_1 < 2.001248439 \\ k = 4 & \text{if } 2.001248439 \leqslant x_1 < 5.985643685 \\ k = 5 & \text{if } 5.985643685 \leqslant x_1 \leqslant 10 \end{cases} \tag{6.32}$$

and $c(k)$ is a function of k and taken as

$$c(k) = \begin{cases} -100 & \text{if } k = 5 \\ 0 & \text{otherwise} \end{cases} \tag{6.33}$$

The function $f_{5.5}$ consists of five segments, and each segment is a shifted Rosenbrock function. However, the depth of the segment 5 is adjusted by subtracting the number 100 from the value of the function. Hence, the global minimum exists only in the segment 5. The global minimum value of $f_{5.5}(x_1, x_2)$ is -100 at

$$(x_1, x_2) = (9, 1)$$

The test problem using the test function 5.5 is now stated as follows:

$$\begin{array}{ll} \text{Find} & (x_1, x_2) \\ \text{such that} & f_{5.5}(x_1, x_2) \leqslant -99.99999 \end{array} \tag{6.34}$$

Test Problem 6.1

The test function 6.1 is a superposition of five shifted camel functions with one of the five functions adjusted in height to give one pair of global minima defined as

$$f_{6.1}(x_1, x_2) := \left(4 - 2.1\left(x_1 - 6(k-1) + 12\right)^2 + \left(x_1 - 6(k-1) + 12\right)^4/3\right)$$
$$\times (x_1 - 6(k-1) + 12)^2 + (x_1 - 6(k-1) + 12) x_2$$
$$+ \left(-4 + 4x_2^2\right) x_2^2 + c(k)$$
$$-15 \leqslant x_1 \leqslant 15$$
$$-2 \leqslant x_2 \leqslant 2$$

(6.35)

where k is decided in accordance with the value of x_1 as follows:

$$\begin{cases} k = 1 & \text{if } -15 \leqslant x_1 < -8.982347639 \\ k = 2 & \text{if } -8.982347639 \leqslant x_1 < -3 \\ k = 3 & \text{if } -3 \leqslant x_1 < 3 \\ k = 4 & \text{if } 3 \leqslant x_1 < 9 \\ k = 5 & \text{if } 9 \leqslant x_1 \leqslant 15 \end{cases} \quad (6.36)$$

and $c(k)$ is a function of k and taken as

$$c(k) = \begin{cases} -10 & \text{if } k = 1 \\ 0 & \text{otherwise} \end{cases} \quad (6.37)$$

The function $f_{6.1}$ consists of five segments, and each segment is the shifted camel function. However, the depth of the segment 1 is adjusted by subtracting the number 10 from the value of the function. Therefore, the global minima exist only in the segment 1. The global minimum value of $f_{6.1}(x_1, x_2)$ is -11.031628 at

$$(x_1, x_2) = \begin{cases} (-12.0898, 0.7126) \\ (-11.9102, -0.7126) \end{cases}$$

The test problem using the test function 6.1 is stated as follows:

$$\begin{aligned} &\text{Find} && (x_1, x_2) \\ &\text{such that} && f_{6.1}(x_1, x_2) \leqslant -11.03162 \end{aligned} \quad (6.38)$$

Test Problem 6.2

The test function 6.2 is also the five shifted camel functions with one pair of global minima defined as

$$f_{6.2}(x_1, x_2) := \left(4 - 2.1(x_1 - 6(k-1) + 12)^2 + (x_1 - 6(k-1) + 12)^4/3\right)$$
$$\times (x_1 - 6(k-1) + 12)^2 + (x_1 - 6(k-1) + 12)x_2$$
$$+ \left(-4 + 4x_2^2\right)x_2^2 + c(k)$$
$$-15 \leqslant x_1 \leqslant 15$$
$$-2 \leqslant x_2 \leqslant 2$$

(6.39)

where k is decided in accordance with the value of x_1 as follows:

$$\begin{cases} k = 1 & \text{if } -15 \leqslant x_1 < -9.017652361 \\ k = 2 & \text{if } -9.017652361 \leqslant x_1 < -2.982347639 \\ k = 3 & \text{if } -2.982347639 \leqslant x_1 < 3 \\ k = 4 & \text{if } 3 \leqslant x_1 < 9 \\ k = 5 & \text{if } 9 \leqslant x_1 \leqslant 15 \end{cases}$$

(6.40)

and $c(k)$ is a function of k and taken as

$$c(k) = \begin{cases} -10 & \text{if } k = 2 \\ 0 & \text{otherwise} \end{cases}$$

(6.41)

The function $f_{6.2}$ consists of five segments, and each segment is the shifted camel function. However, the depth of the segment 2 is adjusted by subtracting the number of 10 from the value of the function. Hence, the global minima exist only in the segment 2. The global minimum value of $f_{6.2}(x_1, x_2)$ is -11.031628 at

$$(x_1, x_2) = \begin{cases} (-6.0898, 0.7126) \\ (-5.9102, -0.7126) \end{cases}$$

The test problem using the test function 6.2 is stated as follows:

$$\begin{array}{ll} \text{Find} & (x_1, x_2) \\ \text{such that} & f_{6.2}(x_1, x_2) \leqslant -11.03162 \end{array}$$

(6.42)

Test Problem 6.3

The test function 6.3 is also the five shifted camel functions with one pair of global minima defined as

$$f_{6.3}(x_1, x_2) := \left(4 - 2.1\left(x_1 - 6(k-1) + 12\right)^2 + \left(x_1 - 6(k-1) + 12\right)^4 / 3\right)$$
$$\times \left(x_1 - 6(k-1) + 12\right)^2 + \left(x_1 - 6(k-1) + 12\right) x_2$$
$$+ \left(-4 + 4x_2^2\right) x_2^2 + c(k)$$
$$-15 \leqslant x_1 \leqslant 15$$
$$-2 \leqslant x_2 \leqslant 2$$

(6.43)

where k is decided in accordance with the value of x_1 as follows:

$$\begin{cases} k = 1 & \text{if } -15 \leqslant x_1 < -9 \\ k = 2 & \text{if } -9 \leqslant x_1 < -3.017652361 \\ k = 3 & \text{if } -3.017652361 \leqslant x_1 < 3.017652361 \\ k = 4 & \text{if } 3.017652361 \leqslant x_1 < 9 \\ k = 5 & \text{if } 9 \leqslant x_1 \leqslant 15 \end{cases}$$

(6.44)

and $c(k)$ is a function of k and taken as

$$c(k) = \begin{cases} -10 & \text{if } k = 3 \\ 0 & \text{otherwise} \end{cases}$$

(6.45)

The function $f_{6.3}$ consists of five segments, and each segment is the shifted camel function. However, the depth of the segment 3 is adjusted by subtracting the number 10 from the value of the function. Hence, the global minima exist only in the segment 3. The global minimum value of $f_{6.3}(x_1, x_2)$ is -11.031628 at

$$(x_1, x_2) = \begin{cases} (-0.0898, 0.7126) \\ (0.0898, -0.7126) \end{cases}$$

The test problem using the test function 6.3 is stated as follows:

Find (x_1, x_2)
such that $f_{6.3}(x_1, x_2) \leqslant -11.03162$

(6.46)

Test Problem 6.4

The test function 6.4 is also the five shifted camel functions with one pair of global minima defined as

$$f_{6.4}(x_1, x_2) := \left(4 - 2.1\left(x_1 - 6(k-1) + 12\right)^2 + \left(x_1 - 6(k-1) + 12\right)^4/3\right)$$
$$\times (x_1 - 6(k-1) + 12)^2 + (x_1 - 6(k-1) + 12) x_2$$
$$+ \left(-4 + 4x_2^2\right) x_2^2 + c(k)$$
$$-15 \leqslant x_1 \leqslant 15$$
$$-2 \leqslant x_2 \leqslant 2$$

(6.47)

where k is decided in accordance with the value of x_1 as follows:

$$\begin{cases} k = 1 & \text{if } -15 \leqslant x_1 < -9 \\ k = 2 & \text{if } -9 \leqslant x_1 < -3 \\ k = 3 & \text{if } -3 \leqslant x_1 < 2.982347639 \\ k = 4 & \text{if } 2.982347639 \leqslant x_1 < 9.017652361 \\ k = 5 & \text{if } 9.017652361 \leqslant x_1 \leqslant 15 \end{cases}$$

(6.48)

and $c(k)$ is a function of k and taken as

$$c(k) = \begin{cases} -10 & \text{if } k = 4 \\ 0 & \text{otherwise} \end{cases}$$

(6.49)

The function $f_{6.4}$ consists of five segments, and each segment is the shifted camel function. However, the depth of the segment 4 is adjusted by subtracting the number 10 from the value of the function. Hence, the global minima exist only in the segment 4. The global minimum value of $f_{6.4}(x_1, x_2)$ is -11.031628 at

$$(x_1, x_2) = \begin{cases} (5.9102, 0.7126) \\ (6.0898, -0.7126) \end{cases}$$

The test problem using the test function 6.4 is stated as follows:

$$\begin{aligned} &\text{Find} && (x_1, x_2) \\ &\text{such that} && f_{6.4}(x_1, x_2) \leqslant -11.03162 \end{aligned}$$

(6.50)

Test Problem 6.5

The test function 6.5 is also the five shifted camel functions with one pair of global minima defined as

$$f_{6.5}(x_1, x_2) := \left(4 - 2.1(x_1 - 6(k-1) + 12)^2 + (x_1 - 6(k-1) + 12)^4/3\right)$$
$$\times (x_1 - 6(k-1) + 12)^2 + (x_1 - 6(k-1) + 12)x_2$$
$$+ \left(-4 + 4x_2^2\right)x_2^2 + c(k)$$
$$-15 \leq x_1 \leq 15$$
$$-2 \leq x_2 \leq 2$$

(6.51)

where k is decided in accordance with the value of x_1 as follows:

$$\begin{cases} k = 1 & \text{if } -15 \leq x_1 < -9 \\ k = 2 & \text{if } -9 \leq x_1 < -3 \\ k = 3 & \text{if } -3 \leq x_1 < 3 \\ k = 4 & \text{if } 3 \leq x_1 < 8.982347639 \\ k = 5 & \text{if } 8.982347639 \leq x_1 \leq 15 \end{cases}$$

(6.52)

and $c(k)$ is a function of k and taken as

$$c(k) = \begin{cases} -10 & \text{if } k = 5 \\ 0 & \text{otherwise} \end{cases}$$

(6.53)

The function $f_{6.5}$ consists of five segments, and each segment is the shifted camel function. However, the depth of the segment 5 is adjusted by subtracting the number 10 from the value of the function. Hence, the global minima exist only in the segment 5. The global minimum value of $f_{6.5}(x_1, x_2)$ is -11.031628 at

$$(x_1, x_2) = \begin{cases} (11.9102, 0.7126) \\ (12.0898, -0.7126) \end{cases}$$

The test problem using the test function 6.5 is now stated as follows:

$$\begin{aligned} &\text{Find} & &(x_1, x_2) \\ &\text{such that} & &f_{6.5}(x_1, x_2) \leq -11.03162 \end{aligned}$$

(6.54)

6.4 Test Results

The fourteen test problems defined above are used to test the node array method.

Table 6.1. Summary of the search results (with Stopping Rule 3).

Problem Number	Number of Function Evaluations	$f(x_1, x_2)$	x_1	x_2	$\hat{m}$		Final Node Address
1	424	0.000000	1.000000	1.000000	2	13	(0.75, 0.5)
2	50	−1.031625	0.090192	−0.713297	0	0	(0, 0)
3	466	0.000000	5.000000	1.000000	2	13	(0.75, 0.5)
4	55	−1.031628	−12.089945	0.712762	0	0	(0, 0)
5.1	11628	−99.999990	−6.996883	1.006245	5	99	(0, 0.09375)
5.2	10839	−99.999998	−3.000648	0.998594	5	75	(0.28125, 0.0625)
5.3	6842	−99.999992	0.997338	0.994568	4	229	(0.5, 0.8125)
5.4	242	−100.000000	5.000000	1.000000	2	13	(0.75, 0.5)
5.5	16055	−99.999992	8.997228	0.994455	5	131	(1, 0.09375)
6.1	55	−11.031628	−12.089945	0.712762	0	0	(0, 0)
6.2	181	−11.031621	−5.908871	−0.713075	2	6	(0.25, 0.25)
6.3	142	−11.031628	0.090103	−0.712834	2	1	(0.25, 0)
6.4	524	−11.031627	6.090516	−0.712746	3	14	(0.625, 0.125)
6.5	246	−11.031627	12.090269	−0.712499	2	9	(1, 0.25)

The test results are summarised in Table 6.1 where the column headings indicate the following quantities. The Problem Number, the Number of Function Evaluations required to find an admissible point, the function value $f(x_1, x_2)$ and the coordinates x_1 and x_2 corresponding to the admissible point found, the integer $\hat{m}$ which indicates the highest density reached by the node array and the Final Node Address (the node from which the admissible point is found) given in two forms, first as the number of the last element of the node sequence and second as its corresponding coordinates in the normalised search space.

The small number δ needed in the stuck detection rule is taken to be 10^{-5} throughout the fourteen problems. As can be seen, the NA&MBP(R) successfully found an admissible point for every one of the fourteen problems. This confirms that the node array method is globally convergent and suggests that the method is also universally globally convergent for all piecewise continuous objective functions.

The seven Test Problems 1, 3 and 5.1-5.5 are all based on the Rosenbrock function, which, as can be seen in Figure 6.1, has been constructed to have an awkward shape so as to make it relatively difficult for a local search method (in this case the MBP(R)) to find an admissible point in a test problem, even when the local search starts within the convergence space (Zakian, Section 1.7) of the function. This difficulty is reflected in the number of function evaluations needed to find an admissible point.

Table 6.2. Summary of the search results (without Stopping Rule 3).

Problem Number	Number of Function Evaluations	$f(x_1, x_2)$	x_1	x_2	$\hat{m}$		Final Node Address
1	7086	0.000003	0.998431	0.996809	1	1	$(0.5, 0)$
2	50	−1.031625	0.090192	−0.713297	0	0	$(0, 0)$
3	12917	0.000003	0.998431	0.996809	1	1	$(0.5, 0)$
4	55	−1.031628	−12.089945	0.712762	0	0	$(0, 0)$
5.1	40380	−99.999991	−7.003067	0.993842	2	15	$(0, 0.75)$
5.2	31400	−99.999997	−3.001127	0.997884	2	6	$(0.25, 0.25)$
5.3	12933	−99.999997	0.998431	0.996809	1	1	$(0.5, 0)$
5.4	39804	−100.000000	5.000000	1.000000	2	13	$(0.75, 0.5)$
5.5	98018	−99.999997	8.998370	0.996774	3	35	$(1, 0.375)$
6.1	55	−11.031628	−12.089945	0.712762	0	0	$(0, 0)$
6.2	808	−11.031621	−6.089866	0.711697	2	1	$(0.25, 0)$
6.3	403	−11.031625	−0.089009	0.712262	1	1	$(0.5, 0)$
6.4	168	−11.031623	5.908957	0.712786	0	1	$(1, 0)$
6.5	126	−11.031627	11.909749	0.713001	0	1	$(1, 0)$

The seven Test Problems 2, 4 and 6.1-6.5 are all based on the camel function that, as can be seen in Figure 6.2, although it has several global and local minima, does not present great difficulties for local searches.

Consequently, as Table 6.1 shows, the test problems based on the Rosenbrock functions require significantly more function evaluations and, in most cases, also higher node array densities, to find an admissible point, than the test problems based on the camel function.

6.5 Effect of Stopping Rule 3

The node array method includes a device, called Stopping Rule 3 (Zakian, Section 1.7), which terminates any local search when it is deemed to have become relatively unproductive. This section is intended to evaluate the extent to which this device is effective. This is done by removing Stopping Rule 3 and applying the resulting node array method to the fourteen test problems. The results are shown in Table 6.2, which has the same form as Table 6.1. The two tables, when contrasted, provide an obvious way of assessing the effect of Stopping Rule 3 on these test problems.

The general conclusion that can be reached is that, for most of the test problems, Stopping Rule 3 is highly effective in significantly reducing the number of function evaluations needed to find an admissible point. This reduction of labour, made possible by the use of Stopping Rule 3, is very marked with the test problems based on the Rosenbrock function, which involve significantly more intricate topography than the problems based on the camel function. It is also seen that, with the problems based on the camel function, Stopping Rule 3 causes either a less dramatic reduction of labour or a slight increase in labour. However, it is noted that, with or without Stopping Rule 3, the problems based on the camel function require much less labour to solve than the problems based on the Rosenbrock function. It should therefore not be expected that a device such as Stopping Rule 3 would bring about significant reductions in the labour needed for solving problems based on the camel function.

6.6 Conclusions

This chapter provides a numerical evaluation of the efficiency of a new global search method proposed by Zakian (Chapter 1) called the node array (NA) method.

Fourteen challenging test problems were constructed for the purpose of evaluating the efficiency of the node array method. These test problems are also used, elsewhere in the book, to evaluate the efficiency of two other globally convergent methods, thus enabling comparisons to be made of the relative efficiencies of all three methods.

The results of numerical tests, with the specific node array method denoted by NA&MBP(R), confirm that the method is globally convergent and show that the method is relatively efficient when its performance is compared to that of the two other known globally convergent methods presented in this book.

To detect when the NA&MBP(R) method becomes stuck, two simple rules are introduced. These rules, which are to be used in conjunction, are derived by considering the detailed operation of the well-known local search method called the moving boundaries process (MBP). Specifically, the MBP employed here uses the Rosenbrock (R) trial generator and is denoted by MBP(R).

Acknowledgement

The author acknowledges that the work presented in this chapter was suggested and guided by Dr. V. Zakian.

6.A Appendix — Moving Boundaries Process with the Rosenbrock Trial Generator

The moving boundaries process (MBP) is a numerical and iterative local search method for solving a system of inequalities developed by Zakian and Al-Naib (1973). It has been extensively used in the design of control systems; see, for example, Gray and Al-Janabi (1975; 1976), Gray and

Katebi (1979), Taiwo (1978a; 1978b; 1979a; 1979b; 1980; 1986), Bollinger et al. (1979), Coelho (1979), Dahshan (1981), Crossley and Dahshan (1982), Bada (1984; 1985; 1987a; 1987b), Dahshan et al. (1986), Inooka and Imai (1987), Katebi and Katebi (1987), Ng (1989; 1991), Ishihara et al. (1990), Singaby (1991a; 1991b), Janabi and Gray (1991), Prabhu and Chidambaram (1991), Rutland (1991; 1992), Lane (1992; 1995), Whidborne (1992; 1993), Whidborne and Liu (1993), Murad et al. (1993), Patton et al (1993; 1994), Postlethwaite et al. (1994), Rutland (1994), Rutland et al. (1994), Whidborne et al. (1994), Whidborne et al. (1995), Satoh et al. (1996). For a general discussion of the moving boundaries process see Zakian, Section 1.7. In this appendix, the MBP(R) is described in detail.

Suppose that the design problem is formulated by a set of inequalities as follows:

$$\phi_i(c) \leqslant \varepsilon_i, \quad i = 1, 2, \ldots, n \tag{6.55}$$

where ϕ_i denote real functions of c representing performance indices and constraint functions, $c \in \mathbb{R}^N$ represents the real vector comprised of design parameters and $\varepsilon_i \in \mathbb{R}$ are given real numbers. Each inequality in (6.55) defines a set as follows:

$$\mathcal{S}_i := \left\{ c \in \mathbb{R}^N : \phi_i(c) \leqslant \varepsilon_i \right\} \tag{6.56}$$

The *boundary* of the set $\mathcal{S}_i$ is given by $\phi_i(c) = \varepsilon_i$. Let $\mathcal{S}$ be the intersection of all the set $\mathcal{S}_i$ ($i = 1, 2, \ldots, n$) defined by

$$\mathcal{S} := \bigcap_{i=1}^{n} \mathcal{S}_i \tag{6.57}$$

A point $c^\star$ is a solution of the system of inequalities (6.55) if and only if it lies inside the set $\mathcal{S}$.

The MBP starts from an arbitrary starting point $c^0 \in \mathbb{R}^N$ in a subset of $\mathbb{R}^N$, called the search space, and proceeds to an admissible point in $\mathcal{S}$ in an iterative way. Let $c^k \in \mathbb{R}^N$ denote the value of $p \in \mathbb{R}^N$ at the kth move. A set $\mathcal{S}_i^k$ is defined by the inequality $\phi_i(c) \leqslant \phi_i(c^k)$ with a boundary given by $\phi_i(c) = \phi_i(c^k)$. A step is taken from the point c^k to a point $\hat{c}^k \in \mathbb{R}^N$ called a *trial point*. A change from c^k to some trial point $\hat{c}^k$ is referred to as a *trial*. If the boundary defined by $\phi_i(c) = \phi_i(\hat{c}^k)$ is closer, or no further from, the boundary of $\mathcal{S}_i^k$ for every $i = 1, 2, \ldots, n$, then the point $\hat{c}^k$ is accepted and becomes the new point c^{k+1}. After a sufficient number of successful trials, the boundary of $\mathcal{S}_i^k$ coincides with the boundary of $\mathcal{S}_i$ for every $i = 1, 2, \ldots, m$, and the problem is solved.

For an exact statement of the MBP, the following notations are introduced:

$$\mathcal{S}^k := \bigcap_{i=1}^{n} \mathcal{S}_i^k \tag{6.58}$$

$$\mathcal{S}_i^k := \{c \in \mathbb{R}^N : \phi_i(c) \leqslant \varepsilon_i^k\}, \quad i = 1, 2, \ldots, n \tag{6.59}$$

$$\varepsilon_i^k := \begin{cases} \varepsilon_i & \text{if } \phi_i(c^k) \leqslant \varepsilon_i, \quad i = 1, 2, \ldots, n \\ \phi_i(c^k) & \text{otherwise} \end{cases} \tag{6.60}$$

If a trial point $\hat{c}^k$ is a success, then the next point is set as $c^{k+1} = \hat{c}^k$; otherwise, another trial is made from $\hat{c}^k$ until a successful trial occurs. A trial point $\hat{c}^k$ is a success if and only if

$$\phi_i(\hat{c}^k) \leqslant \varepsilon_i^k, \quad i = 1, 2, \ldots, n \tag{6.61}$$

When the strict inequality holds in (6.61), all the boundaries of the set $\mathcal{S}^k$ defined in (6.58) are closer to the boundaries of $\mathcal{S}$. The sequential process described above is terminated when the boundaries of $\mathcal{S}^k$ have converged to the boundaries of $\mathcal{S}$; that is, when $\varepsilon_i^k = \varepsilon_i$ ($i = 1, 2, \ldots, n$). The current boundaries ε_i^k ($i = 1, 2, \ldots, n$) are continuously replaced by the new boundaries (see (6.60)). Since the replacement of the boundaries can be viewed as their movement, the process is called the moving boundaries process.

A trial point $\hat{c}^k$ is usually generated by using the scheme provided by Rosenbrock (1960). The MBP with the Rosenbrock (R) trial generator is denoted by MBP(R). Let $V_j^r \in \mathbb{R}^N$ ($j = 1, 2, \ldots, N$) be the orthonormal unit vectors at the rth stage, and let $e_j \in \mathbb{R}$ ($j = 1, 2, \ldots, N$) be corresponding real numbers. Then a trial point $\hat{c}^k$ is generated as follows:

$$\hat{c}^k = c^k + e_j V_j^r \tag{6.62}$$

If $\hat{c}^k$ is not a successful trial, the real number e_j is replaced by βe_j ($-1 < \beta < 0$). If $\hat{c}^k$ is a successful trial, e_j is replaced by αe_j ($\alpha > 1$). Usually, satisfactory results are obtained with $\alpha = 3$ and $\beta = -0.5$. In either case, j is replaced by $j+1$ in (6.62). If $j = N$, then $j + 1 = 1$.

As soon as one successful trial followed by one failure has occurred for every j, the orthonormal unit vectors V_j^r are replaced by V_j^{r+1} as follows. Let d_j be the sum of all successful values of e_j during the rth stage. Now let $a_j \in \mathbb{R}$ ($j = 1, 2, \ldots, N$) denote real numbers computed as

$$\begin{aligned} a_1 &= d_1 V_1^r + d_2 V_2^r + \ldots + d_N V_N^r \\ a_2 &= d_2 V_2^r + \ldots + d_N V_N^r \\ &\vdots \\ a_N &= d_N V_N^r \end{aligned} \tag{6.63}$$

Then the Gram-Schmidt procedure is used to orthogonalise the vectors a_j and obtain V_j^{r+1} in the following way:

$$b_1 = a_1$$
$$V_1^{r+1} = b_1/\|b_1\|$$
$$b_2 = a_2 - \langle a_2, V_1^{r+1}\rangle V_1^{r+1}$$
$$V_2^{r+1} = b_2/\|b_2\|$$
$$\vdots$$
$$b_N = a_N - \sum_{k=1}^{N-1} \langle a_N, V_k^{r+1}\rangle V_k^{r+1}$$
$$V_N^{r+1} = b_N/\|b_N\|$$

(6.64)

where $\langle x, y\rangle$ denotes the inner products of two N-dimensional vectors x and y, and $\|x\|$ is the Euclidean norm of the vector x. At the beginning of the process (i.e., $j = 0$), $e_j \in \mathbb{R}$ and $V_j^r \in \mathbb{R}$ are arbitrarily chosen. However, as r increases, V_j^r becomes oriented along a direction of rapid advance. Therefore, the rate of convergence of $\mathcal{S}^k$ defined in (6.58) towards $\mathcal{S}$ tends to improve.

The MBP(R) is summarised in the following:

Step 1. The real numbers ε_i ($i = 1, 2, \ldots,$) in (6.55) and the initial parameter vector $c^0 \in \mathbb{R}^N$ are given.

Step 2. Set $c = c^0$ and compute $\phi_i(c)$ ($i = 1, 2, \ldots, n$). If $\phi_i(c) \leqslant \varepsilon_i$ for all $i \in \{1, 2, \ldots, n\}$, then the initial point c^0 satisfies all the inequalities and therefore quit; otherwise go to **Step 3**.

Step 3. Set $e_j^0 = 0.1|c_j|$ if $|c_j| \geqslant 0.1$ ($j = 1, 2, \ldots, N$); otherwise, set $e_j^0 = 0.01$ ($j = 1, 2, \ldots, N$). In addition, set

$$V_1 = [1\,0\,0\,\ldots\,0\,0\,0]$$
$$V_2 = [0\,1\,0\,\ldots\,0\,0\,0]$$
$$\vdots$$
$$V_{N-1} = [0\,0\,0\,\ldots\,0\,1\,0]$$
$$V_N = [0\,0\,0\,\ldots\,0\,1\,0]$$

Step 4. Set $\varepsilon_i^\dagger = \phi_i(c)$ if $\phi_i(c) > \varepsilon_i$ ($i = 1, 2, \ldots, n$); otherwise, set $\varepsilon_i^\dagger = \varepsilon_i$ ($i = 1, 2, \ldots, n$). Set $L = 0$ and $r = 0$.

Step 5. Set $e_j = e_j^0$ and $d_j = 0$.

Step 6. Set $j = 1$. Start a new iteration.

Step 7. Generate a trial point $\hat{c} = c + e_j V_j$ ($j = 1, 2, \ldots, N$). Compute $\phi_i(\hat{c})$ ($j = 1, 2, \ldots, N$). Test whether $\phi_i(\hat{c}) \leqslant \varepsilon_i^\dagger$ ($j = 1, 2, \ldots, N$). If success, go to **Step 8**; otherwise, go to **Step 9**.

Step 8. Set $c = \hat{c}$, $d_j = d_j + e_j$ and $e_j = 3e_j$. Set $\varepsilon_i^\dagger = \phi_i(c)$ if $\phi_i(c) > \varepsilon_i$ $(i = 1, 2, \ldots, n)$; otherwise, $\varepsilon_i^\dagger = \varepsilon_i$ $(i = 1, 2, \ldots, n)$. Check whether $\varepsilon_i^\dagger = \varepsilon_i$ for all $i \in \{1, 2, \ldots, n\}$. If so, quit because the problem is solved; otherwise, go to **Step 10**.

Step 9. Discard $\hat{c}$ and set $e_j = -0.5e_j$. Check whether one success followed by one failure has occurred for every current V_j $(j = 1, 2, \ldots, N)$. If so, re-initialise V_j $(j = 1, 2, \ldots, N)$ using (6.63) and (6.64), set $r = r + 1$ and go to **Step 7.**; otherwise, go to **Step 10**.

Step 10. If $j = N$, go to **Step 6**; otherwise, set $j = j + 1$ and go to **Step 7**.

References

Bada, A.T. (1984). Design of delayed control systems with Zakian's method, *International Journal of Control*, 40(4):773–781.

Bada, A.T. (1985). Design of delayed control systems using Zakian's framework, *IEE Proceedings, Pt. D*, 132(6):251–256.

Bada, A.T. (1987a). Criteria: a package for computer-aided control system design, *Computer-Aided Design*, 19:466–474.

Bada, A.T. (1987b). Robust brake control for a heavy-duty truck, *IEE Proceedings, Pt. D*, 134(1):1–8.

Bollinger, K.E., Cook, P.A. and Sandhu, P.S. (1979). Synchronous generator controller synthesis using moving boundary search techniques with frequency domain constraints, *IEEE Transactions on Power Apparatus & Systems*, 98(5):1497–1501.

Coelho, C.A.D. (1979). Compensation of the speed governor of a water turbine by the method of inequalities, *Transactions of the ASME, Journal of Dynamic Systems, Measurement, and Control*, 101(3):205–211.

Crossley, T.R. and Dahshan, A.M. (1982). Design of a longitudinal ride-control system by Zakian's method of inequalities, *Journal of Aircraft*, 19(9):730–738.

Dahshan, A.E., Fawzy, A.S., Aly, A.W. and Singaby, M.E. (1986). An optimal coordination approach for solving control problems using the method of inequalities, *Proceedings of the IEEE Conference on Systems, Man and Cybernetics*, pp. 1525–1529.

Dahshan, A.M. (1981). A comparison between the method of inequalities and other techniques in control systems design, *International Journal of Systems Science*, 12(9):1149–1155.

Dixon, L.C.W. and Szego, G.P. (1978). *Towards Global Optimization II*, North Holland, New York, NY.

Gray, J.O. and Al-Janabi, T.H. (1975). The numerical design of feedback control systems with set containing of saturation element by the method of inequalities, *Proceedings of 7th IFIP Conference on Optimization Techniques*, pp. 510–521.

Gray, J.O. and Al-Janabi, T.H. (1976). Toward the numerical design of nonlinear feedback systems by Zakian's method of inequalities, *Proceedings of the IFAC Symposium on Large Scale Systems Theory and Applications*, pp. 327–334.

Gray, J.O. and Katebi, S.D. (1979). On the design of a class of nonlinear multivariable control systems, *Proceedings of the IFAC Symposium on Computer-Aided Design of Control Systems*, pp. 71–80.

Inooka, H. and Imai, Y. (1987). Design of a single-loop digital controller by the method of inequalities, *International Journal of Control*, 45(5):1505–1513.

Ishihara, T., Sugimoto, K. and Inooka, H. (1990). Computer-aided robust controller design using the method of inequalities, *Proceedings of Multinational Instrumentation Conference*, pp. 127–132.

Janabi, T.H. and Gray, J.O. (1991). Method for the control of a special class of non-linear systems, *International Journal of Control*, 54(1):215–239.

Katebi, S.D. and Katebi, M.R. (1987). Combined frequency and time domain technique for the design of compensators for non-linear feedback control systems, *International Journal of Systems Science*, 18(11):2001–2017.

Lane, P.G. (1992). *Design of control systems with inputs and outputs satisfying certain bounding conditions*, PhD thesis, University of Manchester Institute of Science and Technology.

Lane, P.G. (1995). The principle of matching: a necessary and sufficient condition for inputs restricted in magnitude and rate of change, *International Journal of Control*, 62(4):893–915.

Murad, G., Postlethwaite, I., Gu, D.W. and Whidborne, J.F. (1993). Robust control of a glass tube shaping process, *Proceedings of 2nd European Control Conference*, p.p. 2350–2355.

Ng, W.Y. (1989). *Interactive Multi-Objective Programming as a Framework for Computer-Aided Control Systems Design*, Lecture Notes in Control & Information Sciences, Springer-Verlag, Berlin.

Ng, W.Y. (1991). An interactive descriptive graphical approach to data analysis for trade-off decisions in multi-objective programming, *Information and Decision Technologies*, 17:133–149.

Patton, R.J., Liu, G.P. and Chen, J. (1993). Multivariable control using eigenstructure assignment and the method of inequalities, *Proceedings of 2nd European Control Conference*, pp. 2030–2034.

Patton, R.J., Liu, G.P. and Chen, J. (1994). Multiobjective controller design using eigenstructure assignment and the method of inequalities, *Journal of Guidance, Control and Dynamics*, 17(4):862–864.

Postlethwaite, I., Whidborne, J.F., Murad, G. and Gu, D.W. (1994). Robust control of the benchmark problem using H_∞ methods and numerical optimization techniques, *Automatica*, 30(4):615–619.

Prabhu, E.S. and Chidambaram, M. (1991). Robust control of a distillation column by the method of inequalities, *Journal of Process Control*, 1:171–176.

Rosenbrock, H.H. (1960). An automatic method for finding the greatest or least value of a function, *Computer Journal*, 3:175–184.

Rutland, N.K. (1991). A crash course on the application of a new principle of design to vehicle speed control, *Proceedings of 30th IEEE Conference on Decision and Control*, pp. 1198–1199.

Rutland, N.K. (1992). Illustration of a new principle of design: vehicle speed control, *International Journal of Control*, 55(6):1319–1334.

Rutland, N.K. (1994). Illustration of the principle of matching with inputs restricted in magnitude and rate of change: vehicle speed control revisited, *International Journal of Control*, 60(3):395–412.

Rutland, N.K., Keogh, P.S. and Burrows, C.R. (1994). Comparison of controller designs for attenuation of vibration in a rotor-bearing system under synchronous and transient conditions, *Proceedings of Fourth International Symposium on Magnetic Bearing*, pp. 1198–1199.

Satoh, T., Ishihara, T. and Inooka, H. (1996). Systematic design via the method of inequalities, *IEEE Control Systems*, 16(5):57–65.

Singaby, M.I.E. (1991a). Application of the discrete version of the method of inequalities, *Proceedings of the IEEE International Conference on Systems, Man and Cybernetics*, pp. 513–515.

Singaby, M.I.E. (1991b). A discrete version of Zakian's method of inequalities, *Proceedings of the IFAC Symposium on Design Methods of Control Systems (D. Franke and F. Kraus, Eds.)*, pp. 101–104.

Taiwo, O. (1978a). Design of multivariable controller for a high-order turbofan engine model by Zakian's method of inequalities, *IEEE Transactions on Automatic Control*, 23:926–928.

Taiwo, O. (1978b). Improvement of a turbogenerator response by the method of inequalities, *International Journal of Control*, 27:305–311.

Taiwo, O. (1979a). Design of multivariable controllers for an advanced turbofan engine by Zakian's method of inequalities, *Transactions of the ASME, Journal of Dynamic Systems, Measurement, and Control*, 101(4):299–307.

Taiwo, O. (1979b). On the design of feedback controllers for the continuous stirred tank reactor by Zakian's method of inequalities, *Chemical Engineering Journal*, 17:3–12.

Taiwo, O. (1980). Application of the method of inequalities to the multivariable control of binary distillation columns, *Chemical Engineering Science*, 35:847–858.

Taiwo, O. (1986). The design of robust control systems for plants with recycle, *International Journal of Control*, 43(2):671–678.

Whidborne, J.F. (1992). Performance in sampled data systems, *IEE Proceedings, Pt. D.*, 139(3):245–250.

Whidborne, J.F. (1993). EMS control system design for a maglev vehicle – a critical system, *Automatica*, 29(5):1345–1349.

Whidborne, J.F. and Liu, G.P. (1993). *Critical Control Systems*, Research Studies Press, Somerset.

Whidborne, J.F., Murad, G., Gu, D.W. and Postlethwaite, I. (1995). Robust control of an unknown plant – the IFAC 93 benchmark, *International Journal of Control*, 61(3):589–640.

Whidborne, J.F., Postlethwaite, I. and Gu, D.W. (1994). Robust controller design using H_∞ loop-shaping and the method of inequalities, *IEEE Transactions on Control Systems Technology*, 2(4):455–461.

Zakian, V. and Al-Naib, U. (1973). Design of dynamical and control systems by the method of inequalities, *IEE Proceedings*, 120(11):1421–1427.

7 A Simulated Annealing Inequalities Solver

James F Whidborne

Abstract. Simulated annealing, based on the statistical mechanical annealing of solids, is a probabilistic method for generating and assessing trial points in the search space. The method is well known and has been used to find the minimum of any function. By defining an energy function for a known simulated annealing method, a probabilistic globally convergent search procedure for solving inequalities is obtained. The method is tested on some numerical test problems and some control design problems. It is found that the method has a high rate of success in finding solutions.

7.1 Introduction

The principle of inequalities (Zakian and Al-Naib, 1973) requires the design problem to be formulated as the conjunction of the inequalities

$$\phi_i(p) \leq \varepsilon_i \text{ for } i = 1 \ldots n \tag{7.1}$$

where p is a real vector $(p_1, p_2, \ldots, p_q)$, chosen from a given search space $\mathcal{P}$, and represents the design parameters that characterise the controller or the entire system, ϕ_i are real functions of p, called objective functions, are measures of the way the system performs and ε_i are real numbers that represent the largest tolerable values of the performance measures. The aim of the design is to find a point p that satisfies all the inequalities. Such a value of p is said to be admissible. The search space is generally a hyper-rectangle of dimension q, because each element of the vector p is bounded from above and below. The search space defines the set of all possible designs within the constraints on the vector p.

In the mechanical annealing of metals, the material is heated and then cooled at a controlled rate to attempt to create a perfect crystal lattice structure. This process is known as annealing. A perfect crystal lattice structure has minimal free energy, hence the objective of annealing is to ensure that the free energy state of the system reaches the minimum energy possible and does not get stuck in local minima whereby there are imperfections in the lattice structure.

A function minimisation algorithm that mimics this process was originally proposed by Kirkpatrick *et al* (1983). A 'temperature' variable is introduced into the search procedure. Initially the temperature is high and, as the search progresses, the temperature is slowly reduced. Random trial steps are made in the locality of the current point. If a step results in an improved point, the new point is accepted. If a trial point is an inferior point, it may still be accepted with a certain probability that is dependent on the temperature. When the temperature is high, the probability of an inferior trial point being accepted is close to unity. As the search progresses, the temperature is reduced in stages until at low temperatures, the probability of an inferior trial point being accepted is close to zero. By accepting some inferior points, the search has the chance to climb out of local minima or other traps and find an admissible point.

This chapter presents a simulated annealing inequalities solver (SAIS) first developed by Whidborne et al (1996). The method is based on the function-minimising simulated annealing scheme of Vanderbilt and Louie (1984). By defining an appropriate energy function, the scheme developed by Vanderbilt and Louie (1984) is modified to produce a method for solving inequalities. The SAIS is a global process for solving inequalities, in the sense that, given any starting point in the search space, an admissible point is almost always located.

In the next section, the concept of simulated annealing is introduced. In Section 7.3, the simulated annealing inequalities solver is developed. In Sections 7.4 and 7.5, the effectiveness of the solver is tested by application to the test problems described by Satoh in Chapter 6 and to some control test problems. Conclusions are given in Section 7.6.

7.2 The Metropolis Algorithm

The simulated annealing algorithm has its origins in the statistical mechanical annealing of solids. A simple algorithm was proposed by Metropolis et al (1953) that can be used to simulate the annealing process. The temperature is reduced in stages. At each temperature, the system is perturbed, and the change in energy calculated. The perturbed state is accepted as the new state if the energy has decreased. If the energy has increased, the new state is accepted with a probability, P, given by

$$P(\Delta E, T) = \exp(-\Delta E/(k_b T)) \qquad (7.2)$$

where $-\Delta E$ is the change in energy, T is the temperature and k_b is the Boltzmann constant. This acceptance rule is referred to as the Metropolis criterion. The basic Metropolis algorithm is shown as Algorithm 7.1 below.

Algorithm 7.1. The basic Metropolis algorithm.

Set T = initial temperature
Set M = a number of steps at each temperature
Set k_b = Boltzmann constant
Until terminating condition is true do{
   for i = 1 to M do{
      Make small perturbation on system
      Calculate change in energy ΔE
      If $\Delta E \leq 0$
         accept perturbation
      else
         accept perturbation with probability $P(\Delta E) = \exp(-\Delta E/(k_b T))$
   }
   Reduce T
}

7.3 A Simulated Annealing Inequalities Solver

7.3.1 Development

The Metropolis algorithm can be used in a simulated annealing scheme to solve function minimisation problems. This may be done by replacing the energy with the objective function, using the temperature as a control parameter, and by assuming that the role of the states of a solid is taken by a description of a system configuration. Many practical simulated annealing techniques have been developed for combinatorial minimisation problems, however, Vanderbilt and Louie (1984) first proposed that the basic scheme be used for searching over a continuous variable space. The main difference with combinatorial methods is the perturbation mechanism for searching the parameter space. The scheme suggested by Vanderbilt and Louie (1984) is for scalar minimisation problems. Whidborne *et al* (1996) have extended the scheme of Vanderbilt and Louie (1984) to a multi-objective minimisation scheme that can be used for solving inequalities. For a review of other suitable simulated annealing schemes, see Jones *et al* (2002).

There are six main features of the Metropolis algorithm that are considered in developing the simulated annealing scheme: (i) the description of the system, (ii) the cooling scheme, (iii) the trial point generator, (iv) the energy function, (v) the acceptance criterion and (vi) the algorithm termination criteria.

The description of the system. The system configuration must be represented in some way that describes the space of possible designs. For continuous parameter problems, this is just the search space $\mathcal{P}$ defined previously.

The cooling scheme. For the determination of the perturbation mechanism described next, an exploration of the search space is required and a rough approximation of the "shape" of the space that results in accepted steps is made. Thus, for the scheme suggested here, the cooling scheme differs from the usual Metropolis algorithm described earlier, in that the temperature is reduced after M accepted steps of the algorithm at the temperature T. The new temperature T^* is set to $T^* = \chi_T T$, where $0 < \chi_T < 1$. The choice of χ_T greatly affects the efficiency of the annealing and its ability to climb out of local minima; if χ_T is too small, the temperature will reduce too quickly, and the process will get stuck in a local minimum. If χ_T is too large, the process will become very inefficient. Vanderbilt and Louie (1984) suggest that χ_T is chosen by trial and error. From the tests by Vanderbilt and Louie (1984) and Whidborne et al (1996), a fairly slow anneal, with $0.097 \leq \chi_T \leq 0.99$, seems to result in a solution, but it is at the expense of a large number of iterations.

The efficiency of the scheme and its ability to climb out of local minima is also very dependent on the initial temperature (or the Boltzmann constant k_b). Vanderbilt and Louie (1984) suggest that the initial temperature is chosen on the basis of the variance of the energy $E(p)$ of a random sample of points, p. Whidborne et al (1996) have conducted an investigation into the effect of the initial temperature.

The trial point generator. A vector $u \in \mathbb{R}^q$ of independent random numbers $(u_1, u_2, \ldots, u_q)$ is generated where each u_i is chosen independently from the interval $[-\sqrt{3}, \sqrt{3}]$, so they have zero mean and unit variance. Thus the vector $u = (u_1, u_2, \ldots, u_q)$ occurs with a constant probability density inside a hypercube of volume $(2\sqrt{3})^q$ and zero outside. A step Δp to a trial point $\tilde{p}$ is taken $\tilde{p} = p + \Delta p$ where

$$\Delta p = Qu \tag{7.3}$$

where the matrix $Q \in \mathbb{R}^{q \times q}$ controls the step distribution. Random steps with a desired covariance matrix $s \in \mathbb{R}^{q \times q}$ can be generated from (7.3) and by solving for Q,

$$s = QQ^T \tag{7.4}$$

A procedure for choosing s so that it adapts to the topography of the objective function has been proposed by Vanderbilt and Louie (1984). The excursions of the random walk are used as the measure of the local topology, so that the search adapts itself to the local topology. Firstly, from Vanderbilt and Louie (1984), the available phase space $\mathcal{E}(T)$ is vaguely defined as a function of the temperature T to be

$$\mathcal{E}(T) = \{p : E(p) - E_{\min}(p) \lesssim T\} \tag{7.5}$$

where $E(\cdot)$ is the energy function defined in (7.9). If the axes of s are poorly aligned with the topology of $E(T)$, much time will be wasted exploring fruitless directions of search. So, at the end of the lth set of M steps, the first and second moments of the walk segment are calculated, where

$$A_i^{(l)} = 1/M \sum_{m=1}^{M} p_i^{(m;l)} \tag{7.6}$$

and

$$S_{ij}^{(l)} = 1/M \sum_{m=1}^{M} \left[p_i^{(m;l)} - A_i^{(l)} \right] \left[p_j^{(m;l)} - A_j^{(l)} \right] \tag{7.7}$$

where $p^{(m;l)}$ is the value of p on the mth step of the lth set. Thus S describes the actual shape of the walk over the lth set. To choose s for the next iteration set, $l+1$,

$$s^{(l+1)} = \frac{\chi_s}{\beta M} S^{(l)} \tag{7.8}$$

where $\chi_s > 1$ is a growth factor, and β is based on a geometric average over the random variables Δp. Typically, $\chi_s = 3$ and $\beta = 0.11$ (Vanderbilt and Louie, 1984).

The idea behind this scheme is that the steps are initially small, but grow as the annealing progresses until the walk can cover the phase space $\mathcal{E}(T)$. Steps outside the phase space will then begin to be rejected, and the size of the walk will reduce as the phase space $\mathcal{E}(T)$ gets smaller as the temperature is reduced. The size and shape of S and hence s will thus adapt to the topology of $E(p)$. In this way, the whole possible phase space is covered in a relatively efficient manner.

The energy function. The following energy function E is used to convert the multi-objective problem to a single objective minimax problem

$$E(p) = \max \left\{ \max \left\{ \frac{\phi_i(p) - \varepsilon_i}{w_i}, 0 \right\} : i = 1, \ldots, n \right\} \tag{7.9}$$

where $w_i, i = 1, \ldots, n$ are positive weights chosen *a priori* by the designer to reflect the relative importance of each objective function ϕ_i.

Evidently, any solution of the inequalities (7.1) corresponds to zero energy. Hence, when the energy is minimised to zero, the corresponding point is admissible.

The acceptance criterion. The algorithm uses the Metropolis acceptance criterion. Thus, if the energy change is $\Delta E = E(\tilde{p}) - E(p)$, where $\tilde{p}$ is a trial point, p is the current point and E is as for (7.9), then if $\Delta E \leq 0$, the trial point is accepted; otherwise, the trial point is accepted with a probability given by $\exp(-\Delta E/(k_b T))$.

Algorithm termination criteria. The algorithm will be terminated when either (i) all the inequalities are satisfied, (ii) a predefined minimum temperature $T_{\min}$ is reached or (iii) the following condition is satisfied (Vanderbilt and Louie, 1984)

$$\frac{\bar{E}^{(l)} - E_{\min}^{(l)}}{\bar{E}^{(l)}} < \eta \tag{7.10}$$

where $\bar{E}^{(l)}$ and $E_{\min}^{(l)}$ are respectively the average and minimum energies over the lth walk segment. This condition means that the configuration has become "frozen", that is, further search is unlikely to reduce the energy. Vanderbilt and Louie (1984) suggest that $\eta = 10^{-3}$.

7.3.2 SAIS

The developed SAIS is shown as Algorithm 7.2. Note that a test should be made to check that each trial point, $\tilde{p}$, is in $\mathcal{P}$, and if not, trials are made until a $\tilde{p} \in \mathcal{P}$ is obtained. An alternative is to characterise $\mathcal{P}$ by a set of inequality constraints, with a penalty function if $\tilde{p} \notin \mathcal{P}$.

7.4 Numerical Test Problems

The effectiveness of the SAIS given by Algorithm 7.2 for solving scalar inequality problems is tested by application to the test problems described by Satoh in Chapter 6. For all the problems, the initial temperature was set to $T^{(0)} = 100$ and the minimum temperature to $T_{\min} = 0.1 \times 10^{-3}$. The weighting, w_1 was set to unity and the search parameters were set to $M = 16$, $\chi_T = 0.96$, $\chi_s = 3$, $\beta = 0.11$, $\eta = 0.1 \times 10^{-3}$ and $k_b = 1$. No attempt was made to tune the temperatures or search parameters. For each problem, the search was conducted 100 times from the same set of initial points selected randomly with an even distribution over the whole search space. The mean number of function evaluations including the failed searches and the minimum and maximum number of function evaluations for successful searches are shown in Table 7.1 along with the percentage of failed searches.

For all the tests, the chances of finding a solution is good, with few failed attempts. All the tests except Problems 3 and 4 were solved with a similar number of function evaluations, averaging in the range 11 000 to 15 000. The ability of the algorithm in finding the global solution, despite the presence of local minima, is demonstrated. It is curious that Problems 3 and 4 sometimes require a very large number of function evaluations, the reasons for this are not yet clear.

Algorithm 7.2. The SAIS algorithm.

Define ϕ_i, ε_i and w_i for $i = 1, \ldots, n$
Set M, χ_T, χ_s, β, ν, and k_b
Set initial and minimum temperatures, $T^{(0)}$ and $T_{\min}$
Set initial point $p^{(0)} = (p_1^{(0)}, p_2^{(0)}, \ldots, p_q^{(0)})$
Calculate objective functions $\phi_i(p^{(0)})$, $i = 1, \ldots, n$
Calculate initial energy $E(p^{(0)})$ by (7.9)
Set initial covariance matrix $s^{(0)}$
Calculate $Q^{(0)}$ by solving (7.4)
Set iteration number $k = 0$
Set cooling stage number $l = 0$
Set cooling stage counter $m = 0$
until $\phi_i(p^{(k)}) \leq \varepsilon_i$ for all $i = 1, \ldots, n$ or $T^{(l)} <= T_{\min}$
  or $(\bar{E}^{(l)} - E_{\min}^{(l)})/\bar{E}^{(l)} < \eta$ do {
  Generate random vector $u = (u_1, u_2, \ldots, u_q)$ where
    $u_i = \text{random}[-\sqrt{3}, \sqrt{3}]$
  Generate trial point $\tilde{p} = p^{(k)} + Q^{(k)}u$
  Calculate $\phi(\tilde{p})$ and $E(\tilde{p})$
  Calculate energy change $\Delta E = E(\tilde{p}) - E(p^{(k)})$
  if
    $\Delta E \leq 0$, accept move
  else
    if $\exp(-\Delta E/(k_b T^{(l)})) > \text{random}[0,1)$, accept move
  if move accepted {
    Set new point $p^{(k+1)} = \tilde{p}$
    Increment $k = k + 1$ and $m = m + 1$
    if $m = M$ {
      Calculate $A_i^{(l)}$ and $S_{ij}^{(l)}$ from (7.6) and (7.7)
      Calculate $s^{(l+1)}$ from (7.8) and $Q^{(l+1)}$ by solving (7.4)
      Reduce temperature $T^{(l+1)} = \chi_T T^{(l)}$
      Set $m = 0$ and increment $l = l + 1$
    }
  }
}

7.5 Control Design Benchmark Problems

The effectiveness of the SAIS for solving control design problems posed using the principle of inequalities was presented in Whidborne *et al* (1996). Some of the results are reproduced here.

Table 7.1. Results of numerical test problems

Problem No.	Mean No. of Function Evaluations	Maximum No. of Function Evaluations	Minimum No. of Function Evaluations	Percent Failed Searches
1	13 454	17 285	5 766	0%
2	11 642	14 557	6 975	0%
3	365 197	1 465 106	7 225	0%
4	49 322	163 503	6 871	1%
5.1	14 609	18 630	9 462	2%
5.2	14 662	18 101	9 376	1%
5.3	14 671	19 696	4 647	1%
5.4	14 703	18 177	4 335	2%
5.5	13 997	19 015	8 430	0%
6.1	12 980	15 880	7 470	3%
6.2	13 028	15 553	8 399	4%
6.3	12 930	15 775	8 518	1%
6.4	12 765	15 364	7 759	1%
6.5	12 919	17 116	8 608	4%

7.5.1 Problem 1

The aim is to obtain a good step response for a simple SISO system which has been investigated by several designers including Zakian and Al-Naib (1973) and Whidborne *et al* (1995). The plant is $G(s) = 10/(s(s+1)(s+5))$ and the controller has the form $K(s,p) = p_1(1+p_2 s)/(1+p_2 p_3 s)$.

The problem is to find values for p such that $\phi_i(p) \leq \varepsilon_i, i = 1, \ldots, 4$ where ϕ_1 is the abscissa of stability, ϕ_2 is the rise-time index, ϕ_3 is the overshoot and ϕ_4 is the control effort index. The objective functions are defined in the Appendix. The design goals are $\varepsilon = (-0.001, 1.5, 0.2, 10)$ and the design parameters are constrained such that $0.01 \leq p_1 \leq 50$, $0 \leq p_2 \leq 20$ and $0.01 \leq p_3 \leq 10$. The weighting w is set to $w = (0.001, 10.0, 1.0, 50)$. The values were chosen to reflect the primary importance of the stability function, ϕ_1 and the approximate relative values of the goals $\varepsilon_2, \varepsilon_3, \varepsilon_4$.

7.5.2 Problem 2

This aim is to obtain a good step response with little cross-coupling for a two-input two-output system (Zakian and Al-Naib, 1973). The plant $G(s, p_1)$ is

$$G(s,p_1) \stackrel{s}{=} \left[\begin{array}{cccc|cc} 1.38 & -0.2077 & 6.715 & -5.676 & 0 & 0 \\ -0.5814 & -4.29 & 0 & 0.675 & 5.679 & 0 \\ 1.067 & 4.273 & -6.654 & 5.893 & 1.136 & -3.146 \\ 0.048 & 4.273 & 1.343 & -2.104 & 1.136 & 0 \\ \hline 1 & 0 & p_1 & -p_1 & 0 & 0 \\ 0 & 1 & 0 & 0 & 0 & 0 \end{array} \right], \quad (7.11)$$

where p_1 is a design variable (Munro, 1972), and the controller $K(s,p)$ has the form

$$K(s,p) = \frac{1}{s}\begin{bmatrix} 0 & p_2(1+p_3 s) \\ p_4(1+p_5 s) & 0 \end{bmatrix}. \tag{7.12}$$

The problem is to find values for p such that $\phi_i(p) \leq \varepsilon_i, i = 1, \ldots, 9$, where ϕ_1 is the abscissa of stability and $\phi_2, \ldots, \phi_9$ are functionals of the closed loop step responses which reflect the design aims. The objective functions $\phi_1, \ldots, \phi_9$ are defined in the Appendix. The design goals are $\varepsilon = (-0.001, 0.2, 0.1, 0.2, 0.1, 0.1, 0.1, 10, 10)$. The design parameters are constrained such that $0.1 \leq p_1, p_3, p_5 \leq 10$, $1 \leq p_2 \leq 100$ and $-20 \leq p_4 \leq 20$. The weightings $w_1, \ldots, w_9$ are set to $w = (0.001, 2, 1, 2, 1, 1, 1, 100, 100)$. These values were chosen to reflect the primary importance of the stability function, ϕ_1, and the relative values of the goals $\varepsilon_2, \ldots, \varepsilon_9$.

7.5.3 Algorithm Performance

For each problem, 100 searches were conducted with the initial points selected randomly with an even distribution over the whole search space. The maximum number of function evaluations for *successful* searches are shown in Table 7.2 along with the percentage of failed searches. The test was also run using the moving boundaries process (specifically, the MBP(R)) (see Zakian, Chapter 1), which is a local search method and is therefore likely to get stuck in local minima, in order to demonstrate that the problem has many local minima or other traps and that the SAIS method is relatively impervious to such obstacles. The initial temperature for each test was set at $T_0 = 0.1$, and the other algorithm variables were fixed at $\chi_T = 0.99$, $\chi_s = 3$, $\beta = 0.11$, $\eta = 0.1 \times 10^{-3}$, and $k_b = 1.0$.

The tests show that the SAIS has, as is to be expected, a greater global capacity than the MBP(R). The SAIS always managed to find solutions, although this could be after a large number of function evaluations compared to the number required by the MBP(R).

Table 7.2. Results of control design problems

Algorithm	Maximum No. of Function Evaluations	Percent Failed Searches
Problem 1		
SAIS	13 007	0 %
MBP(R)	1 548	97 %
Problem 2		
SAIS	12 995	0%
MBP(R)	1 562	80%

7.6 Conclusions

This chapter presents an inequalities solver based on simulated annealing and called the SAIS. The global capacity of the SAIS is demonstrated on the numerical test problems described by Satoh in Chapter 6, and on two control design problems.

The SAIS is shown to be effective at obtaining solutions when a well-known local search method gets stuck in various local traps. This is at the expense of a larger number of function evaluations. A minor drawback of the SAIS is the need to define weights for the energy function.

7.A Appendix – the Objective Functions

7.A.1 Problem 1

The abscissa of stability ϕ_1 is defined as

$$\phi_1 = \max\{\operatorname{Re}\{\lambda\} \text{ for all } \lambda \in \Lambda\} \tag{7.13}$$

where Λ denotes the set of all the roots λ of the characteristic polynomial of the system. The rise-time index ϕ_2 is defined as

$$\phi_2 = \min\{t \text{ such that } y(t,h) = 0.9 y(\infty, h)\} \tag{7.14}$$

where $y(t,h)$ is the step response of the plant output. The overshoot ϕ_3 is defined as

$$\phi_3 = \frac{\hat{y}(h) - |y(\infty, h)|}{|y(\infty, h)|} \quad \text{if } \hat{y}(h) > |y(\infty, h)| \tag{7.15}$$

where $\hat{y}(h) = \sup_{t \geq 0} |y(t,h)|$. The control effort index ϕ_4 is defined as

$$\phi_4 = \sup_{t \geq 0} |u(t,h)| \tag{7.16}$$

where $y(t,h)$ is the step response of the plant input.

7.A.2 Problem 2

The abscissa of stability ϕ_1 is defined as

$$\phi_1 = \max\{\operatorname{Re}\{\lambda\} \text{ for all } \lambda \in \Lambda\} \tag{7.17}$$

where Λ denotes the set of all the roots λ of the characteristic polynomial of the system. The plant output response of the closed-loop system at a time t to a reference step demand $\begin{bmatrix} h_1 \\ h_2 \end{bmatrix} h(t)$ is denoted by $y_i(t, \begin{bmatrix} h_1 \\ h_2 \end{bmatrix})$, $i = 1, 2$ where

$h(t)$ is the unit step function. The step response functional indices $\phi_2, \ldots, \phi_7$ are defined as

$$\phi_2 = \min\{t \text{ such that } y_1(t, [\begin{smallmatrix}1\\0\end{smallmatrix}]) = 0.9\} \tag{7.18}$$

$$\phi_3 = \sup_t \{y_1(t, [\begin{smallmatrix}1\\0\end{smallmatrix}]) - 1\} \tag{7.19}$$

$$\phi_4 = \min\{t \text{ such that } y_2(t, [\begin{smallmatrix}0\\1\end{smallmatrix}]) = 0.9\} \tag{7.20}$$

$$\phi_5 = \sup_t \{y_2(t, [\begin{smallmatrix}0\\1\end{smallmatrix}]) - 1\} \tag{7.21}$$

$$\phi_6 = \sup_t \{y_1(t, [\begin{smallmatrix}0\\1\end{smallmatrix}])\} \tag{7.22}$$

$$\phi_7 = \sup_t \{y_2(t, [\begin{smallmatrix}1\\0\end{smallmatrix}])\} \tag{7.23}$$

Similarly denoting the controller output response of the closed loop system by $u_i(t, [\begin{smallmatrix}h_1\\h_2\end{smallmatrix}])$, $i = 1, 2$, the control effort indices ϕ_8 and ϕ_9 are defined as

$$\phi_8 = \sup_t |u_1(t, [\begin{smallmatrix}1\\0\end{smallmatrix}])| \tag{7.24}$$

$$\phi_9 = \sup_t |u_2(t, [\begin{smallmatrix}0\\1\end{smallmatrix}])| \tag{7.25}$$

References

Jones, D.F., S.K. Mirrazavi and M. Tamiz (2002), Multi-objective meta-heuristics: An overview of the current state-of-the-art, *Eur. J. Op. Research*, 137(1):1–9.

Kirkpatrick, S., C.D. Gelatt, Jr. and M.P. Vecchi (1983), Optimization by simulated annealing, *Science*, 220:671–680.

Metropolis, N., A. Rosenbluth, M. Rosenbluth, A. Teller and E. Teller (1953), Equation of state calculations by fast computing machines, *J. Chem. Phys.*, 21:1087–1092.

Munro, N. (1972), Design of controllers for open-loop unstable multivariable system using inverse Nyquist array, *Proc. IEE*, 119(9):1377–1382.

Vanderbilt, D. and S.G. Louie (1984), A Monte Carlo simulated annealing approach to optimization over continuous variables, *J. Comp. Physics*, 56:259–271.

Whidborne, J.F., D.-W. Gu and I. Postlethwaite (1995), Algorithms for the method of inequalities – a comparative study, *Proc. 1995 Amer. Contr. Conf.*, Seattle, WA, pp. 3393–3397.

Whidborne, J.F., D.-W. Gu and I. Postlethwaite (1996), Simulated annealing for multi-objective control system design, *IEE Proc. Control Theory and Appl.*, 144(6):582–588.

Zakian, V. (1996), Perspectives on the principle of matching and the method of inequalities, *Int. J. Control*, 65(1):147–175.

Zakian, V. and U. Al-Naib (1973), Design of dynamical and control systems by the method of inequalities, *Proc. IEE*, 120(11):1421–1427.

8 Multi-objective Genetic Algorithms for the Method of Inequalities

Tung-Kuan Liu and Tadashi Ishihara

Abstract. Global search capability of genetic algorithms (GAs) provides an attractive method for solving inequalities. For multi-objective optimisation, various multi-objective genetic algorithms (MGAs) have been proposed. However, due to the fundamental difference between the method of inequalities and conventional multi-objective optimisation, it is not immediately apparent how MGAs should be used for solving inequalities. In this chapter, the use of MGAs in the method of inequalities is discussed. For the effective use of MGAs, an auxiliary vector performance index, related to the set of inequalities, is introduced. A simple MGA with Pareto ranking is proposed in conjunction with the auxiliary vector index. The performance of the proposed MGA is tested on a special set of test problems and control design benchmark problems.

8.1 Introduction

Search methods play an important role in the new framework for control systems design presented in this book. One of the global search methods is the genetic algorithms (GAs), which has been proposed by Holland (1975) to apply the principle of biological evolution to artificial systems. For conventional control systems design, GAs have been used to find a controller parameter minimising a single scalar performance index, for which standard non-global hill climbing techniques can fail to find the global minimum, when there are several local minima (Gtefenstete, 1986; Krishnakumar and Goldberg, 1992; Kristinsson and Dumont, 1992; Varsek *et al*, 1993; Porter and Jones, 1992). For vector valued performance index, various multi-objective genetic algorithms (MGAs) (*e.g.*, Goldberg, 1989; Schaffer, 1985; Fonseca and Fleming, 1993; Srinivas and Deb, 1995; Coello *et al*, 2002) have been proposed to find a set of Pareto minimisers.

The multi-objective nature of control systems design is explicitly taken into account in the method of inequalities proposed by Zakian (Zakian and Al-Naib, 1973). Unlike conventional multi-objective methods of design, the method of inequalities seeks an admissible solution, which is not necessarily a Pareto minimiser. Due to this fundamental difference, it is not readily apparent how MGAs should be used to solve inequalities formulated according to the method of inequalities (see the principle of inequalities in Chapter 1).

The authors (Liu et al, 1994; Liu, 1997) have proposed a genetic inequalities solver using a simple MGA for an auxiliary vector index related to the set of inequalities that represent the design specifications. The Pareto minimiser of the proposed method reduces to utopian (see below for definition) solutions if the design problem, formulated by the method of inequalities, is solvable. The proposed solver exploits the global search capability of an MGA to solve inequalities.

Fonseca and Fleming (1993) have proposed an MGA and have used it to obtain a Pareto minimiser subject to inequality constraints. Whidborne et al (1996) have applied this method to the method of inequalities and have compared it with existing local search methods. Note that computation of a Pareto minimiser is not required in the method of inequalities. Fonseca and Fleming (1998) have extended their method to handle more general preference expressions. The extended method includes the genetic inequalities solver proposed by the authors as a special case. However, the advantage of the fitness assignment method used by Fonseca and Fleming (1993, 1998) is lost when the method is specialised to solve inequalities. In addition, Fonseca and Fleming (1998) have not discussed simple but important properties discussed by the authors (Liu et al, 1994; Liu, 1997) for the method of inequalities.

In this chapter, the effective use of MGAs in the method of inequalities is discussed. In particular, the result given by the authors (Liu et al, 1994; Liu, 1997) is presented in more detail with new illustrative results.

This chapter is organised as follows: In Section 8.2, the auxiliary vector index related to the design specifications of the method of inequalities is introduced. Some properties of the set of Pareto minimisers for the auxiliary vector index are discussed. A simple genetic inequalities solver proposed by the authors is discussed in Section 8.4. In Section 8.5, the performance of the proposed genetic solver is examined experimentally for the test problems in Chapter 6. The results for the control design benchmark problems in Chapter 7 is given in Section 8.5. Conclusions are given in Section 8.6.

8.2 Auxiliary Vector Index

In this section, the auxiliary vector index is introduced for the use of MGAs to solve the inequalities

$$\phi_i(c) \leq \varepsilon_i \quad i \in \tilde{M} = \{1, 2, \ldots, M\} \tag{8.1}$$

where $\phi_i(c)$ is an objective function, c is a design point and ε_i is a tolerance or bound specified by the designer. The computational problem is to determine any value of c that satisfies (8.1) or, if such a value does not exist, to confirm this fact. This is called the admissibility problem.

Define the objective vector

$$\Phi(c) \triangleq [\phi_1(c), \phi_2(c), \cdots, \phi_M(c)] \tag{8.2}$$

For an effective use of an MGA to solve the inequalities (8.1), the authors (Liu et al, 1994) have introduced scalar indices related to the inequality performance specifications (8.1) as

$$\lambda_i(c, \varepsilon_i) \equiv \begin{cases} 0 & \text{if } \phi_i(c) \leq \varepsilon_i \\ \phi_i(c) - \varepsilon_i & \text{if } \phi_i(c) > \varepsilon_i \end{cases} \tag{8.3}$$

for $i \in \tilde{M}$. These indices are also related to the ordering relations employed by the local search method called the moving boundaries process (Zakian and Al-Naib, 1973).

Define the auxiliary vector index

$$\Lambda(c, \varepsilon) \triangleq [\lambda_1(c, \varepsilon_1), \lambda_2(c, \varepsilon_2), \cdots, \lambda_M(c, \varepsilon_M)] \tag{8.4}$$

where

$$\varepsilon \triangleq [\varepsilon_1, \varepsilon_2, \cdots, \varepsilon_M] \tag{8.5}$$

is a tolerance vector.

The following notations used by Sawaragi et al (1985) for vector inequalities are employed. For any two vectors x and y, the vector inequality $x \leq y$ means that, for each element, $x_i \leq y_i$ holds. In addition, the notation $x \prec y$ is used to mean $x \leq y$ and $x \neq y$.

For the objective vector (8.2), the following definition is introduced using the above notations.

Definition 8.1. Consider an objective vector $\Phi(c)$ with a design c. The design c_0 is said to be a utopian for the objective vector $\Phi(c)$ if and only if $\Phi(c_0) \leq \Phi(c)$ for all c. A vector c is said to be a Pareto minimiser if and only if there exists no vector $\bar{c}(\neq c)$ satisfying $\Phi(\bar{c}) \prec \Phi(c)$. A utopian is a Pareto minimiser. A Pareto minimiser that is not utopian is said to be a strict Pareto minimiser.

In Chapter 1, the relation between the ordering relations induced by the principle of inequalities and the Pareto ordering has been discussed. The following result is a restatement of the result in terms of the auxiliary vector index.

Property 8.1. A design c is superior to $\bar{c}$ in the sense of the principle of the inequalities with respect to the objective vector $\Phi(c)$ if and only if c is superior to $\bar{c}$ in the Pareto sense with respect to the auxiliary vector index $\Lambda(c, \varepsilon)$.

Proof. Assume that $\phi_j(\bar{c}) \leq \varepsilon_j$ for $j \in \tilde{N} \subset \tilde{M}$. Define $\hat{\Phi}(\bar{c})$ as the vector obtained by deleting the elements $\phi_j(\bar{c})$ ($j \in \tilde{N}$) from the objective vector $\Phi(\bar{c})$. Then the superiority of c in the sense of the principle of inequalities holds if and only if $\hat{\Phi}(c) \prec \hat{\Phi}(\bar{c})$ and $\phi_j(\bar{c}) \leq \varepsilon_j \Rightarrow \phi_j(c) \leq \varepsilon_j$ ($j \in \tilde{N}$).

The condition $\hat{\Phi}(c) \prec \hat{\Phi}(\bar{c})$ is equivalent to $\lambda_i(c, \varepsilon_i) \leq \lambda_i(\bar{c}, \varepsilon_i)$ for all $i \in \tilde{M} \setminus \tilde{N}$ with $\lambda_i(c, \varepsilon_i) < \lambda_i(\bar{c}, \varepsilon_i)$ for some $i \in \tilde{M} \setminus \tilde{N}$. In addition, the condition $\phi_j(\bar{c}) \leq \varepsilon_j \Rightarrow \phi_j(c) \leq \varepsilon_j$ is equivalent to the inequality $\lambda_j(c, \varepsilon_j) \leq \lambda_j(\bar{c}, \varepsilon_j)$ with $\lambda_j(\bar{c}, \varepsilon_j) = 0$. Therefore, the two conditions $\tilde{\Phi}(c) \prec \tilde{\Phi}(\bar{c})$ and $\phi_j(\bar{c}) \leq \varepsilon_j \Rightarrow \phi_j(c) \leq \varepsilon_j$ $(j \in \tilde{N})$ are equivalent to $\Lambda(c, \varepsilon) \prec \Lambda(\bar{c}, \varepsilon)$. In the case that $\phi_i(\bar{c}) > \varepsilon_i$ for all $i \in \tilde{M}$, c is superior to $\bar{c}$ in the sense of the principle of the inequalities if and only if $\Phi(c) \prec \Phi(\bar{c})$, which is obviously equivalent to $\Lambda(c, \varepsilon) \prec \Lambda(\bar{c}, \varepsilon)$ with $\lambda_i(c, \varepsilon_i) > 0$ and $\lambda_i(\bar{c}, \varepsilon_i) > 0$ for all $i \in \tilde{M}$. Consequently, the equivalence of the superiority conditions is proved.

For the effective use of MGAs in the method of inequalities, it is proposed to apply an MGA to the auxiliary vector index. It readily follows from the definitions (8.3), (8.4) and (8.5) that the solvability of the inequalities (8.1) can be stated in terms of the auxiliary vector index as follows.

Property 8.2. A design c_0 satisfying all the inequalities (8.1) exists if and only if c_0 is a utopian solution of the auxiliary vector index $\Lambda(c, \varepsilon)$, i.e., $\Lambda(c_0, \varepsilon) = 0$.

The use of the auxiliary vector index has the following advantages:

(a) The global search capability of MGAs can be exploited to find a solution of the inequalities (8.1).
(b) An MGA applied with the auxiliary vector index always generates designs belonging to a strict Pareto set even if no design satisfies the given design specifications (8.1). The Pareto set provides the designer useful information for modifying the design specifications.

It is apparent that the first advantage is the main reason for using an MGA for solving the inequalities (8.1). If the design problem formulated by the method of inequalities has a solution, the Pareto optimal set generated by the MGA applied to the auxiliary vector index eventually reduces to utopians. The second advantage works only for ill-posed problems but it provides information that cannot be obtained by a failure of a local search method. When a local search method fails to find an admissible solution, it is not easy to decide whether the algorithm is trapped in a local region or no solution satisfying design specifications exists.

Although the strict Pareto set for the auxiliary vector index (8.4) appears only for ill-posed problems, it is interesting to compare it with the Pareto set of the objective vector (8.2) related to the design specifications (8.1). For the two dimensional performance index ($M = 2$), the relation between the two Pareto sets in $\phi_1 - \phi_2$ and $\lambda_1 - \lambda_2$ planes can be illustrated as in Figure 8.1, where the Pareto set is indicated by bold lines and the shaded areas show the achievable regions. The case that the tolerance vector ε is outside of the achievable region in $\phi_1 - \phi_2$ is shown in Figure 8.1(a). In this case, no utopian exists and the Pareto set of the auxiliary vector index $\Lambda(c, \varepsilon)$ in the $\lambda_1 - \lambda_2$

8 Multi-objective Genetic Algorithms for the Method of Inequalities   235

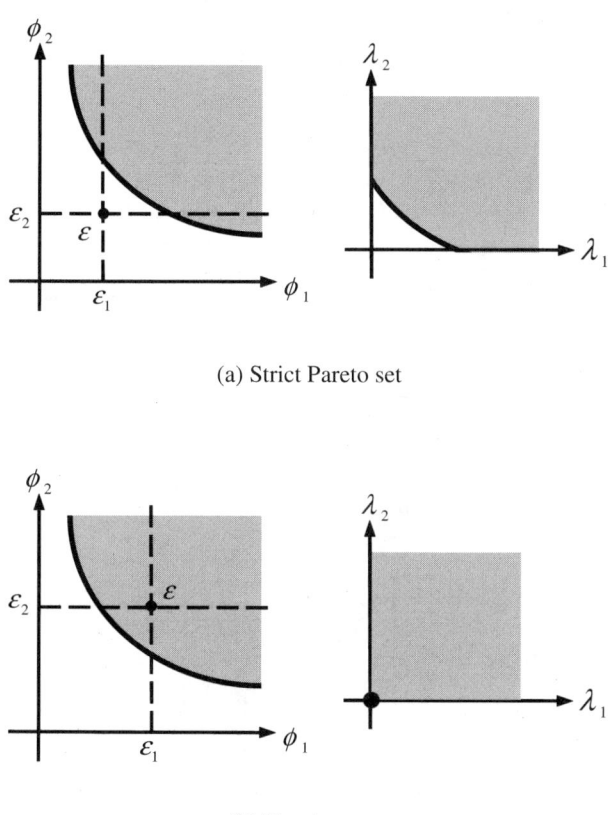

(a) Strict Pareto set

(b) Utopians

Fig. 8.1. The two Pareto sets

plane is a clipped version of that of the vector performance index (8.2). In the case that the tolerance ε is inside of the achievable region, the result is shown in Figure 8.1(b). In the $\lambda_1 - \lambda_2$ plane, the Pareto set is the origin which corresponds to utopian solutions of the auxiliary vector index $\Lambda(c, \varepsilon)$.

In the following, simple properties conjectured from the two dimensional case are given for the general case. The Pareto sets for the objective vector $\Phi(c)$ and the auxiliary vector index $\Lambda(c, \varepsilon)$ are denoted by P and $Q(\varepsilon)$, respectively.

It is known that the principle of inequalities provides a Pareto minimiser for the objective vector (8.2) if the tolerance vectors are properly chosen (Zakian, 1996). For the two dimensional case, this result corresponds to the case that the tolerance vector ε is on the Pareto set as shown in Figure 8.2(a). The following is a restatement of this fact.

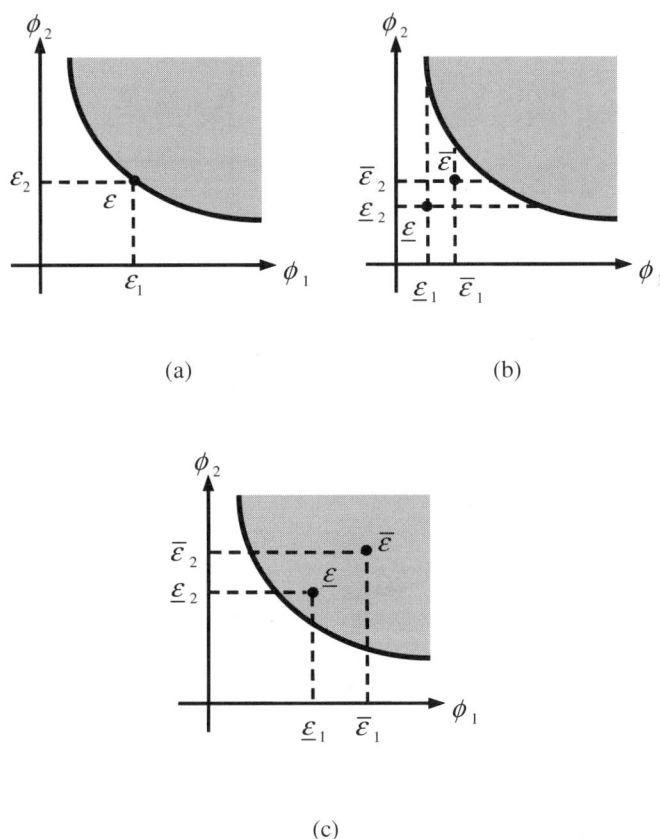

Fig. 8.2. The tolerance vector and the Pareto set

Property 8.3. If the auxiliary vector index $\Lambda(c, \varepsilon)$ has a unique utopian c^* satisfying $\Lambda(c^*, \varepsilon) = 0$, then the utopian c^* is a Pareto minimiser of the objective vector $\Phi(c)$.

Proof. Assume that the unique utopian c^* is not a Pareto minimiser for the objective vector $\Phi(c)$. Then, by the definition of the Pareto minimiser, there exists vector $c^{**} \neq c^*$ satisfying $\Phi(c^{**}) \prec \Phi(c^*)$. Note that $\Lambda(c^*, \varepsilon) = 0$ is equivalent to $\Phi(c^*) \leq \varepsilon$. It follows from $\Lambda(c^*, \varepsilon) = 0$ and $\Phi(c^{**}) \prec \Phi(c^*)$ that $\Phi(c^{**}) \leq \varepsilon$, i.e., $\Lambda(c^{**}, \varepsilon) = 0$. Consequently, c^{**} is also a utopian. This contradicts the uniqueness assumption.

Two properties corresponding to the case shown in Figure 8.2(b) for the two dimensional case are given. The first property is concerned with the inside of the Pareto set $Q(\varepsilon)$ for $\Lambda(c,\varepsilon)$.

Property 8.4. Let $Q_+(\varepsilon)$ denote the Pareto sets consisting of the designs satisfying $\Lambda(c,\varepsilon) > 0$. Then, $Q_+(\varepsilon) \subset P$, where P is the Pareto set for $\Phi(c)$, holds for $\varepsilon > 0$. In addition, for any two tolerance vectors satisfying $\varepsilon \prec \bar{\varepsilon}$, $Q_+(\varepsilon)$ and $Q_+(\bar{\varepsilon})$ satisfy $Q_+(\varepsilon) \supset Q_+(\bar{\varepsilon})$.

Proof. Note that $\Lambda(c,\varepsilon) = \Phi(c) - \varepsilon > 0$ by the assumption. Consider a design $c \in Q_+(\varepsilon)$, for which there exists no design superior to c. Assume that $c \notin P$, which implies that there exist a design c^* such that $\Phi(c^*) \prec \Phi(c)$. Then, for the design c^*, $\Phi(c^*) - \varepsilon \prec \Phi(c) - \varepsilon = \Lambda(c,\varepsilon)$ holds, which is a contradiction to $c \in Q_+(\varepsilon)$. Consequently, $c \in P$. Write the set of designs satisfying $\Phi(c) > \varepsilon$ as $D(\varepsilon)$. The set $Q_+(\varepsilon)$ can be written as $Q_+(\varepsilon) = P \cap D(\varepsilon)$. Since $D(\varepsilon) \supset D(\bar{\varepsilon})$ for $\varepsilon \prec \bar{\varepsilon}$, it is obvious that $Q_+(\varepsilon) \supset Q_+(\bar{\varepsilon})$ holds.

The second property is concerned with the edge of the Pareto set $Q(\varepsilon)$.

Property 8.5. Assume the following two conditions:

(i) There exists $c_0 \in P$ such that $\phi_j(c_0) < \varepsilon_j$ for $j \in \tilde{N} \subset \tilde{M}$.
(ii) For any $c_0 \in P$ satisfying i, there exists $c_1 \in P$ such that $\phi_j(c_0) < \phi_j(c_1) < \varepsilon_j$ for $j \in \tilde{N}$ and $\hat{\Phi}(c_1) \prec \hat{\Phi}(c_0)$ where $\hat{\Phi}(c)$ denotes the vector obtained by deleting the elements $j \in \tilde{N}$ from $\Phi(c)$.

Then, there exists $c_* \in P$ satisfying $\phi_j(c_*) = \varepsilon_j$ for $j \in \tilde{N} \subset \tilde{M}$ and $c_* \in Q(\varepsilon)$,

Proof. Note the existence of $c_1 \in P$ satisfying $\hat{\Phi}(c_1) \prec \hat{\Phi}(c_0)$ for $c_0 \in P$ is possible although there exists no $c \in P$ such that $\Phi(c) \prec \Phi(c_0)$. Define $\hat{\Lambda}(c_0, \hat{\varepsilon}) \triangleq \hat{\Phi}(c_0) - \hat{\varepsilon}$ where $\hat{\varepsilon}$ denotes the vector obtained by deleting the elements $j \in \tilde{N}$ from ε. The assumptions guarantee that, for any $c_0 \in P$ satisfying the first condition, it is possible to find $c_1 \in P$ such that $\hat{\Lambda}(c_1, \hat{\varepsilon}) \prec \hat{\Lambda}(c_0, \hat{\varepsilon})$. Note that $\hat{\Lambda}(c_1, \hat{\varepsilon}) \prec \hat{\Lambda}(c_0, \hat{\varepsilon})$ is equivalent to $\Lambda(c_1, \varepsilon) \prec \Lambda(c_0, \varepsilon)$. By continuing this procedure, it is possible to construct a sequence of the decisions $\{c_k, k = 0, 1, 2, \ldots\}$ which converges to $c_* \in P$ satisfying $\phi_j(c_*) = \varepsilon_j$. Since no decision $c \neq c_*$ exists such that $\hat{\Lambda}(c, \hat{\varepsilon}) \prec \hat{\Lambda}(c_*, \hat{\varepsilon})$, i.e., $\Lambda(c,\varepsilon) \prec \Lambda(c_*, \varepsilon)$, it is concluded that $c_* \in Q(\varepsilon)$.

It can easily be checked that the two assumptions are satisfied at least for the two dimensional case where $N = 1$ and the objective functions $\phi_i(c)$ ($i = 1, 2$) are smooth functions of the design c. Let $Q_*(\varepsilon)$ denote the set of designs satisfying the above property. Then it follows from Property 8.4 and Property 8.5 that the Pareto set $Q(\varepsilon)$ can be written as $Q(\varepsilon) = Q_*(\varepsilon) \cup Q_+(\varepsilon)$.

For non-unique utopians, the following property is obtained. The result corresponds to the two dimensional case shown in Fig. 2 (c).

Property 8.6. Let $Q(\varepsilon)$ denote the Pareto optimal set for the auxiliary vector index $\Lambda(c,\varepsilon)$. Assume that, for the tolerance vectors ε and $\bar{\varepsilon}$ satisfying $\varepsilon \prec \bar{\varepsilon}$, $Q(\varepsilon)$ and $Q(\bar{\varepsilon})$ consist of utopian solutions. Then, $Q(\varepsilon) \subset Q(\bar{\varepsilon})$.

Proof. The result readily follows from the fact that $c \in Q(\varepsilon)$, where the set $Q(\varepsilon)$ consists of utopian solutions, if and only if $\Phi(c) \leq \varepsilon$.

Properties 8.1—8.6 suggest that the use of the auxiliary vector index restricts the Pareto set in the region of interest and reduces the necessary computation burden. In addition, the properties provide some qualitative understanding of the Pareto set of $\Lambda(c,\varepsilon)$. According to the range of the tolerance vector ε, the Pareto set $Q(\varepsilon)$ can be classified into three types:

A. For sufficient small ε, $\Lambda(c,\varepsilon) > 0$ for all designs. Then the Pareto set $Q(\varepsilon)$ coincides with the Pareto set P for $\Phi(c)$.
B. For medium range of ε, designs satisfying $\Lambda(c,\varepsilon) \geq 0$ appear. The Pareto set $Q(\varepsilon)$ includes the Pareto sets $Q_+(\varepsilon)$ and $Q_*(\varepsilon)$ included in the Pareto set P for $\Phi(c)$. As ε increases, the size of $Q(\varepsilon)$ decreases until a utopian appears.
C. For sufficiently large ε, $Q(\varepsilon)$ consists of utopians but includes solutions not contained in P. As ε increases, the size of $Q(\varepsilon)$ increases.

A local parameter search algorithm is not effective for the type A and B since no utopian exists. For this type, an MGA can generate an approximate Pareto set. Noting the above qualitative property, the designer can use the information of the approximate Pareto set to soften the design specifications so that a utopian can be found. On the other hand, for the type C, an MGA may find many utopian solutions. In such a case, the above qualitative properties can be used to tighten the design specifications to reduce the number of the utopian solutions.

8.3 Genetic Inequalities Solver

The method proposed in the previous section can be used with any MGA to construct an inequalities solver. It should be noted that the objective of the method of inequalities is to find a solution satisfying the set of the inequalities while that of an MGA is to characterise the Pareto optimal set. An MGA constructed by taking account of the difference is more appropriate to use for an inqualities solver. In this section, such an MGA is proposed.

8.3.1 Bias Problem of MGAs

It is well known that a simple GA has a bias problem called *genetic drift* for a scalar performance index with multiple minima (*e.g.*, Goldberg, 1989). The GA search has a tendency to concentrate to on one of the minima. The source of the bias is recognised as the stochastic error of the evolution. The effect of the genetic drift can be reduced by preserving population diversity. Various methods including *sharing* and *crowding* have been proposed to reduce the bias (*e.g.*, Goldberg, 1989).

For MGAs, the bias problem is more complex. The vector evaluated genetic algorithm (VEGA) proposed by Schaffer (1985) is well known as a simple MGA. In VEGA, the population is divided into equally sized subpopulations corresponding to the scalar components in the vector performance index. The fitness is assigned by the independent championship in each component of a vector performance index but the mating and the crossover are performed for the whole population. The VEGA has a tendency to ignore middling individuals. Such a bias is apparently undesirable to characterise the whole Pareto set. The genetic drift alone is not a source of the bias of the VEGA.

8.3.2 Ranking Methods for MGAs

MGAs differ from conventional GAs in the fitness assignment. Notice that no Pareto concept is used in the evolution of the VEGA. A part of the bias in the VEGA can be considered due to the bias in the fitness assignment. To reduce the bias, Goldberg (1989) has suggested a fitness assignment using the Pareto (non-dominated) ranking. The procedure is described as follows: First, the Pareto set of the current population is identified. Individuals in the Pareto set are assigned a rank of 1. The Pareto set is removed from the current population. The Pareto set of the remaining population is identified and individuals in the second Pareto set are assigned a rank of 2. This ranking procedure is continued until all individuals in the current population are assigned ranks. The fitness of an individual is assigned based on the rank.

Fonseca and Fleming (1993) have proposed an MGA called the multiple objective genetic algorithm (MOGA) using a ranking method different from the Pareto ranking: The rank of an individual is defined by one plus the number of the other individuals dominating the individual. By this definition, an individual in the non-dominated set is assigned a rank of 1 as in the Pareto ranking. However, the rank of an individual outside of the Pareto set can be different from that obtained by the Pareto sorting. The comparison of the ranking methods is shown in Fig. 3 for the two dimensional case. The ranks of the two points indicated by the arrows in the method of Fonseca and Fleming are different from the ranks obtained by the Pareto ranking.

8.3.3 A Simple MGA with the Pareto Ranking

Fonseca and Fleming (1993) have pointed out that their ranking method is useful to take account of the preference of the decision maker. In addition, they claim that their ranking method makes it easy to analyse some properties mathematically. Using their sorting method, they have proposed an optimiser

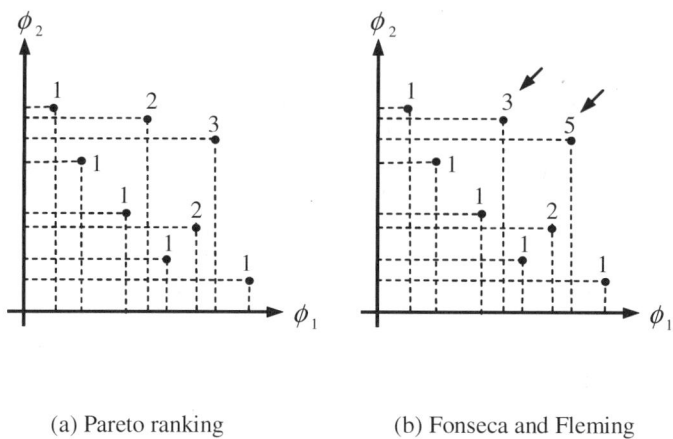

(a) Pareto ranking   (b) Fonseca and Fleming

Fig. 8.3. Comparison of the ranking methods

(Fonseca and Fleming, 1998), which can deal with the method of inequalities as a special case.

All the inequality design specifications in the method of inequalities are considered to have equal importance. Preference ordering of the inequalities in the method of inequalities may introduce unnecessary confusions. It should be noted that the advantage of the fitness assignment method used by Fonseca and Fleming (1993, 1998) is lost when the optimiser (Fonseca and Fleming, 1998) is specialised to the method of inequalities. The fitness assignment based on the Pareto ranking is essential for the use of a MGA in the method of inequalities since it explicitly represents that the method of inequalities does not introduce any preference order among performance indexes.

Goldberg (1989) has pointed out that the Pareto ranking is not sufficient to reduce the bias. He suggested the use of some sharing scheme with the Pareto ranking. Following the suggestion, Srinivas and Deb (1995) have proposed the non-dominated sorting genetic algorithm (NSGA) and have shown its superiority over VEGA by extensive simulations. Fonseca and Fleming (1993) have proposed to use a sharing method with their ranking method.

Independent of the above MGAs, the authors have proposed to use the auxiliary vector performance index with a MGA using the Pareto ranking. The MGA used by the authors does not include any method to reduce the bias other than the Pareto ranking.

8 Multi-objective Genetic Algorithms for the Method of Inequalities 241

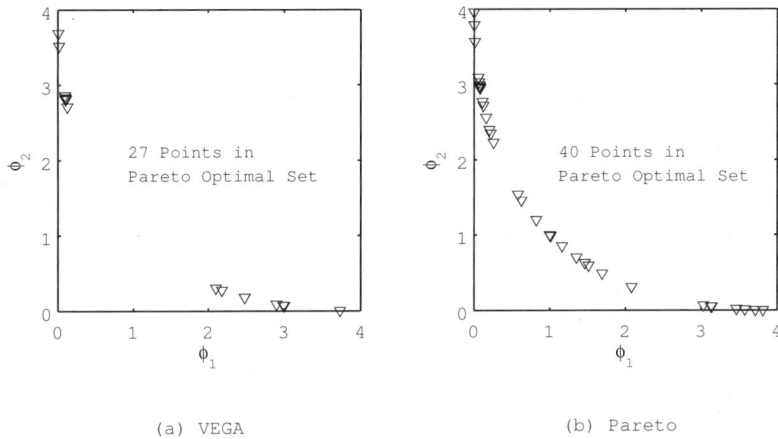

Fig. 8.4. Comparison of Pareto fronts

To show the performance of such a simple MGA, the Pareto set for simple two dimensional functions

$$\phi_1(t) = t^2, \quad \phi_2(t) = (t-2)^2, \quad -6 \le t \le 6 \tag{8.6}$$

is compared with that obtained by VEGA in Figure 8.4. In the comparison, typical GA parameters are used: the tunable parameter t is represented by 16-bit binary code, the population is 100, the crossover rate is 0.8, the mutation rate is 0.001 and the generation number is 3. It is apparent that the use of the Pareto ranking effectively reduces the bias against middling points of the Pareto set.

The primary reason to use an MGA for the method of inequalities is to exploit the global search capability. The appearance of the strict Pareto set means that the method of inequalities problem has no solution. For the application to the method of inequalities, the search capability of an MGA is more important than the characterisation of the Pareto set when the problem has no solution. A sharing method preserves population diversity by reducing fitness of a group of individuals. Intuitively, the search capability is sacrificed by introducing a sharing method. This observation together with the result of Figure 8.4 suggests that a simple MGA with the Pareto ranking is an appropriate choice for the method of inequalities.

8.3.4 An Efficient Computation of the Pareto Ranking

To make the MGA more efficient, it is possible to use an efficient computation method for the Pareto ranking. In MGAs, the brute force sorting is often used. For a set of N points in M dimensional space, the computational

Table 8.1. Comparison of the ranking methods

Number of Points	Computation Time (Sec.)		Improvement Index (%)
	Proposed Method	Brute Force Method	
30	0.0084	0.0067	−20.24
50	0.0211	0.0181	−14.22
80	0.0453	0.0466	3.56
100	0.0552	0.0734	32.97
500	1.3141	1.798	36.82

Calculation condition: MATLAB 6.1 for Windows XP on PC (Intel P4, 2.53GHz, 1GB RAM)

complexity for the brute force sorting is $O(MN^2)$. It is known that the algorithm found by Kung et al (1975) provides the computational complexity $O(N(\log N)^{M-2})+O(N \log N)$, which is smaller than the computational complexity $O(MN^2)$ required for brute force sorting. Kung's algorithm is of great theoretical importance. However, this computational complexity is a worst case estimate that is valid only for sufficiently large N. It has been reported by Gtefenstette (1986) that the population to be generated by GA is usually at most several hundred, sometimes less than hundred. It is doubtful that the Kung's estimate is valid for practical GA computation.

A heuristic computation procedure for non-dominated sorting is proposed. The proposed sorting scheme is based on a simple observation that, if two vectors x and y with non-negative elements satisfy the relation $x \prec y$, then the inequality

$$\sum_{i=1}^{M} x_i \leq \sum_{i=1}^{M} y_i \tag{8.7}$$

holds. Consequently, the inequality (8.7) is a necessary condition for $x \prec y$.

Using the above observation, we propose the following improved ranked-based fitness assignment method which includes the sorting using (8.7). Suppose that a set $\Pi \equiv \{\Theta_1, \Theta_2, \ldots, \Theta_N\}$ with $\Theta_i \equiv \{\theta_{i1}, \theta_{i2}, \ldots, \theta_{iM}\}$ is given. Consider the following procedure.

Step 1: Sort Π from the least to the largest according to $\sum_{j=1}^{M} \theta_{ij}$. Let $K \equiv 1$.
Step 2: Let $Q \equiv \emptyset$, $G \equiv \emptyset$.
Step 3: Let the first point Θ_1 of Π be the criterion.
Step 4: Remove the dominated points Θ_i by checking whether or not the following conditions are satisfied.

$$\max_j g_{ij} \leq 0 \wedge (\exists_j\ )(g_{ij} < 0) \tag{8.8}$$

where g_{ij} is the j-th component of the vector $g_i \equiv \Theta_1 - \Theta_i$, $i = 2, 3, \cdots, N$ and $j = 1, 2, 3 \cdots, m$.

Step 5: Remove the criterion Θ_1 and the dominated points Θ_i from Π. Save Θ_1 to a temporary non-dominated set Q, Θ_i to a temporary dominated set G.

Step 6: If $\Pi \neq \emptyset$ then go to **Step 3**.

Step 7: Rank the points of Q as K. Let $\Pi \equiv G$ and $K \equiv K + 1$.

Step 8: If $\Pi \neq \emptyset$ then go to **Step 2**.

Step 9: The fitness to points is given by the following linear function

$$V_N(i) = \frac{2(f_{\max} - 1)}{N - 1} [N - \text{rank}(i)] \quad (8.9)$$

where rank(i), which is less than or equal to N, are the ranks of points and $f_{\max}$ considered as a GA diversity maintaining parameter (Kristinsson and Dumont, 1992) can be specified by a designer. In this chapter, it is taken as $1 < f_{\max} \leq 2$.

In the worst case, the computational complexity of the proposed sorting procedure is $O(MN^2) + O(N \log N)$ which is not better than $O(MN^2)$ required for the brute force sorting. However, in the most optimistic case, the computational complexity required by the proposed method is $O(MN) + O(N \log N)$. Note that the brute force sorting always requires the computational complexity $O(MN^2)$.

To show the effectiveness of the proposed sorting method, a simple numerical experiment is performed. The trial vector sets are random matrix, made by MATLAB built-in functions, with ten elements uniformly distributed in the interval $[0, 10]$. Each sorting method is implemented 100 times and the mean calculation time is calculated. The result is summarised in Table 8.1 where the improvement index is defined as $R_p = (t_b - t_p)/t_p$, where t_b is the calculation time of the brute force sorting and t_p is that of the proposed sorting. The index R_p increases as the number of points n increases. The result shows that the proposed method provides faster sorting than the brute force method.

8.4 Numerical Test Problems

The ability of the proposed genetic inequalities solver is evaluated by the experiment for the test problems given in Chapter 6. There are 14 problems in

Table 8.2. GA parameters used in the experiment

Population Size	Reproduction Rate	Crossover Rate	Mutation Rate
100	0.8	1.0	0.01

Table 8.3. Summary of the search results

Problem Number	Average Number of Function Evaluations	Maximum Number of Function Evaluations	Minimum Number of Function Evaluations	Number of Failed Searches
1	33036	66340	11140	0
2	21324	72500	1380	0
3	74404	194420	17060	0
4	16860	36820	2500	0
5.1	164284	240020	30660	2
5.2	90884	221060	5380	0
5.3	123012	240020	1780	1
5.4	51668	125700	2580	0
5.5	84892	178100	15860	0
6.1	68156	215940	4180	0
6.2	65068	164740	2740	0
6.3	83092	210820	9940	0
6.4	74724	174180	2980	0
6.5	78796	162020	7460	0

total. For each problem, it is required to find an admissible point satisfying the inequality defined for the scalar-valued function with the two variables. Since the problems are not multi-objective, this experiment evaluates the search capability of the proposed MGA as a simple GA. Many possibilities exist for the choice of the GA parameters used in the experiment. The standard choice of the GA parameters shown in Table 8.2 are used for all the problems. To satisfy the precision requirement of the problems, a string of 60 bits (30 bits for each variable) is used to represent an individual in the GA. For each test problem, ten trials are performed using randomly generated initial populations. The search is terminated when an admissible solution is found or 3000 generations are obtained. The results are summarised in Table 8.3 where the problem numbers are defined in Chapter 6. The MGA can successfully find an admissible solution in all the trials except for the problems 5.1 and 5.3 in which a small number of trials failed among the ten trials. Note that 230020 times function evaluations correspond to 3000 generations. The results of all the trials in the problems 5.1 and 5.3 are shown in Table 8.5. It is seen that the final points of the searches for the failed cases (the third and sixth trials in the problem 5.1 and the sixth trial in the problem 5.3) are very close to the admissible regions, which can be confirmed by the value of the test functions. It is fair to say that the searches are almost successful. The source of the difficulty is not easy to identify. However, it is confirmed experimentally that the improved search results can be obtained by increasing the number of the population to 200.

8 Multi-objective Genetic Algorithms for the Method of Inequalities 245

Table 8.4. Search results for 5.1 and 5.3

Problem Number	Test Times	x_1	x_2	f	Number of Function Evaluations
5.1	1	−7.0030	0.9940	−100.000000	167220
	2	−7.0007	0.9988	−100.000000	147380
	3*	−7.0312	0.9384	−99.999000	240020
	4	−7.0024	0.9951	−100.000000	152260
	5	−6.9970	1.0059	−100.000000	169060
	6*	−7.0157	0.9687	−99.999800	240020
	7	−6.9971	1.0060	−100.000000	123780
	8	−6.9971	1.0057	−100.000000	230020
	9	−7.0027	0.9945	−100.000000	142420
	10	−7.0020	0.9960	−100.000000	30660
5.3	1	0.997460	0.994932	−99.999994	211140
	2	0.998140	0.996484	−99.999993	176820
	3	0.999580	0.999384	−99.999995	73060
	4	1.001464	0.999384	−99.999997	115220
	5	0.999880	1.000004	−99.999994	200980
	6*	1.007760	1.015624	−99.999940	240020
	7	0.999260	0.998592	−99.999999	1780
	8	0.998280	0.996800	−99.999991	4020
	9	1.000420	1.000652	−99.999996	2980
	10	1.002920	1.005968	−99.999990	204100

*failed search

These results suggest that all the test problems are not particularly difficult for the GA in spite of the fact that the results are obtained by the standard choice of GA parameters without special tuning. This property of the GA is very attractive as a global search method. A problem of the GA is that it usually requires a large number of function evaluations. But the progress in computer technology is a great help for the GA to solve practical problems.

8.5 Control Design Benchmark Problems

In this section, the genetic inequalities solver proposed in Section 8.3 is applied to the two control benchmark problems presented in Chapter 7 where the simulated annealing inequalities solver (SAIS) has been compared with the moving boundaries process (MBP(R)).

The first problem is the design of a lead-lag compensator with three design parameters for a third order SISO plant. The aim of the design is to obtain a good step response. Four design specifications are given. The second problem is the design of a decentralised compensator for a loosely coupled

Table 8.5. Results of control design problems

Problem Number	Average Number of Function Evaluations	Minimum Number of Function Evaluations	Maximum Number of Function Evalutaions	Percent Failed Trials (%)
1	5412	1300	11460	0
2	1292	660	3220	0

two-input-two-output plant. The controller includes four design parameters. In addition, one design parameter is included in the plant. The aim of the design is to obtain a good step response with little cross coupling. Nine design specifications are given. For the detailed information on both problems, see Sections 7.5 and 7.A.

The genetic inequalities solver with the standard GA parameters shown in Table 8.2 are used. Each design parameter is represented by a string of 10 bits. Ten trials have been performed for the two problems with different initial populations. Table 5 summarises the results of the experiment. For both problems, the proposed genetic inequalities solver is successful to find an admissible solution for all the trials. But the results suggest that the number of function evaluations required for finding an admissible solution is highly dependent on the initial population. Although the Problem 1 appears simpler than the Problem 2, the proposed genetic inequalities solver requires many more function evaluations for the Problem 1 than are required for the Problem 2. This is a contrast to the SAIS results where almost the same number of maximum number of function evaluations is required for both problems. Since the search mechanisms of the two methods are completely different, it seems difficult to give a simple explanation on the difference.

Note that the results of the SAIS are obtained from 100 initial conditions. On the other hand, the results in Table 8.5 are obtained from ten trials each of which includes 100 initial conditions. The maximum number of function evaluations for the Problem 1 is close to that required for the SAIS reported in the Section 7.5. For the Problem 2, the proposed genetic inequalities solver requires less function evaluations than that required by the SAIS. The proposed genetic inequalities solver does not requie much more function evaluations than the SAIS at least for the two benchmark problems.

8.6 Conclusions

The application of MGAs to control systems design based on the method of inequalities has been discussed. The fundamental difference between the method of inequalities and the conventional multi-objective optimisation has been emphasised. It has been shown that the global search capability of

MGAs can effectively be utilised by introducing the auxiliary vector index related to the design specifications of the method of inequalities. The genetic inequalities solver using a simple MGA with the Pareto ranking has been proposed. The experimental results for the test problems and the design examples have confirmed the global search capability of the proposed solver.

It is an important future work to develop a systematic method for utilising the Pareto set when the problem is ill-posed. Although the proposed MGA utilises only the Pareto ranking to reduce the bias in the Pareto optimal set, the usefulness of the other techniques should be assessed quantitatively.

A large freedom exists to construct other MGAs for the method of inequalities. It seems possible to construct a new MGA by combining the efficient local search capability of a local search method such as the moving boundaries process with the global search capability of GAs.

References

Coello, C.A., D.A. Veldhuizen and G.B. Lamont, (2002) *Evolutionary algorithms for solving multi-objective problems*, Kluwer Academic.

Fonseca, C.M. and P.J. Fleming, (1993) Genetic algorithms for multiobjective optimization: formulation, discussion and generalization. *Proc. of the Fifth International Conference on Genetic Algorithms*, pp. 416–423.

Fonseca, C.M. and P.J. Fleming, (1998) 'Multiobjective optimization and multiple constraint handling with evolutionary algorithms-Part I: Unified formulation.' *IEEE Trans. Systems, Man, and Cybernetics-Part A: Systems and Humans*, 28:26–47.

Goldberg, D.E., (1989) *Genetic algorithms in search, optimization and machine learning*, Addison-Wesley.

Gtefenstette, J.J., (1986) Optimization of control parameters for genetic algorithms. *IEEE Trans. Systems, Man and Cybernetics*, 16:122-128.

Holland, J.H., (1975) *Adaptation in Natural and Artificial Sysmtems*, University of Michigan Press.

Krishnakumar, K. and D.E. Goldberg, (1992) Control system optimization using genetic algorithms. *Journal of Guidance, Control, and Dynamics*, 15:735–740.

Kristinsson, K. and G.A. Dumont, (1992) System identification and control using genetic algorithms. *IEEE Trans. Systems, Man and Cybernetics*, 22:1033–1046.

Kung, H.T., F. Luccio and F.P. Preparata, (1975) On finding the maxima of a set of vectors. *J. of the ACM*, 22:469–476.

Liu, T.K., (1997) *Application of multiobjective genetic algorithms to control systems design*, Ph. D thesis, Tohoku University.

Liu, T.K., T. Ishihara and H. Inooka, (1994) An application of genetic algorithms to control system design. *Proceedings of the first Asian Control Conference*, pp. 701–704.

Porter, B. and A.H. Jones, (1992) Genetic tuning of digital PID Controllers. *Electronics Letters*, 28(9):843–844.

Sawaragi, Y., H. Nakayama and T. Tanino, (1985) *Theory of multiobjective optimization*, Academic Press.

Schaffer, J.D., (1985) Multiple objective optimization with vector evaluated genetic algorithms. *Proc. First Int. Conf. on Genetic Algorithms and Their Applications*, pp. 93–100.

Srinivas, N. and K. Deb, (1995) Multiobjective optimization using nondominated sorting in genetic algorithm. *Evolutionary Computation*, 2:221–248.

Varsek, A., T. Urbancic and B. Filipic, (1993) Genetic algorithms in controller design and tuning. *IEEE Trans. Systems, Man and Cybernetics*, 23:1330–1339.

Whidborne, J.F., D.W. Gu and I. Postlethwaite, (1996) Algorithms for solving the method of inequalities – A comparative study. *Proceedings of American Control Conference, Seattle, WA*, pp. 3393–3397.

Zakian, V., (1996) Perspectives on the principle of matching and the method of inequalities, *Int. J. Contr.*, 65(1):147-175.

Zakian, V. and U. Al-Naib, (1973) Design of dynamical and control systems by the method of inequalities. *Proc. IEE*, 120(11):1421–1427.

Part IV

Case Studies

9 Design of Multivariable Industrial Control Systems by the Method of Inequalities

Oluwafemi Taiwo

Abstract. This chapter reviews the application of the method of inequalities to the multivariable control of important industrial processes such as distillation columns, a turbofan engine and power systems. It is shown that desired performance and stipulated constraints can be expressed in terms of a conjunction of inequalities, any solution of which gives an acceptable design. The method of inequalities is consistently shown to give an efficient solution to the given problem, not only in terms of the ease of problem formulation, but also because of the simple and implementable controllers obtained.

9.1 Introduction

The design of a range of challenging industrial control processes, carried out with the method of inequalities, is reviewed in this chapter. All the designs are in accordance with the conventional definition of control and are subject to standard design criteria expressed in the form of inequalities.

The first process considered is the distillation column, which is used widely (Rademaker *et al* 1975, Shinskey 1977, Skogestad and Postlewaite 1996) in the process industries to separate mixtures with different volatilities. The economic incentives for investigating the feasibility of controlling, simultaneously, more than one product composition of a distillation column become obvious when it is realised that both capital and energy costs are considerably reduced if more than one product meeting specifications, can be obtained from a single column since there is then no need for further processing.

When considering multivariable control of distillation columns, the situation of dual product control is by far the most common in the literature, hence this situation will be considered first. The possible use of decentralised control involving the use of diagonal proportional-plus-integral (PI) controllers to control both product compositions has long been a controversial topic (Rosenbrock 1962, Niederlinski 1971a). Rosenbrock (1962) used somewhat arbitrary transfer functions to demonstrate that de-centralised control will often result in severe interaction. Ridjnsdorp (1965) and Davison (1970) pursued the problem further by outlining the conditions that give rise to interaction. While Ridjnsdorp (1962) and Rademaker *et al* (1975) recommended ratio control, Luyben (1970) proposed non-interacting control (Waller 1974, Taiwo 1980) as a means of eliminating interaction.

Wood and Berry (1973) then compared the merit of both ratio control and non-interacting control experimentally. They concluded that although non-interacting control is to be preferred to ratio control, either scheme is superior to the scheme, which uses two conventional controllers and ignores the interaction present in the plant. Luyben and Vinante (1972) also carried out some experimental investigation of non-interacting control, while others (Waller and Fagervik 1972, Nakanishi et al 1974, Schwanke et al 1977) present only simulation results.

One drawback of non-interacting control is that system performance is very sensitive to the accuracy of the model (Waller and Fagervik 1972) and as noted by Niederlinski (1971a, 1971b) and Foss (1973) complete decoupling is not necessary in a regulator. It has been used in distillation only as an artifice to enable controllers to be designed using well-known single-variable methods. On the other hand, ratio control, although it can be designed without first determining the plant transfer function, ignores the effects of control action at the top of the column on the composition of the bottom. Also, the accurate measurement of the top vapour flow rate might be difficult, as it usually contains some liquid. In any case, a computer will generally be required to implement either scheme.

This has led process control engineers to investigate how the parameters of two conventional controllers can be determined for decentralised control. Most of the methods used to date are heuristic. Typically the gains of the PI controller, which have been independently tuned for the diagonal elements of the plant transfer function matrix, are successively reduced until the entire system is asymptotically stable. However, this approach often fails to give a system with satisfactory transient performace, see for example Wood and Berry (1973), Luyben and Vinante (1972), Waller and Fagervik (1972), Schwanke et al (1977). Another heuristic method, which is slightly more sophisticated and is proposed by Niederlinski (1971a, 1971b), determines the pairing of manipulated and controlled variables such that the gain-bandwidth product for the entire system is maximised. Since this method uses only controllers with integral terms, system steady state performance is generally satisfactory. However, the method has no means of determining controller parameters that ensure that transient interaction is small. In fact, as noted by Niederlinski (1971a) the method may be best suited for determining an initial set of controller parameters which might have to be improved upon using other methods. The progress that has been made in the on-line tuning of decentralised PID controllers has come from three lines of thought. The first method uses relay feedback method (Åström and Hagglund 1984, 1995) in conjunction with sequential loop closure and tuning (Yu 1999, Tan et al 1999, Loh et al 1993, 1994). An extension of this method(Halevi et al 1997, Palmor et al 1995) entails connecting relay to all the feedback loops simultaneously and using ultimate frequency data in PID controller tuning according to Ziegler-Nichols or modified Ziegler-Nichols rules. The second method, called the BLT (biggest log-modulus tuning) method (Monica et al 1988, Luyben and Luyben 1997) detunes the Ziegler-Nichols parameters by certain recommended factors until both stability and intuitive cum empirical performance criteria are satisfied, while for the third category, several design methods have been developed for decentralised PI control systems based principally on Nyquist stability criterion and frequency response information to obtain satisfactory closed loop performance. Hovd and Skogestad (1994) and Skogestad and Postlethwaite (1996) employed sequential loop closure and tuning such that the performance criterion is minimised at each design step. Ho et al (1997) developed a design method for decentralised PID control systems based on shaping the Gershgorin bands, so that the gain and phase margin specifications for the Gershgorin bands are satisfied. Lee et al (1998) extended the iterative continuous cycling method for SISO problems to decentralised PI controller tuning. Chen and Seborg (2003) proposed a method of decentralised PI control based on the idea of independent design, reduction of loop interactions and ensuring system stability. In order to use this method, the plant must be (made) diagonally dominant. The results of applying these methods generally lead to improved controller parameters, although typically, the closed loop systems still suffer from considerable interaction (Balachandran and Chidambaram 1997). It is therefore concluded that reliable design techniques, such as the Ziegler-Nichols method for single-variable systems, have yet to be used in this context. It is also clear that the multivariable model of the plant has to be used for controller design whenever interaction is significant.

Accordingly, the main aim of all the work reviewed here is to apply the computer-aided design method developed by Zakian and known as the method of inequalities (Zakian and Al-Naib 1973, Zakian 1979) to various industrial systems. The effectiveness of the method has been demonstrated in various applications (Taiwo 1978a, 1978b, 1979a, 1979b, 1980,1991, Coelho 1979, Gray and Al-Janabi 1975, 1976, 1991, Bollinger et al 1979). In the present application, both the diagonal PI controller and the full PI controller are designed for a transfer function model of a packed distillation column. The results of implementing the diagonal PI controller are then presented. In order to establish the generality of the results obtained for the packed distillation column, the method of inequalities is also applied to three other distillation columns taken from the literature (Wood and Berry 1973, Schwanke et al 1977, Luyben and Vinante 1972). Designs obtained using the method of inequalities are compared with the non-interacting design. This comparison is made because process engineers commonly employ non-interacting control (Ridjnsdorp and Seborg 1976).

In previous work, investigators (Schwanke et al 1977, Munro and Ibrahim 1973, Pike and Thomas 1974, Davidson 1967, Hu and Ramirez 1972, McGinnis and Wood 1974) have used frequency response methods (Rosenbrock 1974, MacFarlane and Belletrutti 1973), optimal control techniques and modal analysis to design multivariable controllers for the plant. However, these investigations were primarily aimed at obtaining multivariable controllers represented by a full matrix (although Rosenbrock's method can be used more flexibly (Maciejowski, 1989)). Also, the method of inequalities has already been shown (Taiwo 1978, 1979, 1980) to have advantages over these other methods in applications.

Other workers, who have applied the method of inequalities to the control of distillation columns, include Chidambaram and co-workers (Srinivasa and Chidambaram 1991, Balachandran and Chidambaram 1996, 1997). In these cases, both single-variable and multivariable models were considered. For the multivariable situations (Balachandran and Chidambaram 1996, 1997) the emphasis has been on the use of decentralised controllers for two-input two-output and higher dimensional systems. They came up with the conclusion that the method of inequalities outperforms other methods such as BLT method (Luyben and Luyben 1997) and the improved sequential loop tuning method of Hovd and Skogestad (see, for example, Hovd and Skogestad 1994, Skogestad and Postlethwaite 1996), both in terms of the simplicity of problem formulation and system performance.

Another very successful area of application of the method of inequalities is in the control of a 24th order, three-input three-output advanced turbofan engine model (Sain et al 1978, Taiwo 1979). All aspects of desired performance including the rate of change of controller outputs were successfully incorporated into the problem formulation and very simple controllers were obtained meeting all the desired performance and stipulated constraints. Crossley and Dahshan (1982) applied the method of inequalities to design a longitudinal ride-control system for a STOL aircraft. The controller ensures the comfort of passengers and crew by appropriate reduction of normal acceleration. Furthermore Kreisselmeier and Steinhauser (1983), see also, Maciejowski (1989), in a very challenging application, formulated the problem of designing a robust controller capable of ensuring desired performance in all the flight regimes of a fighter aircraft as a conjunction of 42 performance objectives, which were successfully solved. Their control system design formulation is in accordance with the principle and the method of inequalities (see Chapter 1) but they use a different method for solving the inequalities.

The method of inequalities has also been used to design simple multivariable controllers for realistic models of power systems. Taiwo (1978) used the method of inequalities to design simple controllers for the regulation of terminal voltage and load angle of a turbo-alternator. The results were found to be favourable in comparison to those obtained by Ahson and Nicholson (1976), who used the Inverse Nyquist Array method, in terms of the simplicity of the problem formulation, system performance and simplicity of the controllers obtained using the method of inequalities.

Taiwo (1979b) applied the method of inequalities to the multivariable control of an unstable model of the continuous stirred tank reactor whose operating point has been determined using steady state optimisation. By deriving the formulae for the controller outputs, he showed how the method of inequalities can be used to design controllers such that the responses of both the linear and nonlinear models coincide.

Recently, Zhang and Coonick (2000) have used the method of inequalities for the coordinated synthesis of Power System Stabiliser parameters in multi machine power systems in order to enhance overall system small signal stability. They used genetic algorithms to search for the appropriate controller parameters. As usual, the potency of the method in facilitating precise problem formulation as well as the simplicity of controllers was demonstrated. In particular, it was found that the method of inequalities was very effective in designing decentralised controllers for the plant.

9.2 Application of the Method of Inequalities to Distillation Columns

The method of inequalities is used to design controllers for dual quality control of four binary distillation columns whose top and bottom compositions are controlled by manipulating reflux rate and steam. Two types of multivariable proportional plus integral controllers are designed and the performance of the system with these controllers is compared with that of non-interacting control. The results of implementing the simpler controller, which is represented by a diagonal matrix, on a computer-controlled packed distillation column are also presented. In the sequel, we first describe the experimental setup of the packed distillation column.

9.2.1 The Plant and Control System

The packed distillation column used for the experiments is part of the pilot plant in the department of chemical engineering at UMIST (Stainthorp 1970). This continuous fractionating column separates a mixture of methanol and iso-propanol into 97wt.% pure products.

The column is constructed from mild steel 23cm in diameter and packed with 13mm ceramic Raschig rings to a total height of 8.52m. The products taken off at the top and bottom of the column are returned to the appropriate feed tank for recycling (see Figure 9.1). The reboiler is heated by steam at 4.5bar and the total condenser is cooled by water. Both the reflux and the steam flow rates are controlled by electro-pneumatic valve positioners and monitored by orifice plates fitted with differential pressure cells. Temperatures are measured using platinum resistance thermometers which in the case of the top of the column is connected to a continuous transmitter (accuracy $\pm 0.1°C$) and at the bottom is connected to a Kent six-point recorder, which scans each temperature at 30 sec intervals (accuracy$\pm 0.2°C$).

The plant can be controlled either from a conventional analogue instrument panel or by digital computer (the latter has been used in the experiments reported here), which has a control panel, two teleprinters, a paper tape punch and a paper tape reader.

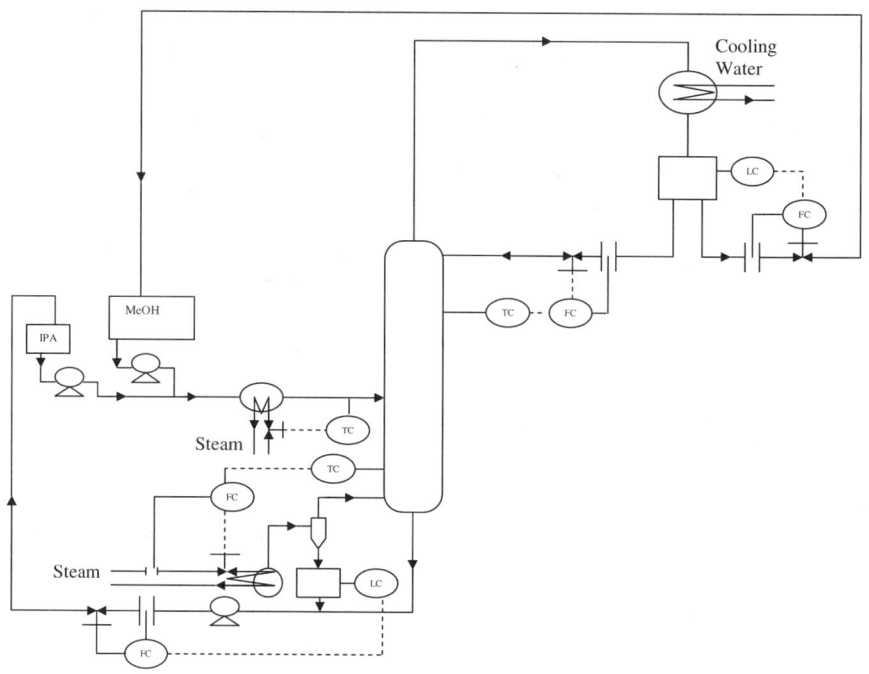

Fig. 9.1. Schematic diagram of the distillation column

The computer is an Argus 308 with 8k core store (24 bit word) and an adequate number of digital to analogue and analogue to digital (10 bits) converters for interfacing to the column and two other similar plants. Organisation of the computer task is achieved by the time-sharing executive program, which also provides facilities for timing and logging operations.

9.2.2 Transfer Function Description of the Packed Distillation Column

The pertinent steady state parameters, which characterise the normal operating point of the plant are given in Table 9.1

It is assumed here that this operating point is based on the economics of the process. The purpose of the control system is then to ensure that

Table 9.1. Steady state operating conditions

	Feed	Distillate	Bottoms	Reflux	Steam
Flow rate (Kg/hr)	61	16	45	48	45
Composition (wt % methanol)	28	97	2	97	-
$T_1 = 68°C$, $T_2 = 77°C$, Feed temperature $= 69°C$					

excursions of system variables from these steady state values are sufficiently small, in spite of the disturbances acting on the system.

Since this is a binary distillation column, composition at a point is uniquely determined by the temperature and pressure at that point. So, by keeping the pressure constant, composition can be inferred from temperature.

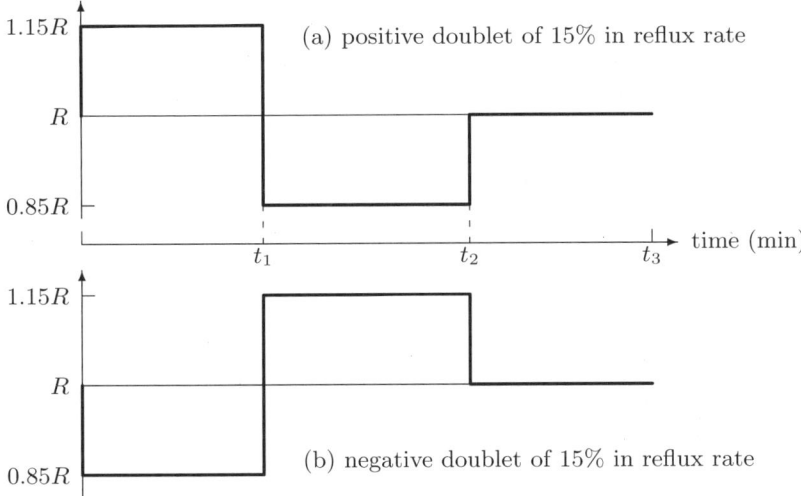

Fig. 9.2. Diagram of positive and negative doublets

In each test the top temperature (T1) and the pressure-corrected bottom temperature (T2) of the plant resulting from "doublet" type (see Figure 9.2) disturbances of reflux and steam rates were recorded every minute. By using this type of disturbance, plant variables are assured to be close to their steady state values, thereby justifying the linearised description; the dangers of driving materials to one end of the column is avoided and plant responses to disturbances in both directions of increasing and decreasing temperatures are combined to elucidate its transfer function in each test.

Preliminary experiments indicated that plant responses were very much obscured by noise when doublet variations were less than ±10%. Consequently, doublet sizes of between ±15% and 20% were adopted. Typical values for t1, t2 and t3 (Figure 9.2) were respectively 15, 30 and 45.

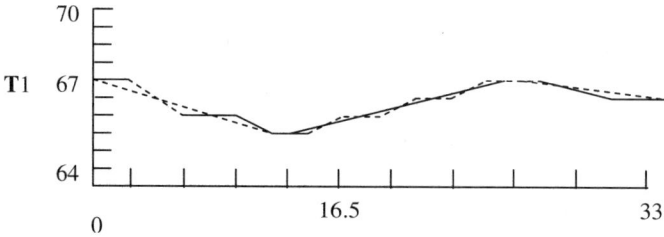

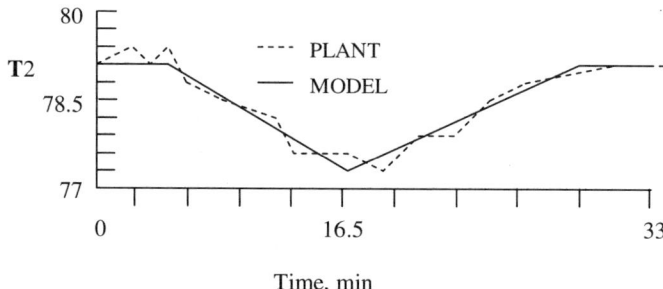

Fig. 9.3. Responses of upper and lower temperatures to a positive doublet of 15% reflux

A least-squares fit employing Rosenbrock's (1960) direct search technique was used to determine the parameters of simple models which describe plant responses. The model adopted for the column is given by

$$\begin{bmatrix} T_1(s) \\ T_2(s) \end{bmatrix} = \begin{bmatrix} g_{11}(s) & g_{12}(s) \\ g_{21}(s) & g_{22}(s) \end{bmatrix} \begin{bmatrix} R(s) \\ S(s) \end{bmatrix} \quad (9.1)$$

where

$$g_{11}(s) = \frac{-0.86e^{-s}}{35.4s+1}, \quad g_{12}(s) = \frac{0.6}{(18.9s+1)(2.6s+1)}$$
$$g_{21}(s) = \frac{-1.5e^{-4s}}{74s+1}, \quad g_{22}(s) = \frac{1.22}{30.6s+1} \quad (9.2)$$

The time constants and time delays are in minutes and the gains have the units Khr/kg. Figure 9.3 shows typical agreement between plant outputs and assumed model.

9 Design of Multivariable Industrial Control Systems by the MoI

The main difficulties encountered during the identification were due to variations in steam quality and the frequent malfunctioning and small accuracy of the Kent six-point recorder. The steam plant which conveyed steam to the plant also supply adjacent offices and factories, and steam pressure varied with the number of other users. For example, the steam quality was particularly low during lunch time. The best results from the pulse tests were generally obtained at night. Also, because of chanelling caused by interfacial tension between the liquid and the packing surface, the column was not always in thermal equilibrium. The temperature recorded by the thermocouples would therefore be slightly different depending on whether they are in contact with vapour or liquid.

9.2.3 Performance Specifications and Constraints

The purpose of this section is to explain why it is valid to specify performance in terms of the functionals of the system step responses. The explanation is provided by the work of Zakian (1978, 1979), which systematises the ideas relating to performance and sensitivity.

In order to behave properly as a regulator or servomechanism, every component $e_i(t)$ of the error vector in the linear, time-invariant ℓ-input, ℓ-output control system shown in Figure 9.4 should remain sufficiently close to zero throughout the time domain in spite of any disturbances acting on the system or any changes in the characteristics of the plant.

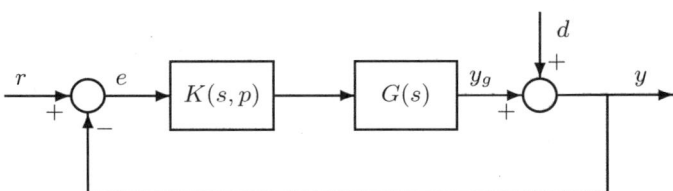

Fig. 9.4. Block Diagram of the feedback structure for the column

Let $J = \{1, 2, \ldots, l\}$, then with regard to Figure 9.4, it can be shown that for every $i \in J$ and $t \geq 0$

$$e_i(t) = \sum_{j=1}^{i} [\delta_{ij} - \sigma_{ij}(t)] f_j(0) + \sum_{j=1}^{l} \int_0^t [\delta_{ij} - \sigma_{ij}(\lambda)] f_j^{(1)}(t-\lambda) d\lambda \quad (9.3)$$

where $f = r - d$, σ_{ij} is the ith system output due to a unit step change in r_j, while δ_{ij} is the Kronecker delta, defined by

$$\delta_{ii} = 1 \text{ and } \delta_{ij} = 0, i \neq j \quad (9.4)$$

Other symbols are defined in the notation.

Clearly, $e_i(t) = 0$ if a controller can be designed such that for all $t \geq 0$

$$\sigma_{ii}(t) = 1 \text{ and } \sigma_{ij}(t) = 0, i \neq j \qquad (9.5)$$

This is not possible in any real system because

$$e_i(0) = f_i(0) \qquad (9.6)$$

which shows that the initial value of the error is independent of the controller parameters.

A further reason why (9.5) may not be satisfied is that control signals must be bounded. However, a system with response σ_{ii} which is sufficiently step-like and σ_{ij} which is small for $i \neq j$ is a satisfactory system. It is noted also from (9.3) that $\lim_{t \to \infty} e_i(t)$ exists for all f which tends to a constant as $t \to \infty$ if, $(\sigma_{ij}(\infty)) = I$, that is to say there is no steady-state error in the system step responses. In practice, a small initial period t_s (which is to be distinguished from the settling time) is allowed for the process to "recover" from initial transients and the design aim is then to determine the controller $K(s,p)$ such that the absolute value of the error is less than a specified quantity after t_s, i.e.,

$$|e_i(t)| \leq k_i > 0, \quad t \geq t_s \qquad (9.7)$$

Experience shows that the settling times of σ_{ii} and σ_{ij}, $i \neq j$ are usually similar even when only the settling time ϕ_{i1} of σ_{ii} is specified in the design. It is therefore sufficient to use only ϕ_{i1} during design in order to reduce computation time. Observe that a design that minimises the maximal value of interaction $\hat{\sigma}_{ij}$ of σ_{ij}, the settling time ϕ_{i1} and the overshoot ϕ_{i2} of σ_{ii} approximates to the ideal situation (9.5). If for the given input f, the matrix (σ_{ij}) is sufficiently close to the condition (9.5) then the desired performance (9.7) thus takes place.

The distillation control problem is often a regulator problem, in which case $f = -d$. Although no effort was made to identify d completely, observations of the steam quality and feed flow rate indicate that $d^{(1)} \neq 0$. Hence the designs given here trade-off between the functionals ϕ_{i1}, ϕ_{i2} and $\hat{\sigma}_{ij}$. It is clear from (9.3) that such a design will in general guarantee relative immunity to disturbances.

Other important factors considered during design are the constraints within which the system has to operate. For example, since linear systems theory is employed in the design, it is important that the linear range of an actuator is not exceeded. Actuator saturation is avoided by constraining the functional

$$U_i = \sup_{0 \leq t < \infty} |u_i(t)| \qquad (9.8)$$

known as the maximum controller output in the ith loop. The constraint on (9.8) should ensure that the signal levels in the control loop are tolerable, that the linearisation approximation is nearly valid and that the actuator satisfies its maximum demand without undue wear. Taiwo (1979b), shows how to determine U_i for the situation when both the nonlinear model and f are known such that the linearisation is valid. However, operating experience with the distillation columns in UMIST suggests that the constraint on (9.8) should be $U_i \leq 15$.

For the system shown in Figure 9.4 let

$$\dot{x}_g(t) = A_g x_g(t) + B_g u(t)$$
$$y_g(t) = C_g x_g(t) \tag{9.9}$$

be a minimal realisation of a rational approximant of $G(s)$. Also let $K(s,p)$ have the realisation

$$\dot{x}_k(t) = A_k x_k(t) + B_k e(t)$$
$$u(t) = C_k x_k(t) + D_k e(t) \tag{9.10}$$

where the matrices A_k, B_k, C_k and D_k depend on the vector p. the closed-loop system therefore has the equations

$$\dot{x}(t) = Ax(t) + Bf(t)$$
$$y(t) = Cx(t) + d(t) \tag{9.11}$$

where

$$A = \begin{bmatrix} A_k & -B_k C_g \\ B_g C_k & A_g - B_g D_k C_g \end{bmatrix}, \quad B = \begin{bmatrix} B_k \\ B_g D_k \end{bmatrix}, \quad C^T = \begin{bmatrix} 0 \\ C_g^T \end{bmatrix}, \quad x = \begin{bmatrix} x_k \\ x_g \end{bmatrix} \tag{9.12}$$

and superscript T denotes matrix transposition. The system is asymptotically stable if and only if all the eigenvalues $\lambda_i(p)$ of A lie in the open left half plane. Thus, one of the inequalities is

$$\phi_1(p) \leq \varepsilon, \quad \varepsilon < 0 \tag{9.13}$$

where

$$\phi_i(p) = \max_i \{\text{Re}[\lambda_i(p)]\} \tag{9.14}$$

and it ensures asymptotic stability of the closed-loop system. If the initial guess p^0 gives rise to a closed-loop unstable system, the MBP is used to find a stable point p by solving the inequalities (9.13). With this as the starting point, the MBP is then used to solve all the inequalities simultaneously.

Each trial of the MBP involves a test for asymptotic stability followed by the computation of $\phi_i(p)$ which are obtained from the step responses of

(9.11). Computation of system time responses is efficiently done by means of Zakian's (1975) I_{MN} approximants. Notice that (9.9) presupposes that the plant description, if expressed in the transfer function matrix form, is rational. A moments method of simplification (Zakian 1978, Taiwo 1995) is used to obtain a rational approximant of the plant transfer function, $G(s)$, whenever this is not rational, in order to facilitate controller design according to the above formulation.

9.2.4 Controller Design

In general, controller design is initiated by choosing a controller form and an arbitrary starting point p^0. Then, with d in (9.11) set equal to zero, system step responses are computed. The aim in controller design is to ensure that all the elements in the matrix $(\delta_{ij} - \sigma_{ij}(t))$ are close to zero for all $t > t_s$ where t_s is sufficiently short and without violating the constraint on U_i.

The functionals ϕ_{i1}, ϕ_{i2} which denote respectively the settling time and overshoot of response σ_{ii}, and $\hat{\sigma}_{ij}$, the maximal interaction, have been used here to specify system performance; the bound on θ_{i1} being largely determined by the performance of the system with non-interacting control while the bounds on θ_{i2} and $\hat{\sigma}_{ij}$ are fixed after a preliminary experimentation with the MBP.

Each element $K_{ij}(s)$ of the matrix $K(s,p)$ is restricted to be of the form of a PI controller. This choice is based on the simplicity and the widespread use of this controller form in the chemical industry.

9.2.5 Design of Controllers for the Packed Distillation Column

Since two terms in $G(s)$ contain pure time delays, the method of moments (Zakian 1978) is used to obtain a rational approximant $G^*(s)$. Minimal realisations of both $G^*(s)$ and $K(s,p)$ are then obtained in order to use the MBP according to the formulation given here.

Formulation 9.1. In this formulation we use the full PI controller (abbreviated FPI) with the transfer function matrix

$$\begin{bmatrix} p_1 + p_2 s^{-1} & p_3 + p_4 s^{-1} \\ p_5 + p_6 s^{-1} & p_7 + p_8 s^{-1} \end{bmatrix} \tag{9.15}$$

A minimal realisation of (9.15) is easily established as

$$A_k = 0_2, \quad B_k = I_2, \quad C_k = \begin{bmatrix} p_2 & p_4 \\ p_6 & p_8 \end{bmatrix}, \quad D_k = \begin{bmatrix} p_1 & p_3 \\ p_5 & p_7 \end{bmatrix} \tag{9.16}$$

The following inequalities are used to specify desired performance and system constraints:

$$\begin{aligned} &\phi_{11} \leq 30, \quad \phi_{12} \leq 0.2, \quad \phi_{21} \leq 30, \quad \phi_{22} \leq 0.2 \\ &\hat{\sigma}_{ij} \leq 0.2, \quad U_i \leq 10, \, i=1,2, \quad \phi_1 \leq -0.001 \end{aligned} \tag{9.17}$$

9 Design of Multivariable Industrial Control Systems by the MoI

A suitable starting point is given by

$$p^0 = (-2, -1, 0.5, 0.5, 0.5, 0.5, 2, 1) \tag{9.18}$$

The performance of the system with $K(s, p^0)$ is expressed by

$$\begin{array}{llll} \phi_{11} = 99, & \phi_{12} = 0.48, & \phi_{21} = 84, & \phi_{22} = 0.22 \\ \hat{\sigma}_{12} = 0.25, & \hat{\sigma}_{21} = 0.65, & U_1 = 6.5, & U_2 = 6.3, & \phi_1 = -0.054 \end{array} \tag{9.19}$$

After 20 iterations, the MBP located

$$p = (-4, -0.968, 0.75, 0.51, 2.244, 0.266, 2.376, 1.1) \tag{9.20}$$

and the functionals of the system with $K(s, p)$ are:

$$\begin{array}{llll} \phi_{11} = 45, & \phi_{12} = 0.2, & \phi_{21} = 45, & \phi_{22} = 0.112 \\ \hat{\sigma}_{12} = 0.216, & \hat{\sigma}_{21} = 0.465, & U_1 = 5.8, & U_2 = 4.7, & \phi_1 = -0.054 \end{array} \tag{9.21}$$

We note from (9.21) that the system has not yet satisfied most of the specification. Hence starting with parameters (9.20) and seeking for ten more iterations, the MBP located

$$p = (-7.2, -1.26, 0.627, 1.97, 3.025, 0.068, 5.4, 2.179) \tag{9.22}$$

The corresponding closed-loop system functionals are now

$$\begin{array}{llll} \phi_{11} = 34, & \phi_{12} = 0.2, & \phi_{21} = 29, & \phi_{22} = 0.114 \\ \hat{\sigma}_{12} = 0.12, & \hat{\sigma}_{21} = 0.28, & U_1 = 7.7, & U_2 = 6, & \phi_1 = -0.054 \end{array} \tag{9.23}$$

Notice the large reduction in p_6 in (9.23), indicating that this element may be set equal to zero. This demonstrates one way in which the method assists the designer to determine controller structure. We therefore stipulate a controller with seven parameters in the next formation.

Formulation 9.2. p_6 is set equal to zero and a PI controller with seven parameters (abbreviated PI7), is used. A minimal realisation of this controller is:

$$A_k = 0_2, \quad B_k = I_2, \quad C_k = \begin{bmatrix} p_2 & p_4 \\ 0 & p_7 \end{bmatrix}, \quad D_k = \begin{bmatrix} p_1 & p_3 \\ p_5 & p_6 \end{bmatrix} \tag{9.24}$$

The design is continued with the parameter values just computed, *i.e.*,

$$p = (-7.2, -1.26, 0.627, 1.97, 3.025, 0.068, 5.4, 2.179) \tag{9.25}$$

The specifications are given by (9.17).

After 20 iterations the MBP located

$$p = (-7.14, -1.76, 1.72, 1.84, 4.4, 7.25, 2.73) \tag{9.26}$$

and the corresponding closed-loop system functionals are:

$$\begin{aligned}&\phi_{11} = 31, \quad \phi_{12} = 0.2, \quad \phi_{21} = 26, \quad \phi_{22} = 0.11\\&\hat{\sigma}_{12} = 0.167, \quad \hat{\sigma}_{21} = 0.24, \quad U_1 = 8.6, \quad U_2 = 7.25, \quad \phi_1 = -0.054\end{aligned} \tag{9.27}$$

We notice from (9.27) that the only system specifications that are violated are the bounds on θ_{11} and $\hat{\sigma}_{21}$. Since these bounds are just marginally exceeded, the design is deemed satisfactory. Observe from (9.26) that the magnitudes of the proportional elements in $k_{ii}(s)$ are distinctly larger than those of $k_{ij}(s)$, $i \neq j$. This suggests that a diagonal controller may control the plant satisfactorily. Therefore we investigate this controller in the next formulation.

Formulation 9.3. Here we use the diagonal PI controller (DPI) given by the transfer function matrix

$$K(s,p) = \begin{bmatrix} p_1 + p_2 s^{-1} & 0 \\ 0 & p_3 + p_4 s^{-1} \end{bmatrix} \tag{9.28}$$

$$p^0 = (-4, -2, 5, 2) \tag{9.29}$$

The performance specification and constraints are given by (9.17).

After 18 iterations the MBP located

$$p = (-9.47, -2.21, 10.7, 3.44) \tag{9.30}$$

and the corresponding closed-loop system functionals are:

$$\begin{aligned}&\phi_{11} = 31.5, \quad \phi_{12} = 0.17, \quad \phi_{21} = 24.5, \quad \phi_{22} = 0.19\\&\hat{\sigma}_{12} = 0.36, \quad \hat{\sigma}_{21} = 0.24, \quad U_1 = 10.4, \quad U_2 = 10.7, \quad \phi_1 = -0.054\end{aligned} \tag{9.31}$$

By slightly relaxing the constraints on U_i the MBP was used to obtain the parameters

$$p = (-12.1, -1.95, 12.9, 4.96) \tag{9.32}$$

The corresponding functionals of the closed-loop system are now

$$\begin{aligned}&\phi_{11} = 36, \quad \phi_{12} = 0.08, \quad \phi_{21} = 21, \quad \phi_{22} = 0.21\\&\hat{\sigma}_{12} = 0.35, \quad \hat{\sigma}_{21} = 0.21, \quad U_1 = 12.1, \quad U_2 = 12.9, \quad \phi_1 = -0.054\end{aligned} \tag{9.33}$$

$\hat{\sigma}_{ij}$ are slightly less than the values given in (9.31).

System responses with the controllers whose parameters are given in (9.26) and (9.32) and non-interacting control (Waller 1974, Taiwo 1980) are shown in Figure 9.5. As expected, the system with controller (9.24) and non-interacting control is slightly superior to that with controller (9.28) with respect to reducing interaction. However, a controller of the form (9.28) can be more cheaply implemented by means of just two conventional analogue controllers, while a computer will generally be required to implement either a controller of the form (9.24) or non-interacting control.

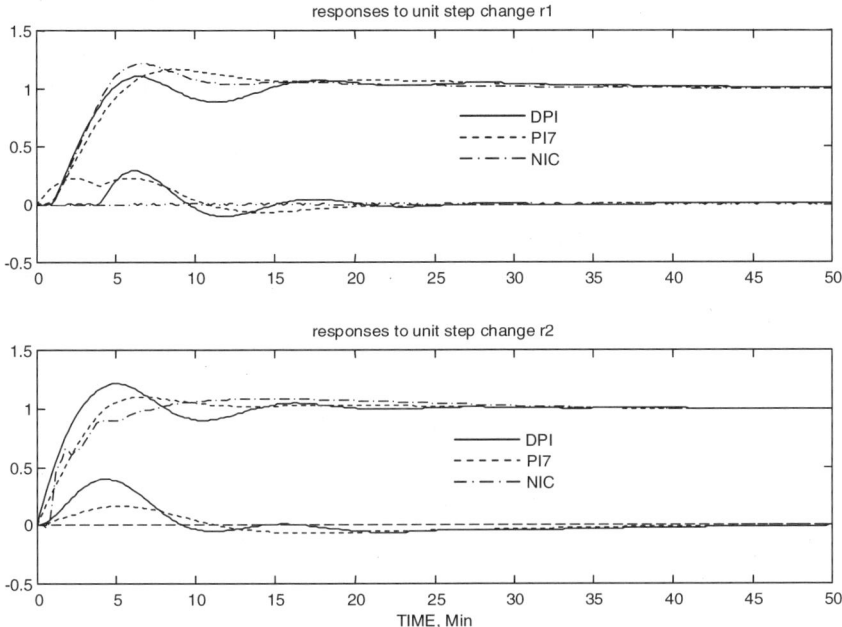

Fig. 9.5. Closed-loop responses of the packed distillation column with diagonal PI controller (DPI), PI controller with seven parameters (PI7) and non-interacting control (NIC)

9.2.6 Experimental Results

Controller (9.28) with parameters (9.32) was implemented on the plant and responses to step changes in T_{1ref} and T_{2ref} were used to compare simulation with actual plant outputs. For plant outputs during test to be distinctly more than the normal noise levels, it was necessary to introduce step changes of $1.5°C$ and $2°C$ in T_{1ref} and T_{2ref} respectively. Output T_1 and T_2 were then normalised before comparing with simulation (see Figure 9.6). Notice that actual plant outputs tend to 'lead" the simulated outputs. Also a great deal of irregularity is evident in the temperatures. These irregularities are more pronounced when there is a step change in T_{2ref}. Also, irregularities are confined to T_2 when there is a step change in T_{1ref}. Most of the irregularity was caused by the relatively small accuracy ($\pm 0.2°C$) and the noise problems encountered with the Kent six-point recorder. These two factors made it relatively difficult to monitor small temperature variations (*e.g.* near the steady state of T_2 and the output of T_2 due to a step change in T_{1ref}) precisely. Unfortunately the poor performance of the bottom temperature recorder also caused erratic action in the top temperature measurement. Note, however, that the simulated responses tend to be the mean of the plant output as would be expected, since the least squares fit criterion was used to obtain the parameters of the plant model. As plant outputs were not very regular, no effort was made to implement the controller with parameters (9.26). It is worth remarking that earlier plans (Abbosh 1973) to implement an adaptive control scheme on this plant were abandoned because of the aforementioned problems.

In this section we have seen (Figure 9.5) that if the multivariable model of the column is considered during design, a diagonal controller may give satisfactory performance. However, since plant output are rather erratic, it is necessary to consider other column models in order to establish the generality of the result.

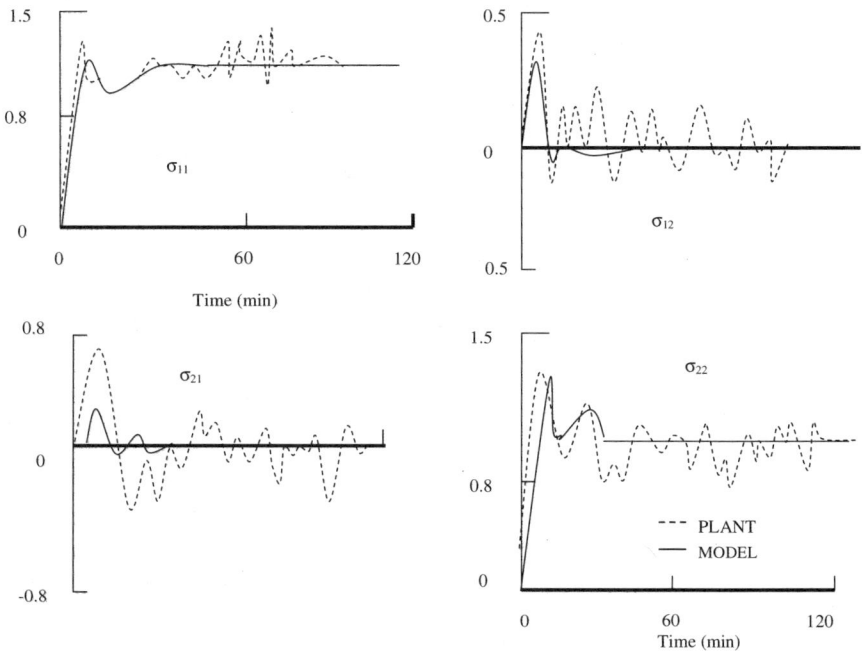

Fig. 9.6. Closed-loop responses of the packed distillation column with the diagonal PI controller whose parameters are given in (9.32)

Other distillation column models. In this section we employ the method of inequalities to design both the full PI controller and the diagonal PI controller for other distillation column models. The main motivation here is to see if the diagonal PI controller can control the column satisfactorily. Previous investigators (Wood and Berry 1973, Luyben and Vinante 1972, Schwanke et al 1977) found that the closed-loop systems were very oscillatory when diagonal control was attempted. The transfer functions of these plants are given in Table 9.2.

Table 9.2. Other distillation column models

Authors	$g_{11}(s)$	$g_{12}(s)$	$g_{21}(s)$	$g_{22}(s)$
Wood and Berry (1973)	$\dfrac{12.8e^{-s}}{16.7s+1}$	$\dfrac{-18.9e^{-3s}}{21s+1}$	$\dfrac{6.6e^{-7s}}{10.9s+1}$	$\dfrac{-19.4e^{-3s}}{14.4s+1}$
Luyben and Vinante (1972)	$\dfrac{-2.16e^{-s}}{8s+1}$	$\dfrac{1.26e^{-0.3s}}{9.5s+1}$	$\dfrac{-2.75e^{-1.8s}}{9.5s+1}$	$\dfrac{4.28e^{-0.35s}}{9.2s+1}$
Schwanke, Edgar and Hougen (1977)	$\dfrac{-10.8(3.08s+1)}{2.13s^2+2.04s+1}$	$\dfrac{0.52(3.125s+1)}{1.78s^2+1.87s+1}$	$\dfrac{-28.14e^{-0.65s}}{1.9s^2+2.21s+1}$	$\dfrac{1.84}{1.87s^2+2.19s+1}$

The details of the designs are similar to those given earlier except that only diagonal PI controllers (Table 9.3) and full PI controllers (Table 9.4) are given. It is obvious in some cases which element of the full PI controllers should be set equal to zero in order to simplify the controller in a way similar to what was done above.

The responses of the system with the various controllers are shown in Figure 9.7 for the case of Wood and Berry's (1973) model. Other responses are available in Taiwo (1980) and have not been repeated here for space economy. As expected non-interacting control gives a system with the least degree of interaction.

In the case of Wood and Berry's (1973) model, the system response σ_{11} with non-interacting control is probably too oscillatory (see Figure 9.7). Better damping might result from further tuning. The controllers designed using the method of inequalities give adequately damped systems, the full PI controller resulting in less interaction. The response σ_{22} is rather sluggish when this latter controller is used. Comparison shows that the performance of the system with diagonal PI controller obtained using the method of inequalities is superior to that achieved with diagonal control in Wood and Berry (1973) and the biggest log-modulus method (Luyben and Luyben 1997) with respect to the level of interaction and adequate damping. The design in Wood and Berry (1973) (the PI controller parameters were not given) was obtained by reducing the gains of the PI controllers which had been designed for $g_{ii}(s)$ independently.

For Luyben and Vinante's model (1972), the biggest difference between the systems with non-interacting control and the full PI controller is in σ_{21}. Non-interacting control gives a system with smaller $\hat{\sigma}_{21}$. Other σ_{ij} are similar. The performances of the systems with these controllers are better than the performance of the system with a diagonal PI controller. The main fault with this latter controller is that it gives a system with larger $\hat{\sigma}_{ij}$ and σ_{11} is sluggish. Nevertheless, this diagonal PI controller gives a better system than the diagonal PI controller designed in Luyben and Vinante (1972). The diagonal PI controller in Luyben and Vinante (1972) was obtained by using the multivariable model of the plant. Nevertheless, θ_{22} and $\hat{\sigma}_{12}$ are too large, while σ_{21} is too oscillatory.

Non-interacting control of Schwanke's model (1977), with the controller parameters given, indicates sustained small amplitude oscillations in σ_{11}. This is probably due to the presence of eight complex conjugate poles in the $g_{ij}(s)$. The system with non-interacting control gives the least amount of interaction σ_{21}, while the full PI controller gives a system which is only slightly better than that with the diagonal PI controller. However, this diagonal PI controller gives a better system, with respect to system damping and interaction, than the diagonal PI controller given in Schwanke et al (1977). That controller was designed using a tuning method due to Hougen (1979). The controller parameters were not given. The characteristics locus design method (MacFarlane and Belletrutti 1973) was also applied by these investigators to this plant with somewhat unsatisfactory results.

Table 9.3. Parameters of the diagonal PI controller

	p_1	p_2	p_3	p_4
Wood's model	0.1644	0.0179	-0.0581	-0.0093
Luyben's model	-1.17	-0.135	2.29	0.206
Schwanke's model	-3.54	-1.56	4.7	2.73

Table 9.4. Parameters of the full PI controller

	p_1	p_2	p_3	p_4	p_5	p_6	p_7	p_8
Wood's model	0.408	0.0835	-0.05557	-0.0176	0.00326	0.0334	-0.023	-0.00967
Luyben's model	-2.14	-1.04	1.12	0.326	0.5	-0.78	1.13	0.44
Schwanke's model	-1.52	-10.11	0.96	0.1	6.59	12.7	6.8	4.6

9.2.7 Conclusion

The method of inequalities has been applied to the multivariable control of four distillation columns. The method is easy to use and gives good results. As design progresses, the designer's understanding of system dynamics improves and the available trade-offs between specifications become more apparent.

The diagonal PI controllers designed by the method of inequalities give systems that are satisfactory, thus indicating that diagonal control may be acceptable with these columns. It is further shown that some previous investigators (Wood and Berry 1973, Luyben and Vinante 1972, Schwanke et al 1977) failed to achieve satisfactory diagonal control, either because designs

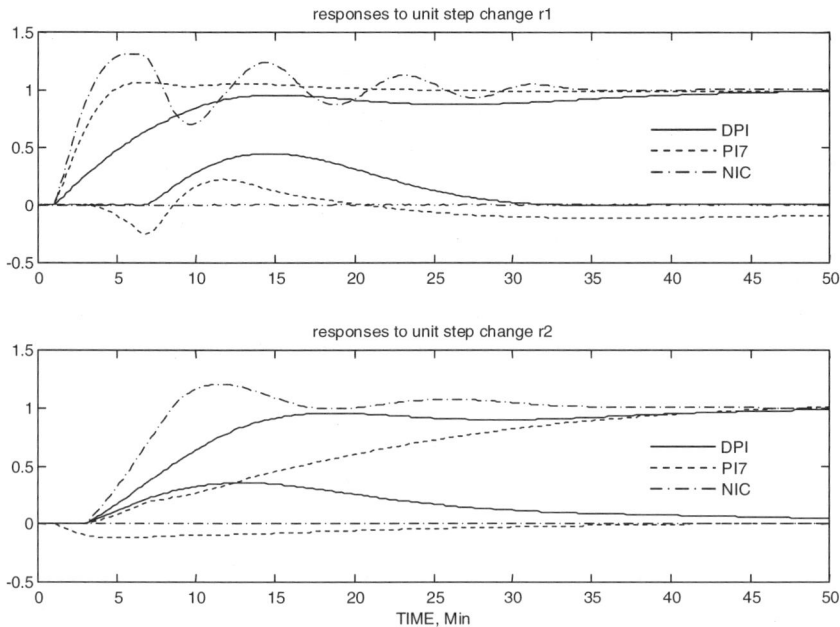

Fig. 9.7. Closed-loop responses of Wood's model with diagonal PI controller (DPI), full PI controller (FPI) and non-interacting control (NIC)

were largely based on the diagonal elements of the plant transfer function matrix, or because of the unavailability of a suitable design method for determining a satisfactory set of controller parameters. The DPI obtained here outperforms the DPI obtained using the BLT method for the Wood and Berry (1973) model. This is consistent with other similar studies (Balachandran and Chidambaram, 1996, 1997) where the DPIs designed using the method of inequalities outperform those obtained using both the BLT and the improved sequential loop closure schemes. Either the full PI controller (FPI) or non-interacting control (NIC) gives rise to somewhat better systems than the systems involving DPI. However, both the FPI and NIC are more expensive to implement and NIC is very sensitive to model inaccuracies.

9.3 Design of Multivariable Controllers for an Advanced Turbofan Engine by the Method of Inequalities

The growing complexity of aircraft engines has been a challenge to control engineers who wish to design controllers for the regulation of system outputs (Skira and De Hoff 1977, De Hoff and Hall 1976, Hackney *et al* 1977). It is now widely accepted that rudimentary hydromechanical control methods are not suitable for these engines. Electronic digital control, therefore is increasingly being used and many new design possibilities have received urgent attention in the industry (Sain *et al* 1978). In this work, we employ the method of inequalities to design simple multivariable controllers for the plant. The models of the engine considered here were provided by the United States National Engineering Consortium to several control groups around the world in 1977 as a form of benchmark to test the potency of the available multivariable design methods at the time. Only a few groups came up with acceptable practical solutions. The one presented here is one of the useful and simple solutions and was single-handedly obtained (Taiwo 1978, 1979). The results reveal some of the inherent difficulties associated with the control of the plant.

9.3.1 Plant Model

The 3-input, 3-output model (Sain *et al* 1978, Peczkowski 1977) of the plant are considered having orders 5 and 24, respectively. The open-loop responses of the latter model are shown in Figure 9.8.For space economy, only the design of controllers for the more difficult full order model is presented here. The reader is encouraged to consult the original work (1979b) for details of controller design for the simplified model.

The 24th Order Model of the Plant Here the plant is considered as being made up of the turbofan engine and actuators. The engine is expressed by the linearised equations

$$\begin{aligned} \dot{x}_e(t) &= A_e x_e(t) + B_e u_e(t) \\ y_e(t) &= C_e x_e(t) + D_e u_e(t) \end{aligned} \qquad (9.34)$$

where the vector y_e has components y_{e1}, y_{e2} and y_{e3} that denote, respectively, engine net thrust level, total engine air flow and engine inlet temperature, while the vector u_e has components u_{e1}, u_{e2}, and u_{e3} that denote, respectively, main burner fuel flow, nozzle jet area and inlet guide vane position. The transfer function matrix of the engine is therefore given by

$$G_e(s) = C_e(sI - A_e)^{-1} B_e + D_e \qquad (9.35)$$

The matrices A_e, B_e, C_e and D_e of dimensions respectively, 16×16, 16×3, 3×16 and 3×3, are chosen such that $G_e(s)$ is the 3×3 transfer function matrix obtained by discarding the last two rows and two columns of the 5×5 transfer function matrix arising from the original 16th order state space model of the engine (Sain *et al* 1978, Peczkowski 1977). The 3×3 transfer function matrix $G_a(s)$ of the actuators, which has a characteristic polynomial of order eight, is obtained by discarding the last two rows and columns of the original 5 × 5 transfer function matrix model of the actuators (Sain *et al* 1978, Peckowski 1977).

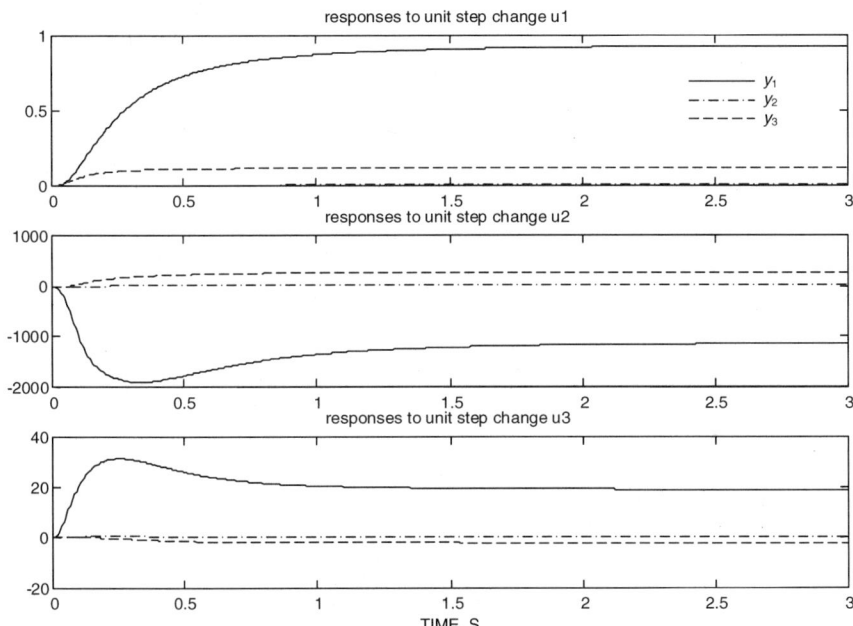

Fig. 9.8. Open-loop unit step responses of the plant

Let
$$\dot{x}_a(t) = A_a x_a(t) + B_a u_a(t)$$
$$u_a(t) = C_a x_a(t) \qquad (9.36)$$

be a minimal realisation of $G_a(s)$ then the plant (engine plus actuators) is given by
$$\dot{x}_c(t) = A_c x_c(t) + B_c u_a(t)$$
$$y_e(t) = C_c x_c(t) \qquad (9.37)$$

where
$$A_c = \begin{bmatrix} A_a & 0 \\ B_e C_a & A_e \end{bmatrix}, \quad B_c = \begin{bmatrix} B_a \\ 0 \end{bmatrix}, \quad C_c = \begin{bmatrix} D_e C_a & C_e \end{bmatrix}, \quad x_c = \begin{bmatrix} x_a \\ x_a \end{bmatrix} \qquad (9.38)$$

and u_a is a vector of controller outputs. The block diagram of the control system is shown in Figure 9.9.

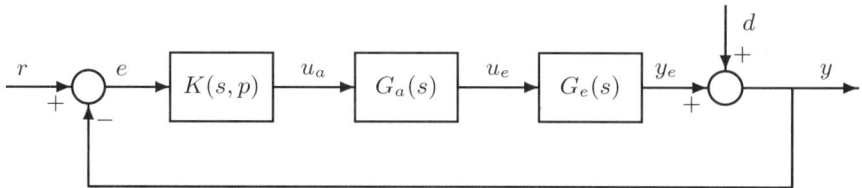

Fig. 9.9. Feedback structure for the 24th order plant

It is assumed here that all the states of the engine are measurable. Measurement lags are neglected. Further details of the work reported here are available (Taiwo 1978b, 1979b).

9.3.2 Performance Specification and Constraints

When in flight, the aircraft pilot manipulates the power lever angle in order to obtain more or less thrust. The power lever angle change is converted by a master engine scheduler into reference input for the closed loop system. The nature of this reference input is not specific, but is assumed here to be a step-like function, in accordance with standard practice (Sain et al 1978, Peczkowski 1977, Zakian 1978). In terms of our linear models this control problem can be summarised thus : with a step change in power lever angle we want to move the engine to the appropriate operating point quickly with the plant variables responding smoothly (Sain et al 1978, Peczkowski 1977, Hackney et al 1977).

The specifications are therefore that the closed-loop system should be approximately decoupled and the net thrust level, y_{e1}, and the inlet temperature, y_{e3}, should respond quickly and with little overshoot. The maximal value of the modulus of the ith input to the engine is

$$U_{ei}(t) = \sup_{0 \leq t \leq \infty} |u_{ei}(t)| \qquad (9.39)$$

This is constrained to be less than 1 ft and 6° for $i = 2$ and 3, respectively, for unit step changes, applied individually, in the reference inputs. Also, the bounds on the maximal rate of change

$$V_{ei}(t) = \sup_{0 \leq t \leq \infty} \left| \frac{du_{ei}(t)}{dt} \right| \qquad (9.40)$$

of these inputs are respectively 15800(lb/hr)/s, 3.6 ft/s and 48°/s for $i = 1, 2, 3$ for unit step changes, applied individually, in the reference inputs.

9.3.3 The Design of Controllers

Formulation 9.4. The controllers designed for the reduced fifth order model give an unstable closed-loop system when implemented on the 24th order model (Taiwo 1979b). Computations using the MBP showed that in order to obtain an acceptable design the multivariable PI controller

$$K(s,p) = \begin{bmatrix} p_1 + p_2 s^{-1} & p_3 + p_4 s^{-1} & p_5 + p_6 s^{-1} \\ 0 & p_7 s^{-1} & 0 \\ p_8 & 0 & p_9 s^{-1} \end{bmatrix} \tag{9.41}$$

which has a minimal realisation

$$A_k = 0_3, \quad B_k = I_3, \quad C_k = \begin{bmatrix} P_2 & P_4 & P_6 \\ 0 & P_7 & 0 \\ 0 & 0 & P_y \end{bmatrix}, \quad D_k = \begin{bmatrix} P_1 & P_3 & P_5 \\ 0 & 0 & 0 \\ P_8 & 0 & 0 \end{bmatrix} \tag{9.42}$$

may be used. Both theoretical considerations and observed system responses indicate that the closed-loop system with controller (9.42) gives correct steady state responses. Therefore, in the subsequent design, only inequalities relating to the settling time, overshoot, U_{ei}, V_{ei} and the interaction functionals were considered.

In the course of the design it became clear that a conflict exists between the requirements of small interaction and adequate damping. The parameters that give a reasonable compromise between these conflicting requirement were found to be

$$p = (-7.74, 80.39, 7.98, 81.6, 47.69, 293.71, 0.03, 0.32, -7.77) \tag{9.43}$$

and the functionals of the closed-loop system with $K(s,p)$ are

$$\begin{array}{llll}
\text{Settling time} & 1.28 & 2.7 & 1.5 \\
\text{Overshoot} & 0.2 & 0 & 0.35 \\
& \hat{\sigma}_{21} = 0.01 & \hat{\sigma}_{12} = 0.13 & \hat{\sigma}_{13} = 0.018 \\
& \hat{\sigma}_{31} = 0.013 & \hat{\sigma}_{32} = 0.13 & \hat{\sigma}_{23} = 0.039 \\
& U_{e1} = 12.6 & U_{e2} = 0.023 & U_{e3} = 2.3 \\
& V_{e1} = 60 & V_{e2} = 0.03 & V_{e3} = 2.8
\end{array} \tag{9.44}$$

where in (9.44) and the rest of the chapter, the numerical values in the first column are functionals of σ_{11}, while those in the second column are functionals of σ_{22}, and those in the third column are functionals of σ_{33}.

The three pairs of complex conjugate poles $(-6.04 \pm j71.37, -5.9 \pm j40.45, -0.86 \pm j6.07)$ cause this system to exhibit some transient oscillation. Also, the overshoots are rather large. This leads to the next formulation.

Formulation 9.5. The controller considered in this formulation is

$$K(s,p) = \begin{bmatrix} (p_1 + p_2 s^{-1})\left(\frac{s+p_{10}}{s+p_{11}}\right) & p_3 + p_4 s^{-1} & p_5 + p_6 s^{-1} \\ 0 & p_7 s^{-1} & 0 \\ p_8\left(\frac{s+p_{10}}{s+p_{11}}\right) & 0 & p_9 s^{-1} \end{bmatrix} \quad (9.45)$$

This structure gives a big improvement in the transients of σ_{11}. An acceptable set of controller parameters was found to be

$$p = (6.58, 402.4, 7.65, 77.78, 66.7, 477, 0.0155, 0.37, -7.4, 2.16, 57.1) \quad (9.46)$$

The closed-loop system with $K(s,p)$ has the following functionals:

$$\begin{array}{llll}
\text{Settling time} & 1.28 & 2.7 & 1.5 \\
\text{Overshoot} & 0.2 & 0 & 0.35 \\
& \hat{\sigma}_{21} = 0.01 & \hat{\sigma}_{12} = 0.13 & \hat{\sigma}_{13} = 0.018 \\
& \hat{\sigma}_{31} = 0.013 & \hat{\sigma}_{32} = 0.13 & \hat{\sigma}_{23} = 0.039 \\
& U_{e1} = 12.6 & U_{e2} = 0.023 & U_{e3} = 2.3 \\
& V_{e1} = 60 & V_{e2} = 0.03 & V_{e3} = 2.8
\end{array} \quad (9.47)$$

This controller succeeds in almost eliminating the overshoot in σ_{33}. However, the overshoot in σ_{11} is rather large. There is also some oscillation in σ_{33}, due mainly to the eigenvalue $-2.97 \pm j46.84$ of the matrix A. As in Formulation 9.4, this design represents a compromise between small iteration and adequate damping. In order to improve σ_{33} also, we now consider another formulation.

Formulation 9.6. The controller considered here is

$$K(s,p) = \begin{bmatrix} (p_1 + p_2 s^{-1})\left(\frac{s+p_{10}}{s+p_{11}}\right) & p_3 + p_4 s^{-1} & (p_5 + p_6 s^{-1})\left(\frac{s+p_{12}}{s+p_{13}}\right) \\ 0 & p_7 s^{-1} & 0 \\ p_8\left(\frac{s+p_{10}}{s+p_{11}}\right) & 0 & p_9 s^{-1}\left(\frac{s+p_{12}}{s+p_{13}}\right) \end{bmatrix} \quad (9.48)$$

The closed-loop system with

$$p = (7.18, 75.74, 62.96, 48.4, 276.7, 0.029, 0.45, -8.55, 8.64, 15.24, 16.45, 22.81) \quad (9.49)$$

has the following functionals:

$$\begin{array}{llll}
\text{Settling time} & 1.6 & 2.0 & 1.75 \\
\text{Overshoot} & 0.54 & 0 & 0.25 \\
& \hat{\sigma}_{21} = 0.03 & \hat{\sigma}_{12} = 0.44 & \hat{\sigma}_{13} = 0.27 \\
& \hat{\sigma}_{31} = 0.04 & \hat{\sigma}_{32} = 0.41 & \hat{\sigma}_{23} = 0.07 \\
& U_{e1} = 12.6 & U_{e2} = 0.023 & U_{e3} = 2.3 \\
& V_{e1} = 49.5 & V_{e2} = 0.03 & V_{e3} = 2.8
\end{array} \quad (9.50)$$

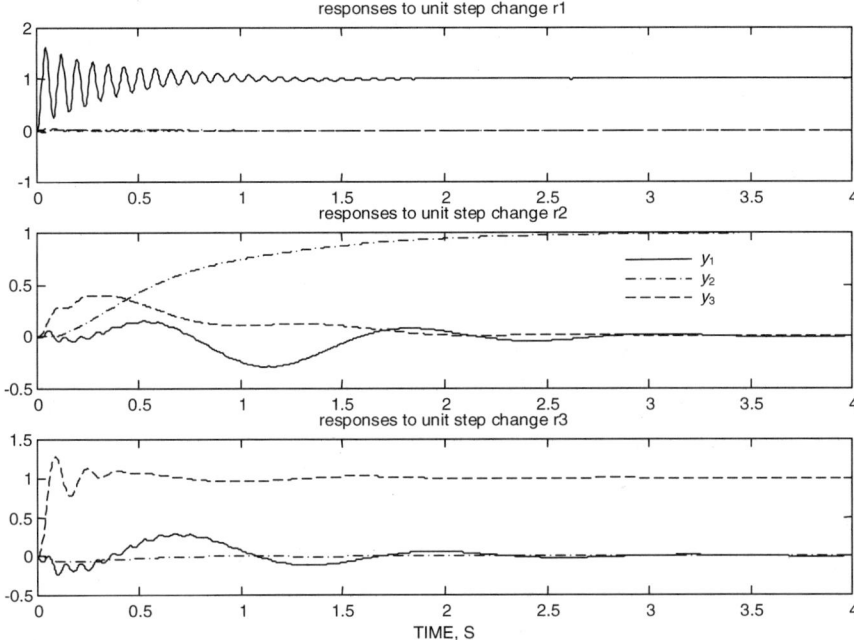

Fig. 9.10. Closed-loop responses of the system with controller parameters (9.49)

The system responses shown in Figure 9.10 clearly demonstrate that the system is lightly damped, though interaction is small. The oscillations are caused by the presence of the eigenvalues $(2.42\pm j81.73, -8.9\pm j41.35, -1.2\pm j5.09$ in the matrix A.

However, with

$$p = (-7.52, 115.3, 9.38, 59.82, 297.7, 0.028, 0.3, -6.52, 2.28, 21.53, 14.52, 40.28) \tag{9.51}$$

the closed-loop system with $K(s,p)$ is reasonably damped. Nevertheless, the interaction level has now increased. The functionals of the closed-loop system are:

$$\begin{array}{llll}
\text{Settling time} & 1.28 & 2.7 & 1.5 \\
\text{Overshoot} & 0.2 & 0 & 0.35 \\
& \hat{\sigma}_{21} = 0.02 & \hat{\sigma}_{12} = 0.78 & \hat{\sigma}_{13} = 0.4 \\
& \hat{\sigma}_{31} = 0.04 & \hat{\sigma}_{32} = 0.93 & \hat{\sigma}_{23} = 0.05 \\
& U_{e1} = 12.6 & U_{e2} = 0.023 & U_{e3} = 2.3 \\
& V_{e1} = 69 & V_{e2} = 0.03 & V_{e3} = 2.7
\end{array} \tag{9.52}$$

The closed-loop system responses are shown in Figure 9.11. The oscillations noticed in the responses are due mainly to the eigenvalues $(-13.8 \pm j73.7, -13.68 \pm j45.28, -1.56 \pm j5.81)$ in the matrix A. Comparisons with the open-loop system clearly show the merits of the various designs.

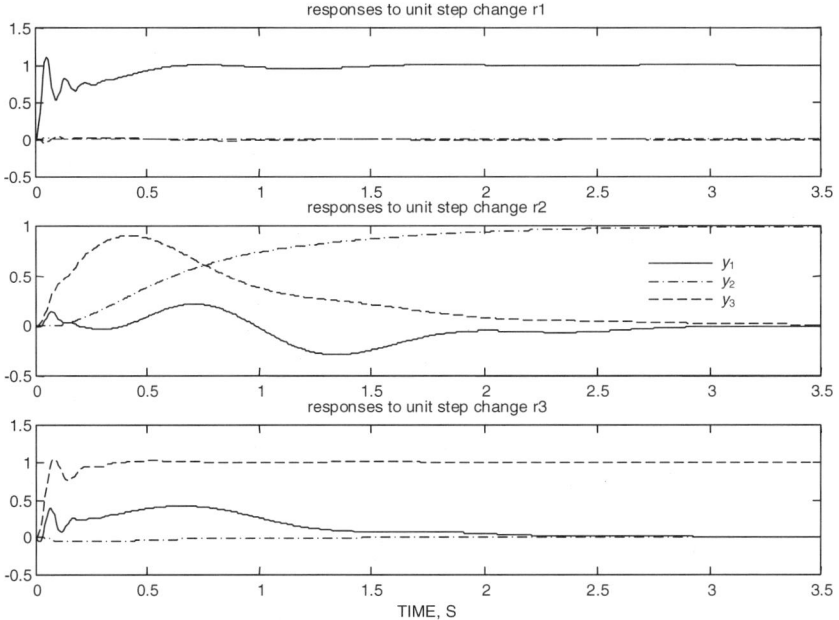

Fig. 9.11. Closed-loop responses of the system with controller parameters (9.51)

9.3.4 Conclusions

The method of inequalities is used to design controllers for the regulation of net thrust level, total air flow and inlet temperature of an F100 turbofan engine. Two models of the plant are considered, having orders 24 and 5, respectively, and simple controllers are designed for each model. As expected, the 24th order model is more difficult to control than the simplified 5th order model. In fact, it is found that the simplified model can give misleading results. Several useful features of the method are illustrated by this application. Most of the difficulty encountered in the design was a result of the sluggish response of the total engine airflow and the interaction which exists when there is a step change in the reference value of the total engine air flow. There is a conflict between little interaction and adequate damping.

9.4 Improvement of Turbo-alternator Response by the Method of Inequalities

The design of a control system for a turbo-alternator connected to an infinite bus through a transmission line has been the subject of several studies aimed at finding simple controllers which meet the usual design specifications. Ahson and Nicholson (1976) applied the Inverse Nyquist Array method (Rosenbrock 1974) to obtain a feedback controller for regulating the terminal voltage and load angle of a turbo-alternator. They employ a ninth-order model of the turbo-alternator (Davison and Rau 1971, Taiwo 1978) which takes the form

$$\dot{x}(t) = Ax(t) + Bu(t)$$
$$y(t) = Cx(t) \tag{9.53}$$

and design a system having a closed-loop transfer function matrix $H(s)$ given by

$$H(s) = [I + L(s)G(s)K(s)F(s)]^{-1}L(s)G(s)K(s) \tag{9.54}$$

where $G(s)$ is the transfer function matrix of the plant, *i.e.*,

$$G(s) = C(sI - A)^{-1}B \tag{9.55}$$

and $K(s)$, $L(s)$, $F(s)$ are the various matrices that make up the controller and called respectively the pre, post and feedback compensators. The results they obtain are expressed by:

$$K(s) = \begin{bmatrix} \frac{10s^2+1001s+100}{s^2+100s} & 0 \\ \frac{0.7s^2+70s+7}{s^2+100s} & \frac{30s+100}{s+100} \end{bmatrix} \tag{9.56}$$

$$L(s) = \begin{bmatrix} 1 & 0 \\ 0 & 0.1 \end{bmatrix}, \qquad F(s) = \begin{bmatrix} 100 & 0 \\ 0 & 0.2 \end{bmatrix} \tag{9.57}$$

Ahson and Nicholson comment upon the advantages of the Inverse Nyquist Array method, and point out that the controller they obtain is simpler than those previously obtained by means of optimal control techniques and can be readily implemented by available hardware.

The present study begins with the observation that the result of Ahson and Nicholson does not quite meet the main performance objectives with respect to steady–state and transient behaviour and stability margin. Furthermore, it is shown that by using the method of inequalities very simple controllers are found which meet the performance specifications. It is well-known that the turbo-alternator problem gives rise to severe difficulties. The results of this section throw some light on the nature of these difficulties.

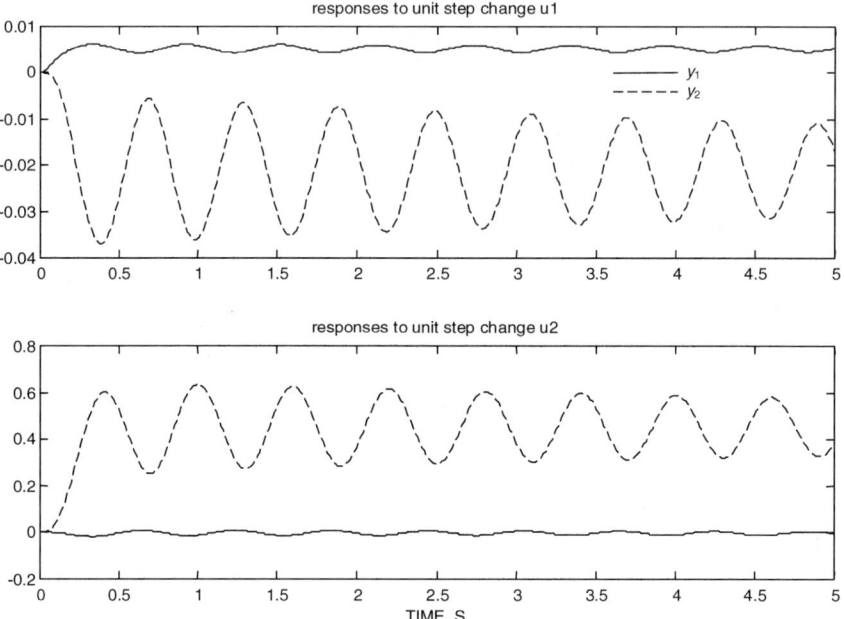

Fig. 9.12. Open-loop responses of the turbo-alternator

9.4.1 Performance

The performance required of a multivariable regulator can be summarised as follows. To a step change applied to any one reference variable, outputs respond by tending smoothly and rapidly to the new values of their respective reference variables. The above statement implies that the closed-loop transfer function matrix $H(s)$, has to satisfy a number of conditions. For correct steady-state performance, $H(s)$ has to satisfy the condition

$$H(0) = I \tag{9.58}$$

The design of Ahson and Nicholson gives

$$H(0) = \begin{bmatrix} 0.01 & 0.0 \\ 0.0021 & 0.0448 \end{bmatrix} \tag{9.59}$$

The transient output changes caused by a step input should not display lightly damped oscillations; or, equivalently, the eigenvalues of the A matrix of the minimal realisation of $H(s)$ should not be located too near the line of imaginaries. In fact the design of Ahson and Nicholson does not satisfy these requirements although, as they demonstrate, their system behaves considerably better in these respects than the uncompensated system. To verify these remarks, the reader should refer to the details in their paper.

9.4.2 Design of a Turbo-alternator Controller

Appropriate inequalities were chosen in accordance with the performance requirements prescribed by Ahson and Nicholson (1976). It may be recalled here that the abscissa of stability of a multivariable system is the largest of the real parts of the eigenvalues of the A matrix of the minimal realisation of the closed loop transfer function matrix $H(s)$.

The controller was chosen to be of the form such that $L(s) = F(s) = I$ and $K(s, p)$ is the simplest matrix having a vector p of the least dimension. The simplest controller which satisfies all the requirements is found to be

$$K(s,p) = \frac{1}{s} \begin{bmatrix} 78.6 & 67.1 \\ 3.8 & 5.1 \end{bmatrix} \qquad (9.60)$$

With this controller the transient responses of the system caused by unit step changes in the reference variables are as shown in Figure 9.13. It is clear from this that the design meets the requirements of transient and steady state performance. The abscissa of stability is -0.2.

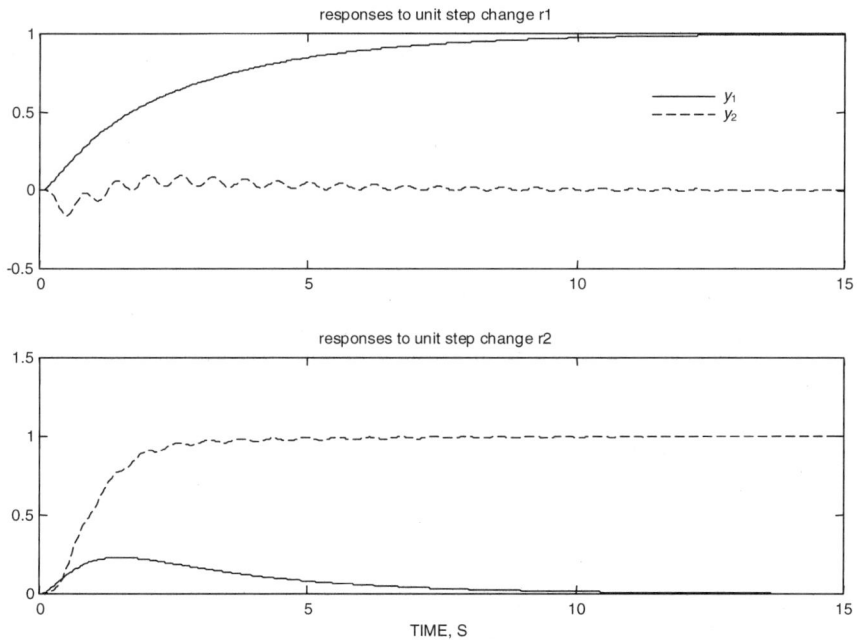

Fig. 9.13. Closed-loop responses of the turbo-alternator with controller (9.60)

For this model of the turbo-alternator, it is found that a conflict exists between the stability margin and interaction. It is also found that for acceptable interaction, increasingly more complicated controllers are required in order to improve stability margin. Controllers of the form

$$K(s,p) = \begin{bmatrix} k_{11}(s) & 0 \\ k_{21}(s) & k_{22}(s) \end{bmatrix} \quad (9.61)$$

was found to give rise to a system with very low interaction. However, this structure was found to produce a system with a relatively small stability margin for small interactions. Even with each $k_{ij}(s)$ in the form of a proportional-plus-integral controller, it was found that unacceptable interaction results for abscissas of stability as low as -0.1. The simplest controller of the form of (9.61) which satisfies the requirements, and with an abscissa of stability of -0.07 is

$$K(s) = \frac{1}{s} \begin{bmatrix} 36 & 0 \\ 1.7 & 0.47 \end{bmatrix} \quad (9.62)$$

Note that the stability margin in this case is identical with that of the design by Ahson and Nicholson (1976). The controller

$$K(s,p) = \frac{1}{s} \begin{bmatrix} 143.7(1+0.3s) & 31.7(1+0.47s) \\ 6.1(1+0.56s) & 5.5 \end{bmatrix} \quad (9.63)$$

gives rise to a system with the better abscissa of stability of -1.25. These designs clearly demonstrate the trade-offs that can take place between controller complexity (*i.e.* cost) and system performance.

9.4.3 Conclusion

This section summarises, very briefly, the work of Taiwo (1978). The method of inequalities has been used to design very simple controllers for the plant. The main difficulty found in designing controllers for the plant lies in the conflict which exists between stability margin and interaction. For an acceptable level of interaction, it is found that a greater stability margin is obtained only at the expense of increasing the complexity of the controller.

Notation

A	closed-loop system stability matrix
d	disturbance variable vector
f	system input $(r - d)$
$f^{(1)}(t)$	$(df(t))/(dt)$
G	transfer function matrix of the plant
I_ℓ	identity $\ell \times \ell$ matrix
J	$\{1, 2, \ldots, \ell\}$
K	transfer function matrix of the controller

ℓ	general designation for the number of inputs (or outputs) of a square system
DPI	diagonal proportional plus integral controller (Equation (9.28))
FPI	full PI controller (Equation (9.15))
PI7	proportional plus integral controller with seven parameters (Equation (9.24))
MBP	Moving Boundaries Process
NIC	non-interacting control
R	change in reflux flowrate (Kg/hr)
r	reference variable vector
S	change in steam flowrate (Kg/hr)
s	Laplace transform variable
T_1, T_{1ref}	upper temperature (or composition), upper temperature (or composition) set point
T_2, T_{2ref}	lower temperature (or composition), lower temperature (or composition) set point
U_i	defined in (9.8)

Greek symbols

δ_{ij}	Kronecker delta, *i.e.*, $\delta_{ij} = 1$ if $i = j$ and $\delta_{ij} = 0$ if $i \neq j$
σ_{ij}	ith component (y_i or T_i) of the closed-loop system output vector due to a unit step change in r_j or T_{jref}
$\hat{\sigma}_{ij}$	$\max_{0 \leq t < \infty} \frac{|\sigma_{ij}(t)|}{\sigma_{jj}(\infty)}, i \neq j, i,j \in J$
ϕ_{i1}	settling time of σ_{ii}, defined as the least value of t_1 such that $|\sigma_{ii}(\infty) - \sigma_{ii}(t)| \leq 0.02 |\sigma_{ii}(\infty)|, \forall\, t > t_1$
ϕ_{i2}	overshoot of σ_{ij}, given by $(\hat{\sigma}_{ii} - |\sigma_{ii}(\infty)|)/|\sigma_{ii}(\infty)|$ if $\hat{\sigma}_{ii} > |\sigma_{ii}(\infty)|$ and zero if $\hat{\sigma}_{ii} \leq |\sigma_{ii}(\infty)|$ where $\hat{\sigma}_{ii} = \max_{0 \leq t < \infty} |\sigma_{ii}(t)|$
λ_i	eigenvalue of A
ϕ_1	defined in (9.13)

References

Abbosh, F.G., (1973) Some experiences in modelling, identification and control of a packed distillation column, *Proc. 5th UKAC Control Convention*, Bath, England

Ahson, S.I. and Nicholson, H., (1976) Improvement of turbo-alternator response using the Inverse Nyquist Array method. *Int. J. Contr.*, 23:657-672.

Åström, K.J. and Hagglund, T., (1984) Automatic tuning of simple regulators with specifications on phase and amplitude margins. *Automatica*, 20(5), 645-651

Åström, K.J. and Hagglund, T., (1995) *PID Controllers: Theory, Design and Tuning*. ISA, Research Triangle Park, NC, USA.

Balachandran, R. and Chidambaram, M., (1996) Comparison of performances of decentralized controllers. *Process Contr. Qual.*, 8(4):123-131.

Balachandran, R. and Chidambaram, M., (1997) Decentralized control of crude unit distillation towers. *Comput. Chem. Eng.*, 21(8):783-786.

Bollinger, K.E., Cook, P.A. and Sandhu, P.S., (1979) Synchronous generator controller synthesis using moving boundaries search techniques with frequency domain constraints. *IEEE Trans. Power Apparatus & Syst.*, 98(5):1497-1501.

Chen, D. and Seborg, D.E., (2003) Design of decentralized PI control systems based on Nyquist stability analysis. *J. Process Contr.*, 13(1):27-39.

Coelho, C.A.D., (1979) Compensation of the speed governor of a water turbine by the method of inequalities. *ASME J. Dyn. Syst. Meas. & Control*, 101(3):205-211.

Crossley, T.R. and Dahshan, A.M., (1982) Design of a longitudinal ride-control system by Zakian's method of inequalities. *J. Aircraft*, 19(9):730–738.

Davison, E.J., (1970) The interaction of control systems in a binary distillation column. *Automatica*, 6(3):447-461.

Davison, E.J., (1967) Control of distillation column with pressure variation. *Trans. I. Chem. Engrs*, 45:229-461.

Davison, E.J. and Rau, N.S., (1971) The optimal output feedback control of a synchronous machine, *IEEE Trans. Pwr. Appar. Syst.*, 90:2123-2134.

De Hoff, R.L. and Hall, W.E. Jr., (1976) Design of a Multivariable Controller for an Advanced Turbofan. *Proc. IEEE Conf. Decision and Control*, pp 1002-1008.

Foss, A. S., (1973) Critique of chemical process control theory. *AIChE Journal*, 19:209-652.

Gray, J.O. and Al-Janabi, T.H., (1975) The numerical design of feedback control systems containing a saturation element by the method of inequalities. *Proc. Seventh IFIP Conf. Optimization Techniques*, Nice, France.

Gray, J.O and Al-Janabi, T.H., (1976) Toward the numerical design of non-linear feedback sytems by Zakian's method of inequalities. *Proc. IFAC Symp. on Large Scale Syst*, Udine.

Hackney, R.D., Miller R.J. and Small, L.L., (1977) Engine criteria and models for multivariable control system design. *Proc. Int. Forum on Alternatives for Multivariable Control*, Peezkowski, J.L., editor, NEC, Chicago, IL.

Halevi, Y., Palmor, Z.J., and Efrati, T., (1997)Automatic tuning of decentralized PID controllers for MIMO processes. *J. Process Contr.* 7(2):119-128.

Ho, W.K., Lee, T.H. and Gan, O.P., (1997) Tuning of multiloop proportional-integral-derivative controllers based on gain and phase margin specifications. *Ind. Eng. Chem. Res.*, 36(6):2231-2238.

Hougen, J.O., (1979) *Measurement and Control Applications*, 2nd Edition, Instrument Society of America, Research Triangle Park, NC.

Hovd, M. and Skogestad, S., (1994) Sequential design of decentralized controllers. *Automatica*, 33(10):1601-1607.

Hu, Y.A. and Raimirez, W.F., (1972) Application of modern control theory to distillation columns. *AIChE Journal*, 18, 479-486.

Janabi, T.H. and Gray, J.O., (1991) Method for the control of a special class of non-linear systems. *Int. J. Contr.*, 54(1):215-239.

Kreisselmeier, G. and Steinhauser, R., (1983) Application of vector performance optimization to a robust control loop design for a fighter aircraft. *Int. J. Contr.*, 37:251-284.

Lee, J.T., Cho, W.H. and Edgar, T.F., (1998) Multiloop PI controller tuning for interacting multivariable processes. *Comp. Chem. Engr.*, 22(11):1711-1724.

Loh, A.P., Hang, C.C., Queck, C.K. and Vasnani, V.U., (1993) Autotuning of multiloop proportional-integral controllers using relay feedback. *Ind. Eng. Chem. Res.*, 32(6):1102-1107.

Loh, A.P. and Vasnani, V.U., (1994) Describing function matrix for multivariable systems and its use in multiloop PI design. *J. Process Contr.*, 4(3):115-120.

Luyben, W.L., (1970) Distillation decoupling. *AIChEJ*, 16:198-203.

Luyben, W.L. and Vinante, C.D., (1972) Experimental studies of distillation decoupling. *Kem. Teollisuus*, 29(8):499-514.

Luyben, W.L. and Luyben, M.L., (1997) *Essentials of Process Control*, MacGraw-Hill, New York, pp 461-471.

MacFarlane, A.G.J. and Belletutti, J.J., (1973) The characteristic locus design method. *Automatica*, 9:575-588.

Maciejowski, J.M., (1989) *Multivariable Feedback Design*. Addison-Wesley, Workingham, U.K.

McGinnis, R.G. and Wood, R.K., (1974) Control of a binary distillation column utilizing a simple control law. *Can. J. Chem. Engng.*, 52:806-809.

Monica, T.J., Yu, C.C. and Luyben, W.L., (1988) Improved Multiloop Single-Input/Single-Output (SISO) Controllers for multivariable Processes, *Ind. Eng. Chem. Res.*, 27(6):969-973.

Munro, N. and Ibrahim, A., (1973) Computer-aided design of multi-variable sample-data systems. *Proc. IEE Conf. Computer-aided Design*, Cambridge, UK.

Nakanishi, E.,Yasuoka, H. and Kunugita, E., (1974) Decoupling feedback control of a binary distillation column, *J. Chem. Eng. Japan*, 118:1298-1301.

Niederlinski, A., (1971a) Two variable distillation control; decouple or not decouple, *AIChE Journal*, 17(5):1261-1263.

Niederlinski, A., (1971b) A heuristic approach to the design of linear multivariable interacting control systems. *Automatica*, 7(6):691-701.

Palmor, Z.J., Halevi, Y. and Krasney, N., (1995) Automatic tuning of decentralized PID controllers for TITO processes. *Automatica*, 31(7):1001-1010

Peczkowski, J.L., (1977) *Proceedings of the International Forum on alternatives for multivariable control*, NEC, Chigaco, Ill.

Pike, D.H. and Thomas, M.E., (1974) Optimal control of a continuos distillation column. *Ind. Engng. Chem. Proc. Des. Dev.*, 13:97-102.

Rademaker, O. Ridjnsdorp, J.E. and Maarleveld, A., (1975) *Dynamics and Control of Continous Distillation Units*. Elsevier, Amsterdam.

Ridjnsdorp, J.E., (1965) Interaction in two variable control systems for distillation columns. *Automatica*, 3(1):15-52.

Ridjnsdorp, J.E., and Seborg, D.E., (1976) A survey of experimental applications of multivariable control to process control problems. *AIChE Journal, Symp. Ser.*, 72:112-123.

Rosenbrock, H.H., (1960) An automatic method for finding the greatest and the least value of a function, *Comp. J.*, 3:175-184.

Rosenbrock, H.H., (1974) *Computer-aided Control System Design*. Academic Press, New York.

Rosenbrock, H.H., (1962) The control of distillation columns, *Trans. I. Chem. Engrs*, 40:35-53.

Sain, M.K., Peczkowski, J.L. and Melsa, J.L., ed, (1978) *Alternatives for Multivariable Control*. Chicago, National Engineering Consortium.

Schwanke, C.O., Edgar, T.F. and Hougen, J.O., (1977) Development of multivariable control strategy for distillation columns. *Trans. ISA*, 16:69-81.

Shinskey, F.G., (1977) *Distillation Control*. McGraw-Hill, New York.

Skira, C. A. and De Hoff, R.L. (1977) A practical approach to linear model analysis for multivariable turbine engine control design. *Proc. Int. Forum on Alternatives for Multivariable Control*, Peezkowski, J.L., editor, NEC, Chicago, IL. pp 29-44.

Skogestad, S. and Postlethwaite, I., (1996) *Multivariable Feedback Control: Analysis and Design*. Wiley, Chichester, England, pp. 432-448.

Srinivasa Prabhu, E. and Chidambaram, M., (1991) Robust control of a distillation column by the method of inequalities. *J. Process Contr.*, 1(3):171-176.

Stainthorp, F.P., (1970) The computer controlled fractionating columns at UMIST. *Brit. Chem. Eng.*, 15:794-796.

Taiwo, O., (1978a) Improvement of turbo-alternator response by the method of inequalities. *Int. J. Contr.*, 2792):305-311.

Taiwo, O., (1978b) Design of multivariable controller for a high-order turbofan engine model by Zakian's method of inequalities. *IEEE Trans. Autom. Control*, 23(5):926-928.

Taiwo, O., (1979a) On the feedback control of continuos stirred tank reactor by Zakian's method of inequalities. *Chem. Engng. J.*, 17:3-12.

Taiwo, O., (1979b) Design of multivariable controllers for an advanced turbofan engine by Zakian's method of inequalities. *ASME J. Dyn. Syst. Meas. & Control*, 101:299-307.

Taiwo, O., (1980) Application of the method of inequalities to the multivariable control of binary distillation columns. *Chem Eng Sci.*, 35(2):847-858.

Taiwo, O. and Krebs, V., (1993) Generalized moment method for the order reduction of multivariable systems. *J. Frank. Inst.*, 330:641-649.

Taiwo, O., (1995) Multivariable system simplification using moment matching and optimization. *IEE Proc. Contr. Theor. Appl.*, 142:103-110.

Tan, K.K., Wang, Q.-G., Hang, C.C. and Hagglund, T.J., (1999) *Advances in PID Control*. Springer, London.

Waller (Toijala) K.V.T., (1974) Decoupling in Distillation. *AIChE Journal*, 20:592-594.

Waller (Toijala) K.V.T. and Fagervik, K., (1972) A Digital simulation study of the two point feedback control of distillation columns. *Kem. Teollisuus*, 29:5-16.

Wood, R.K. and Berry, M.W., (1973) Terminal composition control of a binary distillation column. *Chem. Engng. Sci.*, 28:1707-1717.

Yu, C.C., (1999) *Autotuning of PID Controllers*. Springer, London.

Zakian, V. and Edwards, M.J., (1978) Tabulation of constraint for full grade I_{mn} approximants. *Math. Comp.*, 32, 519-531.

Zakian, V. and Al-Naib, U., (1973) Design of dynamical and control systems by the method of inequalities. *Proc. IEE*, 120(11):1421-1427.

Zakian, V., (1975) Properties of I_{MN} and J_{MN} approximants and applications to numerical inversion of Laplace transforms and initial-value problems, *J. Math. Anal. Appl.*, 50:191-222.

Zakian, V., (1978) The performance and sensitivity of classical control systems, *Int. J. Syst. Sci.*, 9:343-355.

Zakian, V., (1979) New formulation for the method of inequalities. *Proc. IEE*, 126(6):579-584.

Zhang, P. and Coonick, A.H., (2000) Coordinated Synthesis of PSS parameters in multi-machine power systems using the method of inequalities applied to genetic algorithms. *IEEE Trans. Power Systems*, 15(2):811-816.
811-816

10 Multi-objective Control using the Principle of Inequalities

G P Liu

Abstract. This chapter is concerned with three different multi-objective control schemes. These are, respectively, PID control, critical control and eigenstructure assignment. In each case the design problem is formulated in accordance with the principle of inequalities. This is a multi-objective design rule where each design specification is expressed as an inequality. In the PID control problem, a set of frequency-domain performance requirements, such as gain margin, phase margin, crossover frequency and steady-state error are used to construct inequalities. In the critical control problem, the issue of robustness of multivariable critical systems with external and internal uncertainties is addressed. The design inequalities are based on the output performance in the time domain and the robust performance in the frequency domain. The eigenstructure assignment problem considers robustness in multivariable control systems. The performance functions are individual eigenvalue sensitivity functions and the system robustness functions. Based on those performance functions above, the performance criteria for each multi-objective scheme are expressed by a set of inequalities. Some examples demonstrate the operation of the three multi-objective control schemes.

10.1 Introduction

Most control design techniques have only paid attention to optimal solutions for one special performance index, $e.g.$, $\mathcal{H}^\infty$-norm on a closed-loop system transfer function, the eigenvalue sensitivity function or the linear quadratic index. However, many practical control systems are required to have the ability to satisfy simultaneously different and often conflicting performance objectives as best as possible, for instance, closed-loop stability, low feedback gains and insensitivity to model parameter variations. The usual approach to dealing with several design objectives makes use of scalar summation of all weighted objectives in one cost function. Although this method simplifies the approach to optimisation, it is not clear how the weights should be chosen in the summation. This situation was recognised more than thirty years ago (Zakian and Al-Naib, 1973) when it was proposed that each separate design criterion should be expressed as an inequality. This rule of design is now called the principle of inequalities (see Zakian, Chapter 1).

During the last two decades, multi-objective methods have increasingly been considered in the design process (Liu et al, 2003). The control system objectives are described by a set of performance indices which are used to form corresponding inequalities. This type of control problem appears in flight control design, in the control of space structures and in industrial process control. The concept of a multi-objective control problem that involves different types of performance indices is very interesting. For example, the multiple objectives can be considered as different types of norms on transfer functions. In this case, one may take $\mathcal{H}^2$-norms on some of the closed-loop transfer functions, $\mathcal{H}^\infty$-norms on others and $\mathcal{L}^1$-norms on some others. Such a multiple objective problem would naturally arise in considering trade-offs between performance and robust stability. Several multi-objective methods based on the principle of inequalities have been studied. For example, linear quadratic Gaussian control (Tabak et al, 1979; Khargonekar and Rotea, 1991), robust control (Whidborne et al, 1995), identification (Liu, 2001; Liu and Kadirkamanathan, 1999) and fault detection (Chen et al,1996). This chapter discusses three different types of multi-objective schemes. They are, respectively, PID control, critical control and eigenstructure assignment. In each case the design method is based on the principle of inequalities.

10.2 Multi-objective Optimal-tuning PID Control

The proportional, integral and derivative (PID) control algorithm remains the most popular approach for industrial process control despite continual advances in control theory (Liu and Daley, 2000, 2001). This is not only due to the simple structure which is conceptually easy to understand and makes manual tuning possible, but also to the fact that the algorithm provides adequate performance in the vast majority of applications. However, for a variety of reasons optimal setting of PID controller parameters is difficult and as a result many PID design techniques have been developed.

10.2.1 Rule-based PID Controller Design

The design of feedback control systems in industry using frequency-response methods is more popular than any other. This is primarily because the frequency response method provides good designs in the face of uncertainty in the plant model and can easily use experimental information for design purposes. Here, it is assumed that the ideal transfer function of a PID controller is given by

$$K(s) = K_c \left(1 + \frac{1}{T_i s} + T_d s\right) \tag{10.1}$$

where K_c, T_i and T_d are the PID parameters.

A large number of industrial processes can be characterised by a first-order plant with dead time (FOPDT). The transfer function of a FOPDT model is described by

$$G(s) = \frac{K_p e^{-s\tau}}{1+sT} \qquad (10.2)$$

where K_p is the gain, T the time constant and τ the dead time. Based on the FOPDT model, there are a number of PID tuning formulae available. Six PID tuning rules are introduced and assessed in later sections. These tuning rules are Ziegler-Nichols (Ziegler and Nichols, 1942), integral of absolute error (Pessen, 1994), some-overshoot rule (Seborg et al, 1989), no-overshoot rule (Seborg et al, 1989), integral of squared time weighted error (Zhuang and Atherton, 1993), symmetric optimum rule (Kessler, 1958; Voda and Landou, 1995). A summary of the six PID tuning rules is given in Table 10.2.1, where K_u and T_u are the inverse of the system gain and frequency at which the phase is $-180°$. It can be seen from the table that the six PID design methods do not require a parametric transfer function model of the process and only need either one or two frequency response measurements of the process.

Although the traditional PID design methods give simple tuning rules for the controller parameters using either one or two measurement points of the system frequency-response, their control performance may not satisfy the desired requirements. To overcome this disadvantage, a multi-objective optimal PID controller design is needed (Liu et al, 2000).

10.2.2 Multi-objective PID Control Design

In the frequency domain, there are two quantities used to measure the stability margin of the system. One is the gain margin, which is the factor by which the gain is less than the neutral stability value. The other is the phase margin, which is the amount by which the phase of the system exceeds when the system gain is unity. The gain and phase margins are also related to the damping of a system. In addition to the stability of a design, the system is also expected to meet a speed-of-response specification like bandwidth. The crossover frequency, which is the frequency at which the gain is unity, would be a good measurement in the frequency domain for the system's speed of time response. Also, the larger the value of the magnitude on the low-frequency asymptote, the lower the steady-state errors will be for the closed-loop system. This relationship is very useful in the design of suitable compensation. Thus, the following multi-objective performance functions need to be considered during the design of a PID controller (Liu

Table 10.1. Six PID tuning rules

PID Rule	PID Parameters
Ziegler-Nichols (NZ)	$K_c = 0.6K_u$
	$T_i = 0.50T_u$
	$T_d = 0.125T_u$
Integral of absolute error (IAE)	$K_c = 0.70K_u$
	$T_i = 0.4T_u$
	$T_d = 0.150T_u$
Some-overshoot rule (SOR)	$K_c = 0.33K_u$
	$T_i = 0.5T_u$
	$T_d = 0.330T_u$
No-overshoot rule (NOR)	$K_c = 0.20K_u$
	$T_i = 0.5T_u$
	$T_d = 0.330T_u$
Integral of squared time weighted error (ISTWE)	$K_c = 0.509K_u$
	$T_i = 0.05(3.302K_pK_u + 1)T_u$
	$T_d = 0.125T_u$
Symmetric optimum rule (SO)	$K_c = \frac{4+B}{128^{0.5}} \frac{B}{K_{135}}$
	$T_i = \frac{4+B}{B\omega_{135}}$
	$T_d = \frac{B}{(4+B)\omega_{135}}$

and Daley, 1999):

$$\phi_1(K_c, T_i, T_d) = \left\{ \frac{|K(j\omega)G(j\omega)|}{G_M}, \angle K(j\omega)G(j\omega) = -180° \right\} \quad (10.3)$$

$$\phi_2(K_c, T_i, T_d) = \left\{ 2 - \frac{180 + \angle K(j\omega)G(j\omega)}{P_M}, |K(j\omega)G(j\omega)| = 1 \right\} \quad (10.4)$$

$$\phi_3(K_c, T_i, T_d) = \left\{ \frac{\omega}{2\pi f_d}, |K(j\omega)G(j\omega)| = 1 \right\} \quad (10.5)$$

$$\phi_4(K_c, T_i, T_d) = \left\{ \frac{1}{e_{ss}|1 + K(j\omega)G(j\omega)|}, \omega = 0 \right\} \quad (10.6)$$

where $\phi_i(K_c, T_i, T_d)$, for $i = 1, 2, \ldots, 4$ are the normalised gain margin, phase margin, crossover frequency and the steady-state error functions with the desired values G_M, P_M, f_d and e_{ss}, respectively. Thus, the optimal PID controller may be obtained by satisfying the following multi-objective performance criteria.

$$\phi_i(K_c, T_i, T_d) \leq 1, \text{ for } i = 1, 2, 3, 4 \tag{10.7}$$

If the above inequalities are met, then the problem is solved. Clearly, the design problem is to find a PID controller to make (10.7) hold. There are a number of methods to solve the performance criteria problem (10.7). Here, the minimax optimisation method is briefly introduced.

Using the minimax optimisation method, the multi-objective performance criteria (10.7) can be satisfied if

$$\min_{K_c, T_i, T_d} \max_{i=1,2,3,4} \{\phi_i(K_c, T_i, T_d)\} \leq 1 \tag{10.8}$$

Clearly, the above minimises the worst case values of the performance functions. Similar algorithms for solving the problem defined in (10.7) now exist in standard libraries of optimisation software, for example, the Optimisation Toolbox for use with MATLAB (Grace, 1998).

10.2.3 Optimal-tuning PID Control Scheme

When a system has different operating points with widely differing dynamic properties, it is not always possible to exercise control with a fixed parameter controller, even if this is a highly robust controller. For this case, the optimal-tuning PID control scheme shown in Figure 10.1 is proposed. It mainly consists of four parts: frequency response estimation, desired system specifications, optimal-tuning mechanism and PID controller. The frequency response estimated using frequency-domain identification methods provides a non-parametric model for the process. The desired system specifications includes a set of requirements in the frequency domain: gain margin, phase margin, crossover frequency and steady-state error. The optimal-tuning mechanism uses the process frequency-response to find optimal parameters for the PID controller so that the desired system specifications are satisfied.

The operating procedure of the optimal-tuning PID control is as follows. When the system's operating-point or dynamics changes, the new process frequency response is re-estimated by switching on the excitation signal. Then, using this updated frequency-response, the tuning mechanism searches for the optimal parameters for the PID controller to satisfy the desired system specifications. Finally, the PID controller is set to the obtained optimal parameters. In this way, the PID controller may cope with all operating-points of the system and the closed-loop system will have similar optimal control performance.

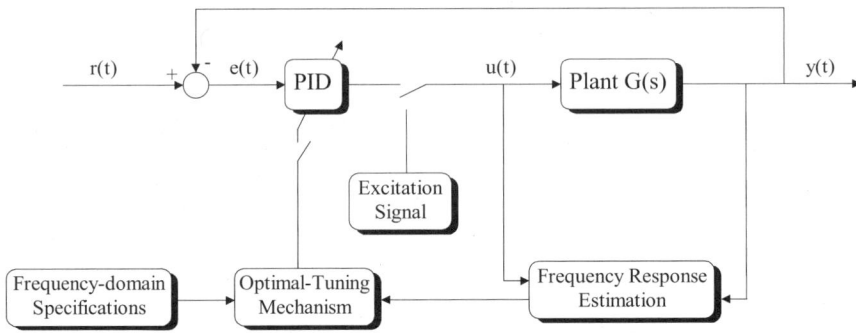

Fig. 10.1. Optimal-tuning PID control structure

10.2.4 Application to a Rotary Hydraulic System

The optimal-tuning PID control scheme is applied to a rotary hydraulic test rig, which is representative of many industrial systems that utilise fluid power. This is a particularly apposite application of the method since hydraulic systems are often very conservatively tuned, due to the fact that the cost of getting the tuning wrong can be highly destructive and costly. To effectively assess the performance of the proposed tuning method the other six tuning rules which were introduced in Section 10.2.1 are also applied to the rig.

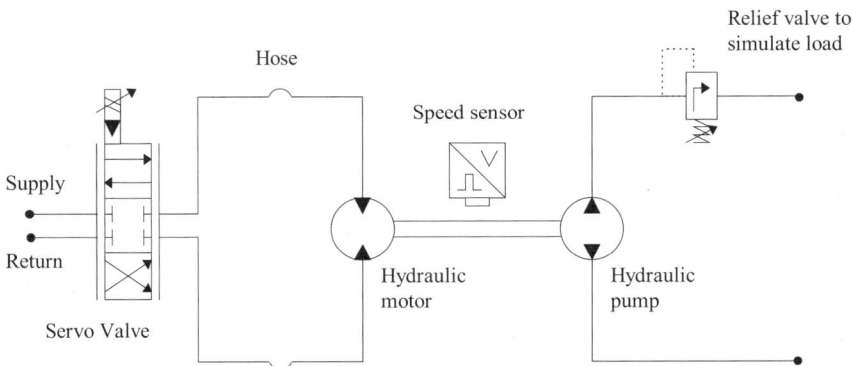

Fig. 10.2. Schematic of hydraulic rig

The rotary hydraulic test rig, as shown in Figure 10.2, comprises an electro-hydraulic servo control valve driving a fixed displacement hydraulic motor up to 8000 rpm with a maximum operating pressure of 21 MPa. The motor is coupled by a rigid shaft to a hydraulic pump of the same displacement as the motor and a solenoid controlled relief valve is used to simulate

10 Multi-objective Control using the Principle of Inequalities

variations in load. This type of hydraulic system is typically applied to mixer drives, centrifuge drives and machine tool drives where accurate speed control with fast response time is required and large changes in load can be expected.

For the sake of simplicity, a periodic multi-sine excitation signal was directly applied to the hydraulic system in an open-loop way. Based on the input-output data, the frequency-response of the system was estimated using frequency-domain identification methods.

Following the six PID tuning rules given in Section 10.2.1 and the multi-objective optimal PID design rule, the speed responses of the hydraulic motor using seven PID controllers are shown in Figures 10.3 and 10.4. From the comparison of seven PID control rules, the experimental results have shown that the multi-objective optimal-tuning PID controller has much better system control performance.

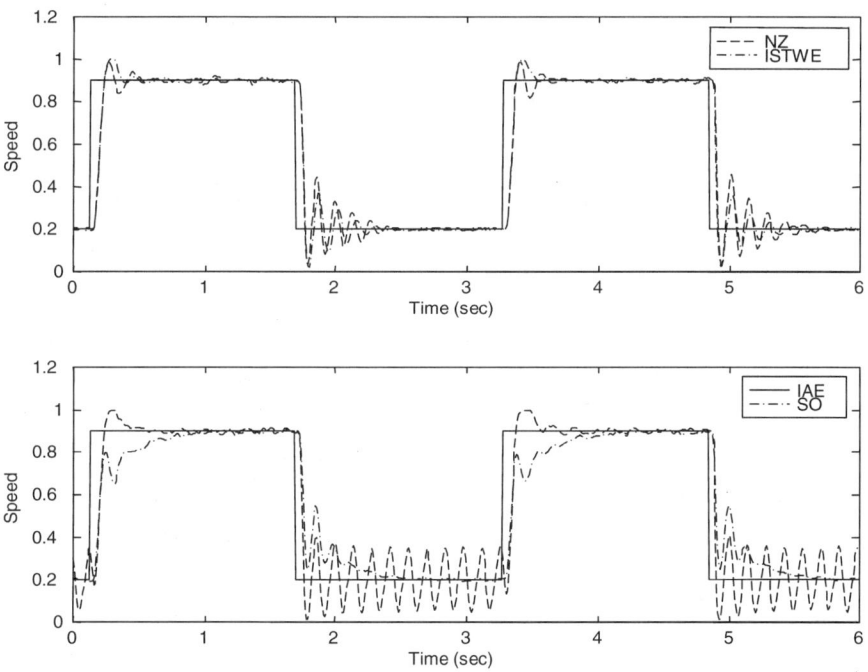

Fig. 10.3. The speed of the hydraulic motor using NZ, ISTWE, IAE and SO controllers

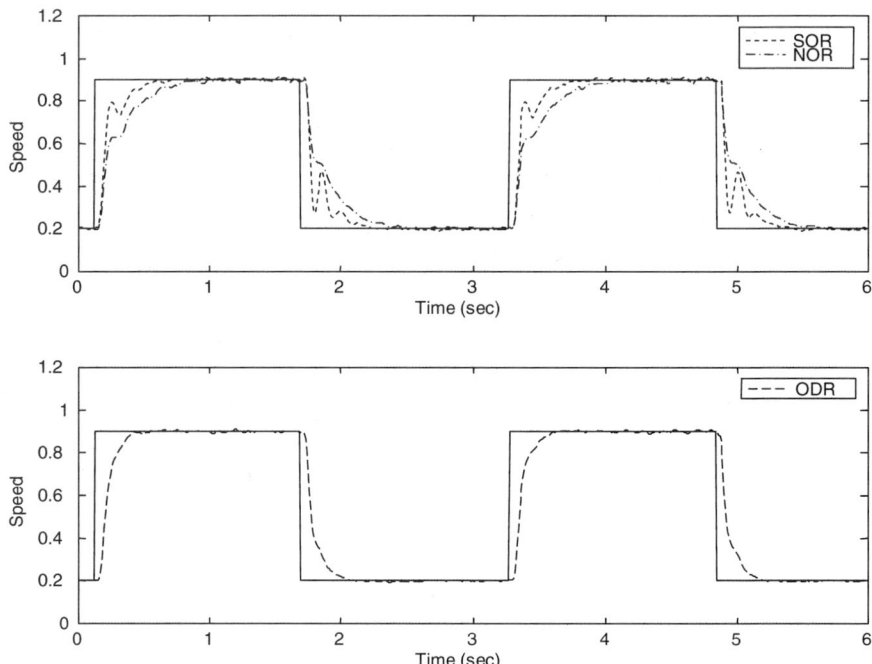

Fig. 10.4. The speed of the hydraulic motor using SOR, NOR and ODR controllers

10.3 Multi-objective Robust Eigenstructure Assignment

In the 1960s Wonham presented the fundamental result on eigenvalue assignment in linear time-invariant multivariable controllable systems (Wonham, 1967). This states that the closed-loop eigenvalues of any controllable system may be arbitrarily assigned by state feedback control. Later, Moore found that degrees of freedom are available over and above eigenvalue assignment using state feedback control for linear time-invariant multi-input multi-output (MIMO) systems (Moore, 1976). Since then, numerous methods and algorithms involving both state and output feedback control have been developed to exercise those degrees of freedom to give the systems some good performance characteristics (Kautsky *et al*, 1985; Liu and Patton, 1998, 1999).

Most eigenstructure assignment techniques have only paid attention to optimal solutions for one special performance index, *e.g.*, the eigenvalue sensitivity function or the linear quadratic index. However, many practical control systems are required to have the ability to fit simultaneously different and often conflicting performance objectives as best as possible. This chapter is concerned with multi-objective robust eigenstructure assignment. The multi-objective performance indices include the individual eigenvalue sensitivities and the sensitivity functions in the frequency domain. The performance criteria are expressed by a set of inequalities on the basis of the multi-objective performance indices. In order to obtain an approximate global optimisation for the multi-objective control system design, a numerical algorithm is outlined by combining eigenstructure assignment, the principle of inequalities and genetic algorithms.

10.3.1 Eigenstructure Assignment

Eigenstructure assignment is a design technique which may be used to assign the entire eigenstructure (eigenvalues and right or left eigenvectors) of a closed-loop linear system via a constant gain full state or output feedback control law.

Consider the following linear time-invariant system

$$\delta x = Ax + Bu \tag{10.9}$$
$$y = Cx \tag{10.10}$$

where $x \in \mathbb{R}^{n \times 1}$ is the state vector, δx represents $\dot{x}(t)$ for continuous systems and $x(t+1)$ for discrete systems, $u \in \mathbb{R}^{r \times 1}$ is the control input vector, $y \in \mathbb{R}^{m \times 1}$ is the output vector, $A \in \mathbb{R}^{n \times n}$, $B \in \mathbb{R}^{n \times r}$ and $C \in \mathbb{R}^{m \times n}$.

A linear output feedback control law applied to the system above is

$$u = Ky \tag{10.11}$$

where the matrix $K \in \mathbb{R}^{r \times m}$ is the output feedback controller gain.

It is well known that $\max\{r,m\}$ eigenvalues are assignable arbitrarily by output feedback. This restricts the choice of eigenvalue and eigenvector pairs. However, if $m+r > n$, the whole spectrum can be assigned, with some restrictions on the eigenvector selection (Kimura, 1975). This also implies that an eigenvector associated with an eigenvalue can be assigned in either the right eigenvector set or the left eigenvector set but not both. In addition, whatever the selection of eigenvectors is, it may be very difficult to assign some spectrum if the system to be designed suffers from pathological structures. For this case or $m+r \leq n$, the dynamical compensator will be helpful (Liu and Patton, 1998). Thus, without loss of generality, the following assumptions are made for system (10.9) and (10.10): rank$(B) = r$, rank$(C) = m$ and $m+r > n$. Then the closed-loop system representation is given by

$$\delta x = (A + BKC)x \tag{10.12}$$

Now, define a closed-loop self-conjugate eigenvalue set $\Lambda = \{\lambda_i : \lambda_i \in \mathbb{C}, i = 1, 2, \ldots, n\}$, i.e., the set of the eigenvalues of the closed-loop matrix $A+BKC$, where n is the number of distinct eigenvalues.

According to the definition of right eigenvectors, then

$$(\lambda_i I - A - BKC)R_i = 0 \tag{10.13}$$

for $i = 1, 2, \ldots, n$. Thus, the right eigenvector matrix is given by

$$R = [R_1, R_2, \ldots, R_n] \in \mathbb{C}^{n \times n} \tag{10.14}$$

where R_i is the right eigenvector associated with the eigenvalue λ_i.

Similarly, the left eigenvectors are defined by

$$L_i^T(\lambda_i I - A - BKC) = 0 \tag{10.15}$$

for $i = 1, 2, \ldots, n$. Then, the left eigenvector matrix is given by

$$L = [L_1, L_2, \ldots, L_n] \in \mathbb{C}^{n \times n} \tag{10.16}$$

where the matrix L_i is the left eigenvector associated with the eigenvalue λ_i.

Therefore, the problem of eigenstructure assignment via output feedback for system (10.9) – (10.11) can be stated as follows: Given the closed-loop eigenvalue set Λ, find a real controller K such that the eigenvalues of the matrix $A + BKC$ are in the set Λ by choosing both the right and left eigenvector matrices R and L properly.

Consider the eigenstructure of the closed-loop system corresponding to eigenvalues described by (10.13) and (10.15), which may be rewritten as

$$(A + BKC)R_i = \lambda_i R_i \tag{10.17}$$
$$L_i^T(A + BKC) = \lambda_i L_i^T \tag{10.18}$$

For the sake of simplicity, it is assumed that $|\lambda_i I - A| \neq 0$. Rearranging the above equations results in

$$R_i = (\lambda_i I - A)^{-1} B W_i \tag{10.19}$$
$$L_i = (\lambda_i I - A^T)^{-1} C^T V_i \tag{10.20}$$

where

$$W_i = KCR_i \tag{10.21}$$
$$V_i = K^T B^T L_i \tag{10.22}$$

are called the right and left parameter vectors, respectively.

For output-feedback eigenstructure assignment, the controller K is given by either (Kwon and Youn, 1987)

$$K = W(CR)^T(CR(CR)^T)^{-1} \tag{10.23}$$

or

$$K^T = V(B^T L)^T(B^T L(B^T L)^T)^{-1} \tag{10.24}$$

if and only if

$$R^T L = I \tag{10.25}$$

If the matrix $C = I$, this is the state-feedback eigenstructure assignment. The controller K can be calculated by the right eigenvector matrix R and the right parameter-vector matrix W, that is,

$$K = WR^{-1} \tag{10.26}$$

The basic controller forms introduced in this section are very useful during the design of systems using eigenstructure assignment.

10.3.2 Multi-objective Eigenstructure Assignment

In most parameter insensitive design methods using eigenstructure assignment the performance indices are given on the basis of the right and left eigenvector matrices. For example, a very common performance index is the overall eigenvalue sensitivity:

$$\phi(R) = \parallel R \parallel_2 \parallel L \parallel_2 \tag{10.27}$$

which gives an overall measure of conditioning of the eigenproblem.

It has been shown that the individual eigenvalue sensitivity ϕ_i of a matrix $A + BKC$ to perturbations for the i-th eigenvalues λ_i is

$$\phi_i(R, L) = \frac{\parallel L_i \parallel_2 \parallel R_i \parallel_2}{|L_i^T R_i|} \tag{10.28}$$

where L_i and R_i are the i-th left and right eigenvectors of $A + BKC$, respectively.

Though it has been shown that a minimisation of $\phi(R)$ reduces a bound on the individual eigenvalue sensitivities and the actual values of $\phi_i(R, L)$ themselves will become small so that the conditioning of the eigenproblem is improved, it is often conservative because the $\phi(R)$ measures the upper bound of all individual eigenvalue sensitivities, *i.e.*,

$$\phi(R) \geq \max_{i=1,2,\ldots,n} \phi_i(R, L) \tag{10.29}$$

Hence, in order to reduce the conservatism the following formulation is considered:

$$\phi_i(R) \leq \varepsilon_i, \text{ for } i = 1, 2, \ldots, n \tag{10.30}$$

where ε_i is a real positive number.

There are also a number of robust performance indices which are considered in the optimisation design of control systems in the frequency domain. First, define the following norms:

$$\| H \|_\infty = \sup\{|H(j\omega)|: \quad \omega \in \mathbb{R}\} \tag{10.31}$$

$$\| H \|_{\infty,\mathcal{F}} = \sup\{|H(j\omega)|: \quad \omega \in \mathcal{F}\} \tag{10.32}$$

where H is a transfer function and $\mathcal{F}$ denotes a frequency range.

In robust control using $\mathcal{H}^\infty$ for state feedback systems, the objectives are expressed in terms of the $\mathcal{H}^\infty$-norm of transfer functions. One of the objectives of the system with $C = I$ is the following:

$$\min_K \left\| \begin{matrix} S \\ C_S \end{matrix} \right\|_\infty \tag{10.33}$$

where

$$S = (I - (sI - A)^{-1} BK)^{-1} \tag{10.34}$$

$$C_S = K((sI - A)^{-1} BK - I)^{-1} \tag{10.35}$$

S is the sensitivity function, C_S is the complementary sensitivity function, and the minimisation is over the whole set of stabilising controllers K.

As the singular value techniques are used to evaluate control system stability and robustness characteristics it becomes apparent that a singular value matrix norm is often conservative in its ability to predict near instability. Though the robust performance formulation (10.33) is widely used in $\mathcal{H}^\infty$ control, it is often conservative because of the following relations:

$$\max\{\| S \|_\infty, \| C_S \|_\infty\} \le \left\| \begin{matrix} S \\ C_S \end{matrix} \right\|_\infty \tag{10.36}$$

This means that the maximum singular value of the matrix $[S^T, (C_S)^T]^T$ gives a measurement of the upper bound of the maximum singular values of the sensitivity function S and the function C_S. Thus, in order to reduce the conservatism above, the following robust control formulation is considered:

$$\| S \|_\infty \le \varepsilon_{n+1} \tag{10.37}$$

$$\| C_S \|_\infty \le \varepsilon_{n+2} \tag{10.38}$$

where ε_{n+1} and ε_{n+2} are positive real numbers.

In practice, it requires that disturbance rejection at low frequencies and enhancement in robust stability to safeguard against modelling errors at high frequencies. Let the low frequency range and the high frequency range be denoted by $\mathcal{F}_L$ and $\mathcal{F}_H$, respectively. Then the following robust control formulation is obtained:

$$\| S \|_{\infty,\mathcal{F}_L} \le \varepsilon_1 \tag{10.39}$$

$$\| C_S \|_{\infty,\mathcal{F}_H} \le \varepsilon_2 \tag{10.40}$$

Thus, the multi-objective performance functions for multivariable control systems may be

$$\phi_i(K) = \phi_i(R, L), \quad i = 1, 2, \ldots, n \tag{10.41}$$
$$\phi_{n+1}(K) = \|S\|_{\infty, \mathcal{F}_L} \tag{10.42}$$
$$\phi_{n+2}(K) = \|C_S\|_{\infty, \mathcal{F}_H} \tag{10.43}$$

In practice, it is usually intended to locate the eigenvalue vector λ in a well-defined set to meet the requirements of the practical control system (*e.g.*, stability, speed of response, *etc.*). This leads to eigenvalue constraints, for example of the form $\underline{\lambda}_i \leq \lambda_i \leq \overline{\lambda}_i$, where $\underline{\lambda}_i \in \mathbb{R}$ and $\overline{\lambda}_i \in \mathbb{R}$ are the lower and the upper bound vectors, respectively. These constraints may be removed by considering the change of variables given by

$$\lambda_i(z_i) = \underline{\lambda}_i + (\overline{\lambda}_i - \underline{\lambda}_i)\sin^2(z_i) \tag{10.44}$$

with $z_i \in \mathbb{R}$.

It has been shown that the right and left eigenvector matrices are determined by closed-loop eigenvalues and two parameter matrices (V, W). Clearly, the controller matrix K is a function of $Z = [z_1, z_2, \ldots, z_n]$, V and W. Thus, the performance functions $\phi_i(K)$ ($i = 1, 2, \ldots, n+2$) can be described by $\phi_i(Z, V, W)$.

If one of the performance functions $\phi_i(Z, V, W)$ ($i = 1, 2, \ldots, n+2$) is minimised individually (single-objective approach), then unacceptably large values may result for other performance functions $\phi_j(Z, V, W)$ ($j \neq i, j = 1, 2, \ldots, n+2$). Generally, there does not exist a solution for all performance functions $\phi_i(Z, V, W)$, for $i = 1, 2, \ldots, n+2$ to be minimised by the same controller K.

Therefore the optimisation can be reformulated into a multi-objective problem as

$$\begin{cases} \min_{Z,V,W} \sum_{i=1}^{n+2} \phi_i(Z, V, W)/\varepsilon_i \\ \phi_i(Z, V, W) \leq \varepsilon_i, \text{ for } i = 1, 2, \ldots, n+2 \end{cases} \tag{10.45}$$

where the positive real number ε_i represents the numerical bound on the performance function $\phi_i(Z, V, W)$ and is determined by the designer.

10.3.3 Controller Design using Genetic Algorithms

As several types of objectives (or cost functions) are considered for control systems, this section introduces a numerical algorithm for the multi-objective optimisation control problem (10.45) by combining eigenstructure assignment, the principle of inequalities and genetic algorithms (Liu and Patton, 1996), based on the formulation in the previous section. The steps to be executed for the GA implementation are as follows:

Step 1 Each solution in the population is represented as a real number string. As the eigenvalues $Z \in \mathbb{R}^{1 \times n}$, the right and left parameter matrices $V = [V_1, V_2, \ldots, V_n]$ and $W = [W_1, W_2, \ldots, W_n]$ then the chromosomal representation may be expressed as an array

$$P = [Z, V_1^T, V_2^T, \ldots, V_n^T, W_1^T, W_2^T, \ldots, W_n^T] \tag{10.46}$$

Step 2 The N (an odd number) sets of parameter strings P for the initial population are randomly generated.

Step 3 Evaluate the performance functions $\phi_i(P_j)$ $(i = 1, 2, \ldots, n+2)$ for all N sets of the parameters P_j and

$$\Delta_j = \max_{i=1,2,\ldots,n+2} \phi_i(P_j)/\varepsilon_i \tag{10.47}$$

$$\Phi_j = \sum_{i=1}^{n+3} \phi_i(P_j) \tag{10.48}$$

for $j = 1, 2, \ldots, N$.

Step 4 According to the fitness of the performance functions for each set of parameters, remove the $(N-1)/2$ weaker members of the population and reorder the sets of parameters. The fitness of the performance functions is measured by

$$F_j = \Delta_j^{-1}, \text{ for } j = 1, 2, \ldots, N \tag{10.49}$$

Step 5 Perform the crossover using an average crossover function to produce the $(N-1)/2$ offspring. The average crossover operator takes two parents which are selected in S4 and produces one child that is the result of averaging the corresponding fields of two parents. Thus, the average crossover function is given by

$$P_{Cj} = (P_{j+1} + P_j)/2, \text{ for } j = 1, 2, \ldots, (N-1)/2 \tag{10.50}$$

Step 6 A real number mutation operator is used. The maximum amount that this operator can alter the value of a field is a parameter of the operator. The mutation operation is defined as

$$P_{Mj} = P_{Cj} + d_m \xi_j, \text{ for } j = 1, 2, \ldots, (N-1)/2 \tag{10.51}$$

where d_m is the maximum to be altered and $\xi_j \in [-1, 1]$ is a random variable with zero mean.

Step 7 To prevent the best parameter set from loss in the succeeding parameter sets, the elitist strategy is used to copy the best parameter set into the succeeding parameter sets. The best parameter set P_b is defined as one satisfying

$$\Phi_b = \min_l \{\Phi_l : \Phi_l \leq \Phi_m - \alpha(\Delta_l - \Delta_m) \text{ and } \Delta_l \leq \Delta_m + \delta\} \tag{10.52}$$

where

$$\Delta_m = \min_{j=1,2,\ldots,n+3}\{\Delta_j\} \tag{10.53}$$

E_m and E_l correspond to Δ_m and Δ_l, $\alpha > 1$ and δ is a positive number, which are given by the designer (e.g., $\alpha = 1.1$ and $\delta = 0.1$).

Step 8 Insert the $(N-1)/2$ new offspring to the population which are generated in a random fashion. Actually, the new offspring are formed by mutating the best parameter set P_b with a probability, i.e.,

$$P_{Nj} = P_b + d_n \xi_j, \text{ for } j = 1, 2, \ldots, (N-1)/2 \tag{10.54}$$

where d_n is the maximum to be altered and $\xi_j \in [-1, 1]$ is a random variable with zero mean. Thus, the next population is formed by the parameter sets P_{Mj} ($j = 1, 2, \ldots, (N-1)/2$), P_{Nj} ($j = 1, 2, \ldots, (N-1)/2$) and P_b.

Step 9 Continue the cycle initiated in Step 3 until convergence is achieved. The population is considered to have converged when Φ_b cannot be reduced any further, subject to

$$\Delta_j - \Delta_b \leq \varepsilon, \text{ for } j = 1, 2, \ldots, N \tag{10.55}$$

where ε is a positive number.

Take the best solution in the converged generation and place it in a second 'initial generation'. Generate the other $N-1$ parameter sets in this second initial generation at random and begin the cycle again until a satisfactory solution is obtained or Δ_b and Φ_b cannot be improved any further.

Consider a linear representation of a distillation column (Kautsky et al, 1985) with a state feedback controller, which is of the form:

$$\dot{x} = Ax + Bu \tag{10.56}$$
$$u = Kx \tag{10.57}$$

where

$$A = \begin{bmatrix} -0.1094 & 0.0628 & 0 & 0 & 0 \\ 1.3060 & -2.1320 & 0.9807 & 0 & 0 \\ 0 & 1.5950 & -3.1490 & 1.5470 & 0 \\ 0 & 0.0355 & 2.6320 & -4.2570 & 1.8550 \\ 0 & 0.0023 & 0 & 0.1636 & -0.1625 \end{bmatrix}$$

$$B = \begin{bmatrix} 0 & 0.0638 & 0.0838 & 0.1004 & 0.0063 \\ 0 & 0 & -0.1396 & -0.2060 & -0.0128 \end{bmatrix}^T$$

The closed-loop eigenvalues are chosen such that

$-10 \leq \lambda_1 \leq -0.01$

$-10 \leq \lambda_{2,re} \leq -0.01$

$-10 \leq \lambda_{2,im} \leq -0.01$

$-10 \leq \lambda_i \leq -0.01$ for $i = 4, 5$

Note that $\lambda_3 = \lambda_2^*$. A state feedback controller is required to assign the eigenvalues in the above regions to satisfy the following performance criteria:

$\phi_i(Z, V, W) \leq 3.0$, for $i = 1, 2, 3, 4, 5$
$\phi_6(Z, V, W) \leq 8.5$
$\phi_7(Z, V, W) \leq 20.0$

The performance functions ϕ_i ($i = 1, 2, \ldots, 5$) are the individual eigenvalue sensitivity functions. The performance functions ϕ_6 and ϕ_7 are the sensitivity function S and the function C_S of the closed-loop system in the frequency domain. The parameters for the numerical algorithm are

parameter length	15
population size N	21
d_m (mutation)	0.1
d_n (new population)	0.2
α(elitism)	1.1
δ(elitism)	0.1
frequency range $\mathcal{F}$	$[0.01, 100]$

Using the eigenstructure assignment toolbox (Liu and Patton, 1999), after 150 generations the optimal results have been found approximately by the numerical algorithm. The performance functions are

$\phi_1(Z, V, W) = 1.4968$ $\qquad$ $\phi_2(Z, V, W) = 1.1875$
$\phi_3(Z, V, W) = 1.1875$ $\qquad$ $\phi_4(Z, V, W) = 1.5144$
$\phi_5(Z, V, W) = 1.0757$ $\qquad$ $\phi_6(Z, V, W) = 1.3056$
$\phi_7(Z, V, W) = 12.7281$

and the eigenvalue set Λ of the closed-loop system is

$\Lambda = \begin{bmatrix} -0.1676 & -2.6299 - 0.9682j & -2.6299 + 0.9682j & -0.0950 & -5.9625 \end{bmatrix}$

The optimal state feedback controller is

$K = \begin{bmatrix} -6.7708 & -4.7671 & 1.2392 & 2.1961 & 2.8328 \\ -2.8498 & 8.6277 & 7.2037 & 3.1976 & 3.8011 \end{bmatrix}$

Clearly, with the above controller K, the required multi-objective performance criteria are satisfied.

10.4 Multi-objective Critical Control

A problem that occurs frequently in control engineering is to control outputs (usually the errors) of a system subjected to random external inputs (reference inputs and/or disturbances) so that the absolute value of the outputs is within prescribed bounds. Any violation of the bounds results in unacceptable, perhaps catastrophic, operation. For example, the electro-magnetic suspension control system for a magnetically levitated (maglev) vehicle. It is necessary that the airgap between the guideway and the levitating magnets is maintained for the effective operation of the system in spite of disturbance resulting from variations in the guideway profile. Such a system is said to be critical (Zakian, 1989; Whidborne and Liu, 1993;).

10.4.1 Critical Control Systems

A critical system in control engineering is to control an output v (usually the error) of a system subjected to random external inputs w (reference inputs and/or disturbances) so that the absolute value of the output is within a prescribed bound ε, that is,

$$|v(t,w)| \leq \varepsilon, \quad \forall t \in \mathbb{R} \tag{10.58}$$

Any violation of the bound results in unacceptable, and perhaps catastrophic, operation (Zakian, 1989). For example, motor vehicles are required to travel within their prescribed lanes and any departure from those can have serious consequences.

In the robust control design of critical systems, one should consider four aspects: external input space (or external uncertainty), internal uncertainty (*e.g.*, modelling error), output performance and robust stability (Liu, 1992; Liu *et al*, 1995).

a) External input space: This is concerned with environmental conditions which the system is subjected to, including reference inputs and disturbances.
b) Internal uncertainty: This is concerned with modelling errors of the plant, including parameter variations and unmodelled dynamics.
c) Output performance: This is concerned with the ability of the system to reach satisfactory outputs.
d) Robust stability: This is concerned with stability of the system in the case of internal uncertainty.

Based on the definition of the critical system and the four aspects above, a possible definition of output performance criteria for critical control systems may be formulated in a set of inequalities of the form

$$\phi_i(K) \leq \varepsilon_i, \text{ for } i = 1, 2, \ldots, n \tag{10.59}$$

$\phi_i(K)$ is a real scalar and represents the supremum of the absolute value of the i-th output for all time, and for all external inputs and all internal uncertainties to which the system is subjected. It may also represent some physical constraints such as actuator or sensor saturation. K is a stabilising controller which is being designed. The finite positive real number ε_i represents the numerical bound on the aspect of dynamic behaviour described by $\phi_i(K)$

10.4.2 Input Spaces of Systems

The definition of an input space plays an important role in critical control systems since it forms an integral part of the control system representation. The choice of the input space is crucial to the definition of performance and directly affects the design method of a system (Liu, 1990). In the literature, the following kinds of external input models have been investigated. First, the external inputs are assumed to be some known standard functions, *e.g.* a unit step, a pulse, sine, etc., which are considered in much of classical control theory. Second, it is assumed that the external inputs are a white Gaussian process with zero mean and covariance not greater than unity, which are considered in the well-known LQG optimisation problem, $\mathcal{H}^2$ optimisation problem and much of stochastic control theory. Third, it is assumed that the external inputs are a square-integrable signal of unity energy, which are considered in $\mathcal{H}^\infty$ optimisation problem. Finally, it is assumed that the external inputs are a persistent bounded casual signal, which is considered in the set theoretic approach and in l^1/L^1 optimisation problems.

In order to design continuous systems, an input space is introduced in the frequency domain (Liu, 1993). This input space is defined as follows.

Suppose that $F_1 \in \mathcal{H}^2(\sigma)$ and $F_2 \in \mathcal{H}^2(1)$, then $F = F_1 F_2$ is said to be an input in the frequency domain. The set of all functions F is known as an input space denoted by $\mathcal{F}(\sigma)$, *i.e.*

$$\mathcal{F}(\sigma) = \{F_1 F_2 : F_1 \in \mathcal{H}^2(\sigma), F_2 \in \mathcal{H}^2(1)\} \tag{10.60}$$

where the space $\mathcal{H}^2(\sigma_i)$ ($\sigma_i = 1, \sigma$) is the set of $\mathcal{H}^2$ Hardy space consisting of functions $H(s)$ satisfying that $H \in \mathcal{H}^2$ and $\|H\|_2 \leq \sigma_i$, for $\sigma_i \geq 0$.

Here one can think of the input as being generated by the given input F_1 through the transfer function F_2. Similarly, one can think of the input space $\mathcal{F}(\sigma)$ as being generated by a set of given inputs F_1 with $\mathcal{H}^2$-norm no larger than σ through a class of variable transfer functions F_2 with $\mathcal{H}^2$-norm no larger than 1.

For the sake of simplicity, it assumes that the reference input is set to zero and the disturbance F is modelled by the space $\mathcal{F}(\sigma)$. Of course, if it assumes that the reference input is also modelled by the space $\mathcal{F}(\sigma)$, then the results for this case are also similar to the results to be given in this chapter but the output should be replaced by the error in the following discussion.

10.4.3 Robust Critical Control

Several robust control methods for systems are studied on the basis of the use of singular values. It was recognised that singular values are good indicators of matrix size. Using the small gain theorem, conditions for guaranteed stability can be obtained in terms of the maximum singular values of perturbing transfer functions. Now, consider a standard feedback system with the controller $K(s)$ and the plant $G(s) + \Delta(s)$, where $G(s)$ is a known nominal model and $\Delta(s)$ is an unknown perturbation representing parameter variations and unmodelled dynamics. Thus, the following bound may be established for the maximum singular value of the perturbation of the system in order to guarantee robust stability:

$$\|\Delta\|_\infty < \|K(I+GK)^{-1}\|_\infty^{-1} \tag{10.61}$$

where $\|\cdot\|_\infty$ is the $\mathcal{H}^\infty$-norm of a matrix, *i.e.* the maximal singular value of a matrix.

It is well-known that the robust stability of systems with internal uncertainty can be guaranteed by (10.61). Let us define a dynamical modelling error space which models the internal uncertainty in the frequency domain. The modelling error space $\mathcal{D}(\delta)$ is a set of functions $\Delta(s)$ satisfying that

$$\mathcal{D}(\delta) = \{\delta : \|\Delta\|_\infty < 1/\delta, \text{ for } \delta > 0\} \tag{10.62}$$

Then from the stability formula (10.61) the robust performance criterion of systems can be formulated as

$$\psi(K) \leq \delta \tag{10.63}$$

where

$$\psi(K) = \|K(I+GK)^{-1}\|_\infty \tag{10.64}$$

Therefore, after combining the output performance criteria (10.59) and the robust performance criterion (10.63) a formulation of the robust control design of critical systems will be expressed by the following form (Liu *et al*, 1995):

$$\begin{cases} \psi(K) \leq \delta \\ \phi_i(K) \leq \varepsilon_i, \text{ for } i = 1, 2, \ldots, n \end{cases} \tag{10.65}$$

Obviously, the design problem is to find a controller such that (10.65) is satisfied. For this design problem, firstly, simplify the output performance functions $\phi_i(K)$ by some analytical methods (Liu, 1993); secondly, find a controller K satisfying (10.65) employing some numerical optimisation methods.

10.4.4 Multi-objective Critical Control

Now we consider multivariable systems with disturbance

$$F = [F_1, F_2, \ldots, F_n]^T \tag{10.66}$$

Let each disturbance F_i belong to a corresponding space $\mathcal{F}(\sigma_i)$, and the modelling error $\Delta \in \mathcal{D}(\delta)$. Also, define the following Cartesian spaces:

$$\mathcal{F}(\sigma) = \mathcal{F}(\sigma_1) \times \mathcal{F}(\sigma_2) \times \ldots \times \mathcal{F}(\sigma_n) \tag{10.67}$$

where $\sigma = [\sigma_1, \sigma_2, \ldots, \sigma_n]$.

The output performance of the system is defined by

$$\phi_i(K) = \sup\{|y_i(t, F, \Delta, K)| : t \in \mathbb{R}, F \in \mathcal{F}(\sigma), \Delta \in \mathcal{D}(\delta)\}, \text{ for } i = 1, 2, \ldots, n \tag{10.68}$$

where $y_i(t, F, \Delta, K)$ is the ith output of the system and the function of the disturbance F, the modelling error Δ and the controller K. Thus the output performance criteria are defined by

$$\phi_i(K) \leq \varepsilon_i, \text{ for } i = 1, 2, \ldots, n \tag{10.69}$$

where ε_i is a real positive number.

The robust performance criterion is still defined by

$$\psi(K) \leq \delta \tag{10.70}$$

where

$$\psi(K) = \left\| K(I + GK)^{-1} \right\|_\infty \tag{10.71}$$

and δ is given by the external uncertainty space $\mathcal{D}(\delta)$.

A relationship between the input space $\mathcal{F}(\sigma)$ and the controller K in the frequency domain, and the output performance in the time domain is given below (Liu et al, 1995). Assume that $(I+(G+\Delta)K)^{-1} \in \mathcal{H}^\infty$ and $F \in \mathcal{F}(\sigma)$, then

$$\phi_i(K) = \sum_{j=1}^n \left\| e_i^T (I + (G + \Delta)K)^{-1} e_j \right\|_\infty \sigma_j \tag{10.72}$$

where $e_i \in \mathbb{R}^{n \times 1}$ is an identity vector, e.g. $e_2 = [0, 1, 0, \ldots, 0]^T$. If all $F \in \mathcal{F}(\sigma)$, all $\Delta \in \mathcal{D}(\delta)$ and $\psi(K) \leq \delta$, then for $i = 1, 2, \ldots, n$

$$\phi_i(K) \leq \frac{\delta \gamma}{\delta - \psi(K)} \phi^+(K) \tag{10.73}$$

where

$$\gamma = \sum_{j=1}^n \sigma_j \tag{10.74}$$

$$\phi^+(K) = \left\| (I + GK)^{-1} \right\|_\infty \tag{10.75}$$

Thus, the control design problem of multi-objective critical systems can be simplified as

$$\begin{cases} \psi(K) \leq \delta \\ \phi^+(K) \leq \varepsilon^+ \end{cases} \tag{10.76}$$

where

$$\psi(K) = \| K(I+GK)^{-1} \|_\infty \tag{10.77}$$
$$\phi^+(K) = \| (I+GK)^{-1} \|_\infty \tag{10.78}$$
$$\varepsilon^+ = \frac{\delta - \psi(K)}{\delta\gamma} \max_{i=1,2,\ldots,n} \{\varepsilon_i\} \tag{10.79}$$

It is clear that the output performance criteria (10.69) and the robust performance criterion (10.70) can be satisfied by the solution of the inequalities (10.76). Therefore, the control design of multi-objective critical systems is largely simplified.

10.4.5 An Example

The following example illustrates the control design of multi-objective critical systems. Here, consider a two-input two-output plant given by

$$G(s) = \begin{bmatrix} \frac{-45s+1}{s^2+5s+6} & \frac{24s}{s^2+5s+6} \\ \frac{-35s}{s^2+5s+6} & \frac{58s+1}{s^2+5s+6} \end{bmatrix} \tag{10.80}$$

The external uncertainty (the disturbance) F and the internal uncertainty (the modelling error) Δ which the system is subjected to are assumed to be

$$F \in \mathcal{F}(\sigma), \quad \sigma = \begin{bmatrix} 0.01 & 0.01 \end{bmatrix}$$
$$\Delta \in \mathcal{D}(2.5)$$

The output performance functions $\phi_1(K)$ and $\phi_2(K)$ in time domain are required to satisfy

$$\phi_1(K) \leq 0.13$$
$$\phi_2(K) \leq 0.13$$

It is easy to know that $\gamma = \sum_{i=1}^{2} \sigma_i = 0.02$, $\delta = 2.5$ and $\max_{i=1,2}\{\varepsilon_i\} = 0.13$.
According to the result (10.76) of Section 10.4.4, the robust control design problem of system (10.80) can be simplified to find a stabilising controller such that

$$\psi(K) = \| K(I+GK)^{-1} \|_\infty \leq 2.5 \tag{10.81}$$
$$\phi^+(K) = \| (I+GK)^{-1} \|_\infty \leq \varepsilon^+ = \frac{0.13(2.5 - \psi(K))}{0.05} \tag{10.82}$$

are satisfied. Combining the method of inequalities (Whidborne and Liu, 1993) and the $\mathcal{H}^\infty$ optimisation techniques, a stabilising controller that satisfies both the output and the robust performance requirements (10.81) and

(10.82) is found to be

$$K(s) = \begin{bmatrix} \frac{-0.0386s^2-0.0056s-0.0001}{s^3+0.1918s^2+0.0104s+0.0002} & \frac{-0.0108s^2+0.0022s+0.0001}{s^3+0.1918s^2+0.0104s+0.0002} \\ \frac{-0.0182s^2-0.0012s+0.0001}{s^3+0.1918s^2+0.0104s+0.0002} & \frac{-0.0179s^2+0.0107s+0.0005}{s^3+0.1918s^2+0.0104s+0.0002} \end{bmatrix}$$

which gives the following values of the output and robust performance functions

$$\psi(K) = 1.9574 \tag{10.83}$$
$$\phi^+(K) = 1.3309 \tag{10.84}$$

It is easily obtained from (10.82) and (10.83) that $\varepsilon^+ = 1.4108$ which is greater than 1.3309. From (10.73), It can be calculated that $\phi_1(K) \leq 0.1226$ and $\phi_2(K) \leq 0.1226$, which are less than 0.13. Therefore, both the output and the robust performance requirements are satisfied.

References

Chen, J., R.J. Patton and G.P. Liu, (1996) Optimal residual design for fault diagnosis using multi-objective optimisation and genetic algorithms, *International Journal of Systems Science*, 27(6):567-576.

Grace, A., (1998) *Optimisation Toolbox for Use with MATLAB*, The MathWorks Inc.

Kautsky, J., N.K. Nichols and P. Van Dooren, (1985) Robust pole assignment in linear state feedback. *International Journal of Control*, 41(5):1129-1155.

Kessler, C., Das symmetrische optimum. (1958) *Regelungstetechnik*, 6(11):395-400.

Khargonekar, P.P. and M.A. Rotea, (1991) Multiple objective optimal control of linear systems: the quadratic norm case, *IEEE Trans. Automat. Contr.*, 36(1):4-24.

Kimura, H., (1975) Pole assignment by gain output feedback. *IEEE Transactions on automatic Control*, 20(4):509-516.

Kwon, B.H. and M.J. Youn, (1987) Eigenvalue-generalized eigenvector assignment by output feedback. *IEEE Transactions on Automatic Control*, 32(5):417-421.

Liu, G.P., (1990) Frequency-domain approach for critical systems. *International Journal of Control*, 52(6):1507-1519.

Liu, G.P., (1992) *Theory and Design of Critical Control Systems*, Ph.D thesis, Control Systems Centre, University of Manchester Institute of Science and Technology.

Liu, G.P., (1993) Input space and output performance in the frequency domain for critical systems, *IEEE Transactions on Automatic Control*, 38(1):152-155.

Liu, G.P., (2001) *Nonlinear Identification and Control: A Neural Network Approach*, Springer-Verlag.

Liu, G.P. and S. Daley, (1999) Optimal-tuning PID controller design in the frequency-domain with application to rotary hydraulic systems, *Control Engineering Practice*, 7:821-830.

Liu, G.P. and S. Daley, (2000) Optimal-tuning nonlinear PID control for hydraulic systems, *Control Engineering Practice*, 8(9):1045-1053.

Liu, G.P. and S. Daley, (2001) Optimal-tuning PID control for industrial systems, *Control Engineering Practice*, 9(11):1185-1194.

Liu, G.P., R. Dixon and S. Daley, (2000) Multiobjective optimal-tuning proportional-integral controller design for ALSTOM gasifier problem, *Proceedings of ImechE Part I*, 214:395-404.

Liu, G.P. and V. Kadirkamanathan, (1999) Multiobjective criteria for nonlinear model selection and identification with neural networks, *IEE Proceedings, Part D*, 146(5):373-382.

Liu, G.P. and R.J. Patton, (1996) Robust control design using eigenstructure assignment and multiobjective optimization. *Int. Journal of Systems Science*, 27(9):871-879.

Liu, G.P. and R.J. Patton, (1998) *Eigenstructure Assignment for Control System Design*, John Wiley.

Liu, G.P. and R.J. Patton, (1999) *Eigenstructure Assignment Toolbox for Use with MATLAB*, Pacilantic International, Oxford.

Liu, G.P., H. Unbehauen and R.J. Patton, (1995) Robust control of multivariable critical systems, *International Journal of Systems Science*, 26(10):1907-1918.

Liu, G.P., J.B. Yang and J.F. Whidborne, (2003) *Multiobjective Optimisation and Control*, Research Studies Press.

Moore, B.C., (1976) On the flexibility offered by state feedback in multivariable systems beyond closed-loop eigenvalue assignment. *IEEE Transactions on Automatic Control*, 21(6):689-692.

Pessen, D.W., (1994) A New look at PID-Controller tuning, *Transactions of ASME, Journal of Dynamic Systems, Measurement, and Control*, 116:553-557.

Seborg, D.E., T.F. Edgar and D.A. Mellichamp, (1989) *Process Dynamics and Control*. Wiley, New York.

Tabak, T., A.A. Schy, D.P. Giesy, and K.G. Johnson, (1979) Application of multiple objective optimization in aircraft control systems design, *Automatica*, 15:595-600.

Voda, A. and I.D. Landau, (1995) A method for the auto-calibration of PID controllers. *Automatica*, 31(1):41-53.

Whidborne, J. F., G. Murad, D.-W. Gu and I. Postlethwaite, (1995) Robust control of an unknown plant – the IFAC 93 benchmark. *International Journal of Control*, 61(3):589-640,

Whidborne, J. F., D.-W. Gu and I. Postlethwaite, (1996) Simulated annealing for multi-objective control system design, *IEE Proc. Control Theory and Appl.*, 144(6):582-588.

Whidborne, J.F. and G.P. Liu, (1993) *Critical Control Systems: Theory, Design and Applications*, Research Studies Press and John Wiley.

Wonham, W.M., (1967) On pole assignment in multi-input, controllable linear systems. *IEEE Transactions on Automatic Control*, 12:660-665.

Zakian, V., (1989) Critical systems and tolerable inputs. *International Journal of Control*, 49(5):1285-1289.

Zakian, V. and U. Al-Naib, (1973) Design of dynamical and control systems by the method of inequalities, *Proc. Inst. Elec. Eng.*, 120:1421-1427.

Zhuang, M. and D.P. Atherton, (1993) Automatic tuning of optimum PID controllers. *IEE Proceedings-D*, 140(3):216-224.

Ziegler, J.G. and N.B. Nichols, (1942) Optimum settings for automatic controllers. *Trans. ASME*, 64:759-768.

11 A MoI Based on $\mathcal{H}^\infty$ Theory — with a Case Study

James F Whidborne

Abstract. Control system design methods based on $\mathcal{H}^\infty$ theory are able to produce controllers that are both robust to model uncertainty and that have good performance. Furthermore, the method is applicable to multivariable systems. However, specifications for control system designs are rarely expressed explicitly in terms of the $\mathcal{H}^\infty$-norm. Thus it is often not clear how to choose the weighting functions required in $\mathcal{H}^\infty$-based design in order to meet the specifications. By posing the problem using the principle of inequalities (PoI), a method of inequalities (MoI) is proposed that can overcome some of the limitations of $\mathcal{H}^\infty$-based control design. The efficacy of the method is demonstrated by a case study of the control of a distillation column.

11.1 Introduction

Over the last two decades, there have been very significant developments in the analysis of robustness for control systems and a parallel development in methods for synthesising robust controllers. One of the better known approaches is based on $\mathcal{H}^\infty$ theory. The $\mathcal{H}^\infty$-norm of a single-input single-output (SISO) transfer function is essentially the maximum magnitude of the frequency response. Control design using $\mathcal{H}^\infty$ theory is based on minimising the $\mathcal{H}^\infty$-norm of some input-output transfer function(s). It can be easily shown that by making certain $\mathcal{H}^\infty$-norms sufficiently small, robust stability can be guaranteed.

A number of relatively simple algorithms have been developed for synthesising $\mathcal{H}^\infty$-optimal and sub-optimal controllers (Zhou et al, 1996; Skogestad and Postlethwaite, 1996); these are readily available in MATLAB toolboxes (Chiang and Safonov, 2001; Balas et al, 2001). However, the procedure for designing controllers is more involved. Simply minimising the relevant $\mathcal{H}^\infty$-norms will not result in good control, because of the fundamental conflicts between the various performance and stability requirements of the control system. Hence these conflicts need to be resolved. This is done by introducing frequency-domain "weighting functions" into the problem and the process of the design is to adjust these weighting functions in an iterative manner until a satisfactory design is achieved. The procedure is usually to inspect various frequency responses and, using the experience of the designer, adjust the weighting functions accordingly.

The attractions of the $\mathcal{H}^\infty$-based approaches are several:

- the method extends in a natural and simple manner to multivariable control systems,
- the design procedure is based in the frequency domain, which control designers are well-versed in, and
- the resulting control systems are guaranteed to have certain levels of robustness.

The method also has a number of disadvantages:

- the resulting controllers tend to be high-order, which can cause problems with implementation,
- the design procedure is heuristic and hence requires experience to obtain good designs, and
- the performance measures used are fairly arbitrary, in that, being based on the $\mathcal{H}^\infty$-norm, they rarely correspond to the specified requirements for the system.

The method of inequalities (MoI) (Zakian and Al-Naib, 1973) described in this chapter is able to overcome the second and third disadvantages, but not the first.

The basic idea is simple, and that is to use search methods (see Satoh, Chapter 6; Whidborne, Chapter 7; Liu and Ishihara, Chapter 8) to automate the iterative procedure of designing the weighting functions to achieve the performance specifications for the system. By formulating the problem using the principle of inequalities (PoI), the procedure for designing controllers using $\mathcal{H}^\infty$-theory is made easier and less heuristic. Furthermore, the procedure for meeting the specifications for the control system requirements is methodical rather than *ad-hoc*.

In this chapter, a MoI is proposed that utilises the two degree-of-freedom $\mathcal{H}^\infty$ method of Limebeer *et al* (1993). The MoI is demonstrated by the design of a controller for a distillation column (Limebeer, 1991). A MoI utilising the two degree-of-freedom $\mathcal{H}^\infty$ method was originally proposed by Whidborne *et al* (1995), but the approach proposed here makes use of some additional design freedoms that were originally ignored. The first MoI based on $\mathcal{H}^\infty$ theory was originally proposed by Whidborne *et al* (1994). This MoI used the robust stabilisation of normalised coprime factorisations of Glover and McFarlane (1989) (also see McFarlane and Glover, 1990; McFarlane and Glover, 1992). The main limitation of this approach is that it is a one degree-of-freedom method. Other MoI's based on $\mathcal{H}^\infty$ and $\mathcal{H}^2$ theory can be found in Postlethwaite *et al* (1994), Whidborne *et al* (1995) and Dakev *et al* (1997).

11.2 Preliminaries

Notation. The transpose of a matrix X is denoted by X^T. The maximum singular value of a matrix X is denoted by $\bar{\sigma}(X)$. To define a state-variable description of a system $F(s) = C(sI - A)^{-1}B + D$, the following notation is used

$$F(s) \stackrel{s}{=} \left[\begin{array}{c|c} A & B \\ \hline C & D \end{array}\right] \tag{11.1}$$

The $\mathcal{H}^\infty$-norm. For a stable system $F(s)$, the $\mathcal{H}^\infty$-norm is defined as

$$\|F\|_\infty = \sup_\omega \{\bar{\sigma}(F(j\omega))\} \tag{11.2}$$

The $\mathcal{H}^\infty$-norm of a system gives the largest steady-state gain for any sinusoidal input at any frequencies. It is also equal to the largest energy gain of the system (for further details, see Skogestad and Postlethwaite, 1996, pp 153).

The small gain theorem. The robust stability of a system can be analysed by means of the small gain theorem.

Theorem 11.1. *The system shown in Figure 11.1 is stable if and only if*
a) $\|F\|_\infty < \gamma$ *and for all Δ such that* $\|\Delta\|_\infty \leq \frac{1}{\gamma}$,
b) $\|F\|_\infty \leq \gamma$ *and for all Δ such that* $\|\Delta\|_\infty < \frac{1}{\gamma}$.

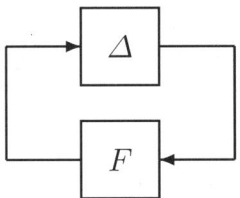

Fig. 11.1. General system for the small gain theorem

The small gain theorem is well-known (see, for example, Skogestad and Postlethwaite, 1996, pp 150-151). The theorem applies for multi-input multi-output as well as single-input single-output systems. Thus $\mathcal{H}^\infty$ theory is used for robust multivariable control system design. From the small gain theorem, stability conditions for a variety of models of system uncertainty can be obtained. For further details, see, for example, Skogestad and Postlethwaite (1996, pp 291-309).

11.3 A Two Degree-of-freedom $\mathcal{H}^\infty$ Method

In control system design, there is a well-known conflict between the objectives of reference tracking, disturbance rejection and sensitivity to parameter variations (governed by the sensitivity function), and the objectives of measurement noise attenuation and robustness to plant uncertainty (governed by the complimentary sensitivity function). The conflict between the reference tracking requirements and its competing objectives can be partially resolved by use of two degree-of-freedom controllers. The idea is to pre-filter the reference signal outside the closed-loop and so partially decouple the reference response of the overall system from the other closed-loop responses. The basic two degree-of-freedom schematic is shown in Figure 11.2 from which the output response to references can be readily determined as $(I - GK_y)^{-1}GK_r$. This means the pre-filter K_r can be used to decouple the response from the sensitivity function $S = (I - GK_y)^{-1}$ and overcome the well-known constraint of $|S| + |T| \geq 1$ where $T = -(I - GK_y)^{-1}GK_y$ is the complimentary sensitivity function.

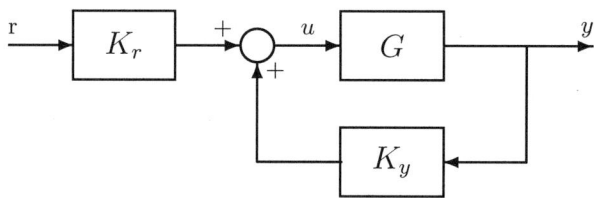

Fig. 11.2. Two degree-of-freedom control system schematic

One method of imposing a desired closed-loop response is to formulate the problem as a model-matching problem, whereby the difference between some desired model response and the closed-loop transfer function is made small. Let T_0 be the desired reference-following model. If

$$\left\| T_0 - (I - GK_y)^{-1}GK_r \right\|_\infty \tag{11.3}$$

is made small, then clearly the output response will closely follow the reference filtered by T_0. This approach is taken in the two degree-of-freedom $\mathcal{H}^\infty$ method developed by Hoyle et al (1991) and Limebeer et al (1993) (see also Green and Limebeer, 1995, pp 441–444). The closed-loop robustness and disturbance rejection properties of the feedback controller K_y are established by a variation of the normalised coprime factorisation approach of Glover and McFarlane (1989).

11.3.1 Robust Stability

A plant model G can be factorised into two stable transfer function matrices (M, N) so that

$$G = M^{-1}N \tag{11.4}$$

The stability of M and N implies that N should contain all the right-half-plane zeros of G, and M should contain all the right-half-plane poles of G as zeros. Such a factorisation is called a left coprime factorisation of G if there exist stable transfer function matrices (V, U) such that

$$MV + NU = I \tag{11.5}$$

The coprime property means that M and N do not have any common right-half-plane zeros, i.e., G has no unstable pole cancellation. A left coprime factorisation is normalised if and only if

$$NN^\sim + MM^\sim = I \text{ for all } s \tag{11.6}$$

where $N^\sim(s) = N^T(-s)$ and $M^\sim(s) = M^T(-s)$.

Let

$$G(s) \stackrel{s}{=} \left[\begin{array}{c|c} A & B \\ \hline C & D \end{array}\right] \tag{11.7}$$

then a normalised coprime factorisation of G is given by

$$[N \; M] \stackrel{s}{=} \left[\begin{array}{c|cc} A + HC & B + HD & H \\ \hline R^{-1/2}C & R^{-1/2}D & R^{-1/2} \end{array}\right] \tag{11.8}$$

where $H := -(BD^T + ZC^T)R^{-1}$, $R := I + DD^T$, and the matrix $Z \geq 0$ is the unique stabilising solution to the algebraic Riccati equation

$$(A - BQ^{-1}D^TC)Z + Z(A - BQ^{-1}D^TC)^T$$
$$- ZC^TR^{-1}CZ + BQ^{-1}B^T = 0 \tag{11.9}$$

where $Q := I + D^TD$.

To obtain a degree of robustness, plant uncertainty is represented by stable additive perturbations on each of the factors of the coprime factorisation of the plant. The perturbed plant model G_p is

$$G_p = (M + \Delta_M)^{-1}(N + \Delta_N) \tag{11.10}$$

where Δ_M, Δ_N are stable unknown transfer functions which represent the uncertainty in the nominal plant model G. The objective of robust stabilisation it to ensure internal stability of not only the nominal model G, but the set of plants defined by

$$\mathcal{G} = \{(M + \Delta_M)^{-1}(N + \Delta_N) : \left\| [\Delta_M \; \Delta_N] \right\|_\infty < 1/\gamma \} \tag{11.11}$$

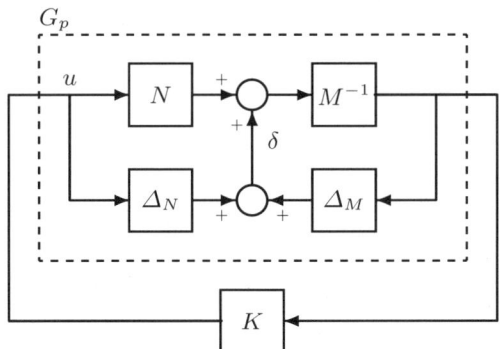

Fig. 11.3. Coprime factor robust $\mathcal{H}^\infty$ stability problem

To maximise the robust stability is the problem of robust stabilisation of normalised coprime factor plant descriptions (Glover and McFarlane, 1989). From the small gain theorem (Theorem 11.1), the perturbed feedback system shown in Figure 11.3 is robustly stable for the family of plants $\mathcal{G}$ defined by (11.11), if and only if

$$\left\| \begin{bmatrix} I \\ K \end{bmatrix} (I - GK)^{-1} M^{-1} \right\|_\infty \leq \gamma \tag{11.12}$$

If $X \geq 0$ is the unique solution of the algebraic Riccati equation

$$(A - BQ^{-1}D^T C)^T X + X(A - BQ^{-1}D^T C) \\ - XBQ^{-1}B^T X + C^T R^{-1} C = 0 \tag{11.13}$$

and Z is the solution to (11.9), then the lowest achievable value of γ for all stabilising controllers is given by

$$\gamma_0 = (1 + \rho(XZ))^{1/2} \tag{11.14}$$

where ρ denotes the spectral radius (maximum eigenvalue magnitude). A controller that achieves γ_0 is given by Glover and McFarlane (1989).

11.3.2 Robust Performance — a Two Degree-of-freedom Formulation

It is clear that if $\left\| T_0 - (I - GK_y)^{-1} GK_r \right\|_\infty$ is made small then there is accurate model-following and subsequently good performance for the nominal plant G. In fact, robust performance is needed, so it is required that

$$\left\| T_0 - (I - G_p K_y)^{-1} G_p K_r \right\|_\infty \leq \gamma \text{ for all } G_p \in \mathcal{G} \tag{11.15}$$

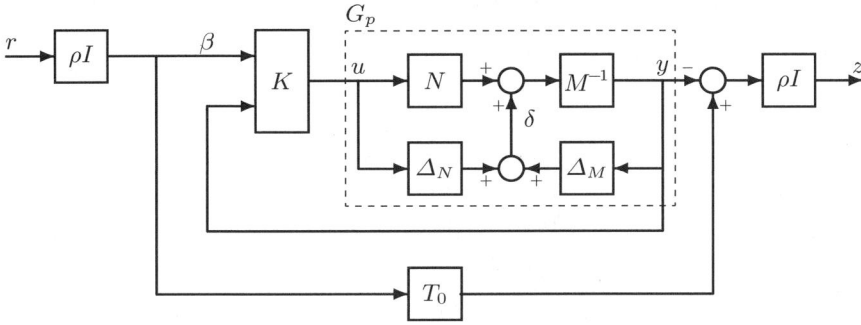

Fig. 11.4. Two degree-of-freedom design configuration

It is shown in Limebeer *et al* (1993) that for the configuration shown in Figure 11.4, both robust stability and robust performance can be achieved. The control signal is given by

$$u = \begin{bmatrix} K_y & K_r \end{bmatrix} \begin{bmatrix} \beta \\ y \end{bmatrix} \tag{11.16}$$

in which K_r is the pre-filter and K_y is the feedback controller. The signals β and y are the scaled reference variable and the measured variable respectively. The transfer function T_0 represents some desired closed-loop transfer function and it is explicitly chosen by the designer to introduce time-domain specifications into the design process. The scalar ρ is a scaling factor to provide a means of trade-off between robustness and model-following properties. Simple algebra from Figure 11.4 shows that

$$\begin{bmatrix} z \\ y \\ u \end{bmatrix} = \begin{bmatrix} \rho^2(T_0 - (I - GK_y)^{-1}GK_r) & -\rho(I - GK_y)^{-1}M^{-1} \\ \rho G(I - K_yG)^{-1}K_r & (I - GK_y)^{-1}M^{-1} \\ \rho(I - K_yG)^{-1}K_r & K_y(I - GK_y)^{-1}M^{-1} \end{bmatrix} \begin{bmatrix} r \\ \delta \end{bmatrix} \tag{11.17}$$

$$=: \Gamma \begin{bmatrix} r \\ \delta \end{bmatrix} \tag{11.18}$$

It can be observed that the $(2,2)$ partition of (11.17) is associated with the robust stability criterion of (11.12) and the $(1,1)$ partition is used to match the closed-loop to the model T_0. In Limebeer *et al* (1993) it is shown that if $\|\Gamma\|_\infty \le \gamma$ then

- The system will remain stable for all $G_p \in \mathcal{G}$.
- The robust performance criterion $\left\| T_0 - (I - G_pK_y)^{-1}G_pK_r \right\|_\infty \le \gamma\rho^2$ is satisfied for all $G_p \in \mathcal{G}$.
- If ρ is set to zero, the two degree-of-freedom problem reduces to the ordinary robust stability problem described in Section 11.3.1.

The problem may be posed as the following optimisation problem.

Problem 11.1. Obtain a stabilising controller $K = \begin{bmatrix} K_y & K_r \end{bmatrix}$ such that

$$\gamma_{\min} = \min_K \| \Gamma(K) \|_\infty \qquad (11.19)$$

To solve Problem 11.1, the problem can be set up in the generalised regulator framework (Skogestad and Postlethwaite, 1996, pp 98–108) with the standard augmented plant given by (Limebeer et al, 1993)

$$\begin{bmatrix} z \\ u \\ y \\ \beta \\ y \end{bmatrix} = \left[\begin{array}{cc|c} \rho^2 T_0 & -\rho M^{-1} & \rho G \\ 0 & 0 & I \\ 0 & M^{-1} & G \\ \hline \rho I & 0 & 0 \\ 0 & M^{-1} & G \end{array}\right] \begin{bmatrix} r \\ \delta \\ u \end{bmatrix} \qquad (11.20)$$

$$=: \left[\begin{array}{c|c} P_{11} & P_{12} \\ \hline P_{21} & P_{22} \end{array}\right] \begin{bmatrix} r \\ \delta \\ u \end{bmatrix} \qquad (11.21)$$

If

$$T_0 \stackrel{s}{=} \left[\begin{array}{c|c} A_0 & B_0 \\ \hline C_0 & D_0 \end{array}\right] \qquad (11.22)$$

T_0 is stable and

$$G \stackrel{s}{=} \left[\begin{array}{c|c} A & B \\ \hline C & D \end{array}\right] \qquad (11.23)$$

then a state-space realisation for (11.20) gives

$$P = \left[\begin{array}{ccc|cc} A & 0 & 0 & (BD^T + ZC^T)R^{-1/2} & B \\ 0 & A_0 & B_0 & 0 & 0 \\ -\rho C & \rho^2 C_0 & \rho^2 D_0 & -\rho R^{1/2} & -\rho D \\ C & 0 & 0 & R^{1/2} & D \\ 0 & 0 & 0 & 0 & I \\ \hline 0 & 0 & \rho I & 0 & 0 \\ C & 0 & 0 & R^{1/2} & D \end{array}\right] \qquad (11.24)$$

and Problem 11.1 can be solved in MATLAB using, for example, the Robust Control Toolbox (Chiang and Safonov, 2001).

Note that the pre-filter, K_r, should be scaled after the optimisation so that the closed-loop transfer function $(I - GK_y)^{-1} GK_r$ matches the unit matrix at steady state. To do this, replace K_r by $K_r K_r^{-1}(0) K_y(0)$.

11.3.3 Robust Design Procedure

For the actual design of robust controllers posed as Problem 11.1, it is necessary for the plant to be augmented by a loop-shaping weighting function, W. The nominal plant G is then replaced by the augmented plant $G_w = WG$ in (11.17) and (11.20). The choice of W may be done by consideration of the open-loop frequency response of G_w using guidelines given by McFarlane and Glover (1992) (see also McFarlane and Glover (1990, pp 98–131) and Skogestad and Postlethwaite (1996, pp 380–385)). An interactive graphical software toolbox has been developed by Whidborne *et al* (2002) to aid the procedure.

The designer also needs to choose the target model, T_0, based on the desired closed-loop response. The scaling factor ρ needs to be chosen to give the correct amount of trade-off between robust stability and performance. The optimal $\mathcal{H}^\infty$ controller, K, is then calculated with the plant G replaced by the weighted plant GW in (11.24). The weighting is then transferred from the weighted plant to the controller, so that the final controller $K = WK_{\min}$ acts on the original unweighted plant, G. Details of the suggested procedure are given by Limebeer *et al* (1993).

11.4 A MoI for the Two Degree-of-freedom Formulation

As discussed in Section 11.3.3, the designer has a number of different design parameters to select, including the structure and parameters of W and T_0 as well as the scaling factor ρ. There are additional design choices possible that are not usually considered.

There is a degree of freedom in the choice of the nominal plant G. Usually, for robust design, G will be chosen in $\mathcal{G}$ so as to minimise the maximum possible size of the perturbation norm. However, this assumes that the chosen uncertainty model is accurate. This is rarely the case, and the uncertainty model is only an approximation of the actual uncertainty that will occur on the real system. Furthermore, the uncertainty model in the two degree-of-freedom formulation is weighted by W. Thus it is possible that another choice of the nominal plant will result in improved control for the real system.

In actual $\mathcal{H}^\infty$ controller design, it is known that there is little advantage in using an optimal controller (Skogestad and Postlethwaite, 1996, p. 384) and designers will usually use a sub-optimal $\mathcal{H}^\infty$ controller. The degree of sub-optimality suggested by Skogestad and Postlethwaite (1996) for the design method of McFarlane and Glover (1992) is 10%, but there does not seem to be any clear theoretical justification for any particular value. The degree of sub-optimality could be utilised by the designer to improve the controller. Thus the designer can specify a desired value of $\| \Gamma \|_\infty =: \gamma_c$, and if

$$\min_{\text{stabilising} K} \| \Gamma \|_\infty \leq \gamma_c \qquad (11.25)$$

a controller that achieves $\|\Gamma\|_\infty = \gamma_c$ can be obtained by the solution of just two algebraic Riccati equations (Doyle et al, 1989) and no γ-iteration is required.

The MoI originally formulated by Whidborne et al (1995) for use with the two degree-of-freedom $\mathcal{H}^\infty$ method of Limebeer et al (1993) proposed that only the parameters of the weighting function W be used as the design parameters. In this chapter, that approach is extended to include the additional design parameters discussed above. The design problem is now stated below.

Problem 11.2. For the system of Figure 11.4, find a $c \in \mathcal{C}$, hence a $(G, T_0, W, \rho, \gamma_c)$ and controller K such that

$$\phi_i(c) \leq \varepsilon_i \text{ for } i = 1 \ldots m \qquad (11.26)$$

where $\phi_i(c)$ are objective functions based on the closed-loop responses, ε_i are real numbers representing desired tolerances on ϕ_i respectively, and $(G, T_0, W, \rho, \gamma_c)$ are parameterised by the design vector c.

Design Procedure

(a) Define a parameterised plant set $\mathcal{G}$ and the objective functions ϕ_i, $i = 1 \ldots m$.
(b) Define the values of ε_i.
(c) Select the structure and order of the nominal model, G, such that $G \in \mathcal{G}$ and the structure and order of the loop-shaping weight, W, and reference model, T_0. Define $\mathcal{C}$ by placing bounds on the relevant c_i to ensure that
  1. $G \in \mathcal{G}$,
  2. W is stable and minimum phase,
  3. T_0 is stable,
  4. ρ is within some range (typically $1 \leq \rho \leq 3$),
  5. and γ_c is within some range (typically $\gamma_0 \leq \gamma_c \leq 5\gamma_0$).
(d) Select initial values of c_i.
(e) Use a numerical search method, such as the Moving Boundaries Process (Zakian and Al-Naib, 1973), to find a $c \in \mathcal{C}$ that satisfies inequalities (11.26). If an admissible solution is found, the design is satisfactory. If no solution is found, either increase the order of W or T_0, relax one or more of the values ε_i, or try again with different initial values of c_i.

11.5 Example — Distillation Column Controller Design

In this section, the proposed MoI is used to design a control system for the high purity distillation column described in Limebeer (1991). The column is considered in just one of its configurations, the so-called LV configuration, for which the following model is relevant

$$G_p(s, k_1, k_2, \tau_1, \tau_2) = \frac{1}{75s+1} \begin{bmatrix} 0.878 & -0.864 \\ 1.082 & -1.096 \end{bmatrix} \begin{bmatrix} k_1 e^{-\tau_1 s} & 0 \\ 0 & k_2 e^{-\tau_2 s} \end{bmatrix} \quad (11.27)$$

where $0.8 \leq k_1, k_2 \leq 1.2$ and $0 \leq \tau_1, \tau_2 \leq 1$, and all time units are in minutes.

The design specifications are to design a controller which guarantees for all $0.8 \leq k_1, k_2 \leq 1.2$ and $0 \leq \tau_1, \tau_2 \leq 1$:

(a) Closed-loop stability.
(b) The output response to a step demand on reference r_1 satisfies $y_1(t) \leq 1.1$ for all t, $y_1(t) \geq 0.9$ for all $t > 30$ and $y_2(t) \leq 0.5$ for all t.
(c) The output response to a step demand on reference r_2 satisfies $y_1(t) \leq 0.5$ for all t, $y_2(t) \leq 1.1$ for all t and $y_2(t) \geq 0.9$ for all $t > 30$.
(d) Zero steady state error.
(e) The frequency response of the closed-loop transfer function between plant output disturbance input and plant input is gain limited to 50 dB.
(f) The open-loop unity-gain cross-over frequency of the largest singular value should be less than 150 rad/min.

A set of closed-loop performance functionals $\{\psi_i(G_p, K): i = 1, 2, \ldots, 8\}$, based on these specifications are defined. Functionals ψ_1 to ψ_6 are measures of the step response specifications. Functionals ψ_1 and ψ_4 are measures of the overshoot; ψ_2 and ψ_5, are measures of the rise-time, and ψ_3 and ψ_6 are measures of the cross-coupling. Denoting the i_{th} output response of the closed loop system with a plant G_p and controller K at a time t to a unit reference step demand on the j_{th} input, r_j, by $y_{ij}(G_p, K, t)$, the step response functionals are defined as

$$\psi_1(G_p, K) = \max_t y_{11}(G_p, K, t) \quad (11.28)$$

$$\psi_2(G_p, K) = -\min_{t>30} y_{11}(G_p, K, t) \quad (11.29)$$

$$\psi_3(G_p, K) = \max_t y_{21}(G_p, K, t) \quad (11.30)$$

$$\psi_4(G_p, K) = \max_t y_{22}(G_p, K, t) \quad (11.31)$$

$$\psi_5(G_p, K) = -\min_{t>30} y_{22}(G_p, K, t) \quad (11.32)$$

$$\psi_6(G_p, K) = \max_t y_{12}(G_p, K, t) \quad (11.33)$$

The steady state specifications are satisfied automatically by the use of integral action in the controller.

From the gain requirement in the design specifications, ψ_7 is the $\mathcal{H}^\infty$-norm (in dB) of the closed-loop transfer function between the plant output and the plant input,

$$\psi_7(G_p, K) = \sup_\omega \bar{\sigma}\left([I - K_y(j\omega)G_p(j\omega)]^{-1} K_y(j\omega)\right) \quad (11.34)$$

From the bandwidth requirement in the design specifications, ψ_8 is defined (in dB) as

$$\psi_8 = \max_{\omega \geq 150} \bar{\sigma}\left(G_p(j\omega)K_y(j\omega)\right) \tag{11.35}$$

To ensure robust performance, the objective functions, ϕ_i, are defined as the worst case functionals ψ_i over a set of plant models, $\mathcal{G}_p$, consisting of some plants at some extremes of the parameter range. Thus the objective functions are

$$\phi_i(K) = \max_{G_p \in \mathcal{G}_p} \psi_i(G_p, K) \text{ for } i = 1, \ldots, 8 \tag{11.36}$$

and the set of extreme plants is $\mathcal{G}_p = \{G_j : j = 1, 2, 3, 4\}$ where plants G_j have actuator time delays and gains shown in Table 11.1. These extreme plant models were chosen because they were judged to be the most difficult to simultaneously obtain good performance. The time delay is approximated using a fifth order Padé approximant.

Table 11.1. Extreme plants G_j, $j = 1, 2, 3, 4$

	τ_1	τ_2	k_1	k_2
G_1	0	0	0.8	0.8
G_2	1	1	0.8	1.2
G_3	1	1	1.2	0.8
G_4	1	1	1.2	1.2

Design variables need to be parameterised and their permissible values defined. The scaling factor ρ and the sub-optimal $\mathcal{H}^\infty$-norm parameter γ are parameterised as c_1 and c_2 respectively, where $1.0 \leq c_1 \leq 3$ and $1.0 \leq c_2 \leq 15$ as suggested by Limebeer et al (1993).

The target model, T_0, is parameterised as

$$T_0(c_3) = \frac{1}{c_3 s + 1} I_2 \tag{11.37}$$

The nominal plant G is parameterised as

$$G(c_9, c_{10}) = \frac{1}{75s + 1} \begin{bmatrix} 0.878 & -0.864 \\ 1.082 & -1.096 \end{bmatrix} \begin{bmatrix} c_{10}\,\text{pade}\,(c_9) & 0 \\ 0 & c_{10}\,\text{pade}\,(c_9) \end{bmatrix} \tag{11.38}$$

where $\text{pade}\,(\tau)$ is defined as the first-order Padé approximant of $e^{-\tau s}$, and $0 \leq c_9 \leq 1$ and $0.8 \leq c_{10} \leq 1.2$. The weighting function, W, is parameterised as

$$W(c_4, c_5, c_6, c_7) = \begin{bmatrix} c_4 & 0 \\ 0 & c_5 \end{bmatrix} \frac{(c_6 s + 1)(c_8 s + 1)}{s(c_7 s + 1)(0.001 s + 1)} \tag{11.39}$$

An integrator term is included to ensure that the final controller has integral action and the steady state specifications are satisfied. The fixed pole at $s = -1000$ is to ensure low high-frequency controller gain. The parameters c_i, $i = 3, \ldots, 8$ are constrained to be positive to prevent unstable controllers and undesirable pole zero cancellations. Based on the design specification listed earlier, the vector of tolerances, ε, on the objective functions, $\phi_i(c)$, $i = 1, \ldots, 7$ is set as

$$\varepsilon = \begin{bmatrix} 1.1, & -0.9, & 0.5, & 1.1, & -0.9, & 0.5, & 50, & 0.0 \end{bmatrix} \tag{11.40}$$

The design procedure described in Section 11.4 was implemented in MATLAB using both the moving boundaries process (Zakian and Al-Naib, 1973) and the Nelder-Mead dynamic minimax approach of Ng (1989). The following design parameters vector was obtained

$$c = \begin{bmatrix} 1.31095, & 4.07825, & 6.43616, & 43.23042, & 43.19785, & 1.54280, & 0.28901, \\ & & & & 0.00113, & 0.60991, & 1.20000 \end{bmatrix} \tag{11.41}$$

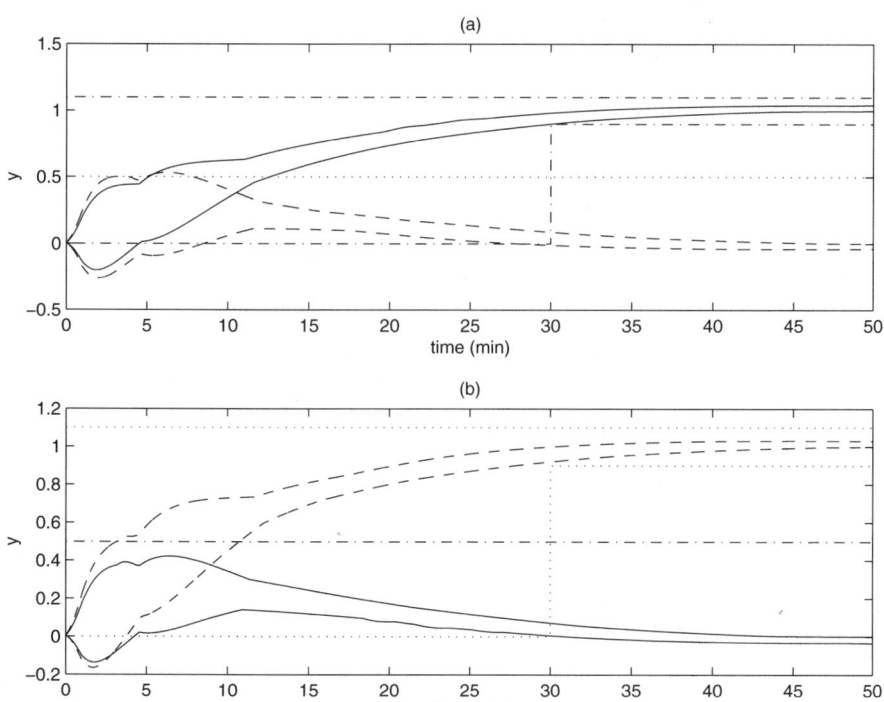

Fig. 11.5. Responses of y_1 (———) and y_2 (- - -) envelopes of all extreme plants to (a) unit step input on reference r_1 and (b) unit step input on reference r_2

which gives a performance

$$\phi = [1.04329, -0.90000, 0.39021, 1.03083, -0.92578, 0.50001,$$
$$56.31846, 0.00506] \quad (11.42)$$

All the step response criteria were satisfied, but ε_6, the 50 dB gain limit, was exceeded.

The envelopes of the 16 possible extreme plants step responses are shown in Figure 11.5. The non-rational transfer function responses are solved using a fifth-order fixed-step Dormand-Prince method in SIMULINK. The prescribed bounds on the responses are also shown in the plots. Over all the extreme plants, the overshoot, rise-time and cross-coupling in the simulations are not significantly worse than for the four extreme Padé-approximant plants used for the design.

The envelopes for the maximum singular values of $[I - K_y(j\omega)G_p(j\omega)]^{-1} K_y(j\omega)$ and $G_p(j\omega)K_y(j\omega)$ for the 16 possible extreme plants, using fifth-order Padé approximants, are shown in Figure 11.6. It is seen that the re-

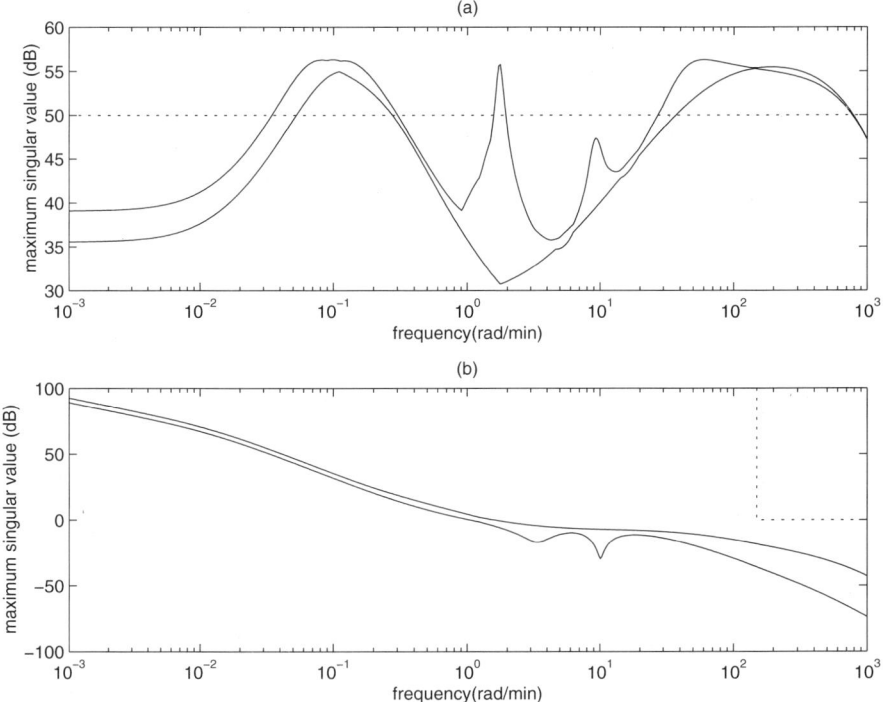

Fig. 11.6. Envelopes of maximum singular values of all extreme plants for (a) $\bar{\sigma}\left([I - K_y(j\omega)G_p(j\omega)]^{-1} K_y(j\omega)\right)$ and (b) $\bar{\sigma}\left(G_p(j\omega)\right) K_y(j\omega)$

quired specifications on the plant output to controller output gain are not met.

11.6 Conclusions

A MoI based on $\mathcal{H}^\infty$ theory was originally proposed by Whidborne et al (1994). The basic approach of using search methods to obtain parameters for weighting functions is appealing and hence other MoIs and variations based on the idea have been developed (for example, see Man et al (1999, pp 133-153) or Jamshidi et al (2003, pp 153-170)). In this chapter, another $\mathcal{H}^\infty$-based MoI is proposed. This MoI uses the two degree-of-freedom $\mathcal{H}^\infty$ method of Limebeer et al (1993). Although a MoI based on this approach was given by Whidborne et al (1995), in this chapter it is proposed that additional design variables are utilised along with the usual weighting function parameters, namely the parameters of the nominal model and the target model and the degree of sub-optimality.

The approach is illustrated on a distillation column benchmark problem. Although not all the design specifications were met, the results compare favourably with other designs for the same problem (Hoyle et al, 1991; Yaniv and Horowitz, 1991).

References

Balas, G.J., J.C. Doyle, K. Glover, A. Packard and R. Smith (2001), *μ-Analysis and Synthesis Toolbox: User's Guide, version 3*, The MathWorks, Inc., Natick, MA.

Chiang, R.Y. and M.G. Safonov (2001), *Robust Control Toolbox: User's Guide, version 2.08*, The MathWorks, Inc., Natick, MA.

Dakev, N.V., J.F. Whidborne, A.J. Chipperfield and P.J. Fleming (1997), Evolutionary H_∞ design of an electromagnetic suspension control system for a maglev vehicle, *Proc. IMechE, Part I: J. Syst. & Contr.*, 311(4):345–355.

Doyle, J.C., K. Glover, P.P. Khargonekar and B.A. Francis (1989), State space solutions to the standard H_2 and H_∞ control problems, *IEEE Trans. Autom. Control*, 34(8):831–847.

Glover, K. and D.C. McFarlane (1989), Robust stabilization of normalized coprime factor plant descriptions with H_∞-bounded uncertainty, *IEEE Trans. Autom. Control*, 34(8):821–830.

Green, M. and D.J.N. Limebeer (1995), *Linear Robust Control*, Prentice-Hall, Englewood Cliffs, N.J.

Hoyle, D.J., R.A. Hyde and D.J.N. Limebeer (1991), An H_∞ approach to two degree of freedom design, *Proc. 30th IEEE Conf. Decision Contr.*, Brighton, U.K., pp. 1579–1580.

Jamshidi, M., L.D.S. Coelho, R.A. Krohling and P.J. Fleming (2003), *Robust Control Systems with Genetic Algorithms*, CRC Press, Roca Baton, Florida.

Limebeer, D.J.N. (1991), The specification and purpose of a controller design case study, *Proc. 30th IEEE Conf. Decision Contr.*, Brighton, U.K., pp. 1579–1580.

Limebeer, D.J.N., E.M. Kasenally and J.D. Perkins (1993), On the design of robust two degree of freedom controllers, *Automatica*, 29(1):157–168.

Man, K.F., K.S. Tang and S. Kwong (1999), *Genetic Algorithms: Concepts and Designs*, Control and Signal Processing, Springer, London, U.K.

McFarlane, D.C. and K. Glover (1990), *Robust Controller Design Using Normalized Coprime Factor Plant Descriptions*, Vol. 138 of *Lect. Notes Control & Inf. Sci.*, Springer-Verlag, Berlin.

McFarlane, D.C. and K. Glover (1992), A loop shaping design procedure using H_∞ synthesis, *IEEE Trans. Autom. Control*, 37(6):759–769.

Ng, W.Y. (1989), *Interactive Multi-Objective Programming as a Framework for Computer-Aided Control System Design*, Vol. 132 of *Lect. Notes Control & Inf. Sci.*, Springer-Verlag, Berlin.

Postlethwaite, I., J.F. Whidborne, G. Murad and D.-W. Gu (1994), Robust control of the benchmark problem using H_∞ methods and numerical optimization techniques, *Automatica*, 30(4):615–619.

Skogestad, S. and I. Postlethwaite (1996), *Multivariable Feedback Control: Analysis and Design*, John Wiley, Chichester, U.K.

Whidborne, J.F., G. Murad, D.-W. Gu and I. Postlethwaite (1995), Robust control of an unknown plant – the IFAC 93 benchmark, *Int. J. Control*, 61(3), 589–640.

Whidborne, J.F., I. Postlethwaite and D.-W. Gu (1994), Robust controller design using H_∞ loop-shaping and the method of inequalities, *IEEE Trans. Control Syst. Technology*, 2(4):455–461.

Whidborne, J.F., P. Pangalos, Y.H. Zweiri and S.J. King (2002), A graphical user interface for computer-aided robust control system design, *Proc. Engineering Design Conf. 2002*, London, pp. 383–392.

Yaniv, O. and I.M. Horowitz (1991), Ill-conditioned plants : A case study, *Proc. 30th IEEE Conf. Decision Contr.*, Brighton, U.K., pp. 1596–1600.

Zakian, V. and U. Al-Naib (1973), Design of dynamical and control systems by the method of inequalities, *Proc. IEE*, 120(11):1421–1427.

Zhou, K., J.C. Doyle and K. Glover (1996), *Robust and Optimal Control*, Prentice Hall, Upper Saddle River, NJ.

12 Critical Control of the Suspension for a Maglev Transport System

James F Whidborne

Abstract. A problem that occurs frequently in control engineering is to control the output of a system so that the output is maintained within strict bounds. Violation of the bounds results in unacceptable and perhaps catastrophic operation. A system of this kind is said to be critical and can be dealt with by the design framework built upon the principles of matching and inequalities. In this chapter, the design of the controller for a critical control system, which is the active magnetic suspension for a maglev transport system, is described. The design illustrates the power of the framework for designing critical control systems.

12.1 Introduction

A common problem in control is to maintain the system error and other responses and outputs within certain prescribed bounds. For such systems, the aim of the control is to satisfy the following condition

$$|z_i(t,w)| \leq \varepsilon_i, \ i=1,\ldots,m \text{ for all } t \geq 0 \qquad (12.1)$$

where $z_i(t)$, $i=1,\ldots,m$ are the system outputs, $w = (w_1(t),\ldots,w_n(t))$ is the system input vector consisting of disturbance and reference signals, t is time and ε_i are positive real numbers representing the prescribed tolerances (bounds) on the outputs. For certain systems, violation of the tolerances by one or more of the output results in unacceptable, or even catastrophic operation. Such systems are known as critical control systems (Zakian, 1989; see also Whidborne and Liu, 1993).

The need for high-speed transport links in modern economies has generated great interest in magnetically levitated (maglev) vehicles. The two most effective suspension methods are electrodynamic suspension (EDS) and electromagnetic suspension (EMS). EDS requires super-conducting materials in order to produce sufficient repulsive force to levitate a vehicle over a conducting track. On the other hand, EMS employs the attractive forces of sets of electromagnets acting upwards to levitate the vehicle towards the tracks. Maglev technology is not new. Working maglev vehicles have been operational at the Transrapid[1] test facility at Emsland, Germany for nearly three

---
[1] http://www.transrapid.de/

decades. In 1985, Japanese Air Lines had a 55 km/h system operating on a 400 m track at Tsukuba Expo 85. From 1984 to 1995, a 42 km/h maglev ran on a 600 m track between Birmingham International rail station and airport in the United Kingdom.

Interest in maglev has recently been boosted by a 1998 decision by the US Congress to run a competition that will lead to the first substantial maglev demonstration line in the US. In 2001, two finalists from seven competing projects were selected (Holmer, 2003). Pennsylvania plans to run a 76-km line connecting Pittsburg to its airport and eastern suburbs, while Maryland plans a 60-km stretch of guide-way from Baltimore to Union Station in Washington via the Baltimore-Washington International Airport. Meanwhile, the first commercial high-speed maglev system has been constructed in Shanghai, between Pudong International Airport and the city centre. Trains take only eight minutes to travel the 30-km track with a maximum speed of 430 km/h (Ross, 2003; Holmer, 2003).

The EMS of a maglev vehicle is an example of a critical control system. EMS employs the attractive forces of sets of electro-magnets acting upwards to levitate the vehicle towards steel tracks. EMS vehicles are inherently unstable and thus require active control. Moreover, it is necessary to maintain an air-gap between the magnet and supporting guide-way in order to avoid undesirable contact between them. Thus it is a critical control system.

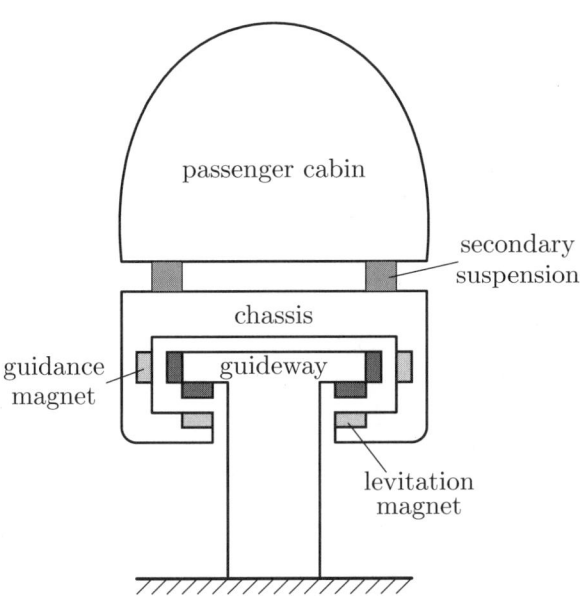

Fig. 12.1. Maglev vehicle cross-section

12 Critical Control of the Suspension for a Maglev Transport System

In this chapter, the method of inequalities (MoI) (Zakian and Al-Naib, 1973) is applied to the critical control problem of designing the suspension controller for a 140 m/s maglev vehicle consisting of a chassis supporting a passenger cabin by means of a secondary suspension of air-springs and hydraulic shock absorbers. The vehicle is shown in cross-section in Figure 12.1. The chassis is levitated by means of direct current electro-magnets under active control providing an attractive force to the guide-way. Skids under the chassis provide some protection in the event of a power failure. The aim of the control is to provide stability to an inherently unstable system, to ensure that the air-gap is strictly maintained and to provide sufficient quality of ride for the passengers. This chapter draws on earlier published work (Whidborne, 1993) but the design is for a larger maglev vehicle. Furthermore, the passive secondary suspension damping factor and spring constant parameters are included as design variables. Additional simulation results are also presented.

12.2 Theory

A critical system has to satisfy (12.1). For the maglev system, the control of three critical outputs will be considered. It is assumed that the input, w, can be decomposed into the sum of two components, so that

$$w = w_1 + w_2 \tag{12.2}$$

where $w_1 \in \mathcal{W}_1$, $w_2 \in \mathcal{W}_2$ and the function spaces $\mathcal{W}_1$ and $\mathcal{W}_2$ are defined below.

Definition 12.1. The input space $\mathcal{W}_1$ is the class of all functions $w_1 : \mathbb{R} \to \mathbb{R}$ where $w_1(t) = 0$ for all $t \leq 0$, $\dot{w}_1(t)$ is piece-wise continuous and there is a positive real number M_1 such that

$$M_1 = \sup\{|\dot{w}_1(t)| : t > 0, w_1 \in \mathcal{W}_1\} \tag{12.3}$$

Definition 12.2. The input space $\mathcal{W}_2$ is the class of all functions $w_2 : \mathbb{R} \to \mathbb{R}$ where $w_2(t) = 0$ for all $t \leq 0$, $\ddot{w}_2(t)$ is piece-wise continuous and there is a positive real number M_2 such that

$$M_2 = \sup\{|\ddot{w}_2(t)| : t > 0, w_2 \in \mathcal{W}_2\} \tag{12.4}$$

The input space $\mathcal{W}_1$ is characterised by a bound on the derivative of the input, and the input space $\mathcal{W}_2$ is characterised by a bound on the second derivative of the input.

Defining

$$\hat{z}_i := \sup\{|z_i(t, w)| : t \in \mathbb{R}, w_1 \in \mathcal{W}_1 w_2 \in \mathcal{W}_2\} \tag{12.5}$$

following Zakian (1986; 1987), for a linear time-invariant system,

$$\hat{z}_i = M_1 \int_0^\infty |z_i(t, h)| \, \mathrm{d}t + M_2 \int_0^\infty |z_i(t, r)| \, \mathrm{d}t \tag{12.6}$$

where the unit step responses, $z_i(t,h)$, and the unit ramp responses, $z_i(t,r)$, are piece-wise continuous ($z_i(t,r)$ are the integrals of the unit step responses, $z_i(t,h)$). Clearly if

$$\hat{z}_i \leq \varepsilon_i, \; i = 1, \ldots, m \tag{12.7}$$

then (12.1) is satisfied.

12.3 Model

The design model (Kortüm and Utzt, 1984; Sinha, 1987) is for a single electro-magnet considering only the vertical movement of the chassis and passenger cabin. The primary electro-magnetic suspension is described by a nonlinear first-order differential equation for the vertical force derived from the magnet-force law and the current/voltage relation. The secondary suspension is modelled as a linear mass-spring-damper system. The model configuration is shown in Figure 12.2.

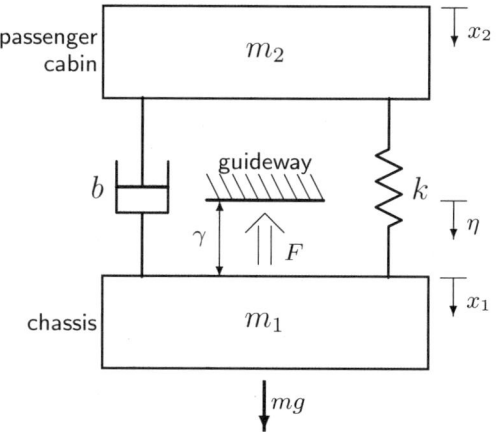

Fig. 12.2. Maglev vehicle model

The force, F, exerted by the magnet is

$$F(i, \gamma, t) = \frac{K_m}{2}\left[\frac{i(t)}{\gamma(t)}\right]^2 \tag{12.8}$$

where i is the current, γ is the gap between magnet and guide-way, and K_m is a constant. By consideration of the system at equilibrium, $K_m = 2mg(\gamma_0/i_0)^2$ where i_0 is the steady-state (nominal) current, γ_0 is the steady-state (nominal) air-gap, m is the total mass of the vehicle and g is the gravitational constant. If a voltage, V, is applied across the electro-magnet winding, the excitation current, i, is determined by

$$V(t) = Ri(t) + K_m \frac{\mathrm{d}}{\mathrm{d}t}\left[\frac{i(t)}{\gamma(t)}\right] \quad (12.9)$$

where R is the total resistance of the circuit (including the amplifier output resistance and the magnet winding resistance).

The chassis has mass m_1 and the passenger cabin has mass m_2. The secondary suspension has a spring constant k and damping constant b. The relationship between the two bodies is assumed linear and so, from Newton's law,

$$m_1\ddot{x}_1 + b(\dot{x}_1 - \dot{x}_2) + k(x_1 - x_2) = mg - F \quad (12.10)$$
$$m_2\ddot{x}_2 + b(\dot{x}_2 - \dot{x}_1) + k(x_2 - x_1) = 0 \quad (12.11)$$

where x_1 is the absolute position of the chassis, and x_2 is the absolute position of the passenger cabin.

The air-gap is related to the absolute chassis position by

$$\gamma(t) = x_1(t) - \eta(t) \quad (12.12)$$

where $\eta(t)$ is the disturbance resulting from variations in the guide-way profile.

Typical values of the parameters are (Kortüm and Utzt, 1984)

$R = 1.8$ Ohms, $\quad i_0 = 50$ A, $\quad \gamma_0 = 12$ mm
$m = 5333$ Kg, $\quad m_1 = 1767$ Kg, $\quad m_2 = 3550$ Kg
$g = 9.807$ m/s^2

Linearising (12.8) and (12.9) about the nominal operating values of F_0 and V_0 respectively where $V_0 = i_0/R$ and $F_0 = mg$, and combining with (12.10), (12.11) and (12.12), a linear state-space nominal model is obtained as

$$\begin{bmatrix} \dot{f} \\ \dot{x}_1 \\ \dot{x}_2 \\ \ddot{x}_1 \\ \ddot{x}_2 \end{bmatrix} = \begin{bmatrix} -\left(\frac{i_0}{\gamma_0}\right)^2 R & \frac{-R}{L} & 0 & 0 & 0 \\ 0 & 0 & 0 & 1 & 0 \\ 0 & 0 & 0 & 0 & 1 \\ \frac{-1}{m_1} & \frac{-k}{m_1} & \frac{k}{m_1} & \frac{-b}{m_1} & \frac{b}{m_1} \\ 0 & \frac{k}{m_2} & \frac{-k}{m_2} & \frac{b}{m_2} & \frac{-b}{m_2} \end{bmatrix} \begin{bmatrix} f \\ x_1 \\ x_2 \\ \dot{x}_1 \\ \dot{x}_2 \end{bmatrix} + \begin{bmatrix} \frac{i_0}{\gamma_0} & \left(\frac{i_0}{\gamma_0}\right)^2 R \\ 0 & 0 \\ 0 & 0 \\ 0 & 0 \\ 0 & 0 \end{bmatrix} \begin{bmatrix} v \\ \eta \end{bmatrix} \quad (12.13)$$

$$\begin{bmatrix} e \\ \ddot{x}_2 \\ v \end{bmatrix} = \begin{bmatrix} 0 & 1 & 0 & 0 & 0 \\ 0 & \frac{k}{m_2} & \frac{-k}{m_2} & \frac{b}{m_2} & \frac{-b}{m_2} \\ 0 & 0 & 0 & 0 & 0 \end{bmatrix} \begin{bmatrix} f \\ x_1 \\ x_2 \\ \dot{x}_1 \\ \dot{x}_2 \end{bmatrix} + \begin{bmatrix} 0 & -1 \\ 0 & 0 \\ 1 & 0 \end{bmatrix} \begin{bmatrix} v \\ \eta \end{bmatrix} \quad (12.14)$$

where $L = K_m/\gamma_0$ is the nominal inductance, $e = \gamma - \gamma_0$ is the air-gap error, $v = V - V_0$ is the control voltage and $f = F - F_0$ is the control force.

12.4 Design Specifications

An electromagnet excited by a constant voltage will either clamp to the rail or fall because the attractive forces decrease with an increasing air-gap. Thus the electromagnet suspension is open-loop unstable and so the primary aim of the controller is to provide stability. As well as ensuring the stability of the system, there are three other important considerations that are required when assessing the effectiveness of a control scheme.

(a) In order to avoid undesirable contact between the guide-way and electromagnet, the air-gap must be maintained between 7 and 17 mm with a nominal value of 12 mm. Contact leads to vibration, noise, friction and possible damage to the vehicle and/or guide-way. The nominal air-gap cannot not be too large as the lifting capacity decreases with a growing air-gap and power consumption and magnet weights would be infeasible if the gap were too large. A minimum air-gap of 7 mm assures a sufficient safety margin to accommodate the failure of a single magnet (Kortüm and Utzt, 1984) and therefore the maximum permissible error between the actual air-gap and nominal air-gap is 5 mm.
(b) Vertical acceleration experienced by passengers may be used as a measure of ride comfort. Typically, this should not exceed 500 mm/s^2 in either direction (Kortüm and Utzt, 1984). However, this is not a rigid requirement and it may be allowed to increase to as much as 1000 mm/s^2 (Armstrong, 1984).
(c) The applied voltage should be within the feasible limits of the electromagnets and power amplifiers. Kortüm and Utzt (1984) recommend that the control voltage should be within ±600 V. The power electronics will be cheaper and easier to implement if the voltage, V, is constrained to be positive. A reasonable limit is to maintain the voltage V between 0 and $2V_0$. Thus, the control voltage v should be within $\pm V_0$, that is ±90 V.

12.5 Performance for Control System Design

The performance measures used must reflect the objective of the control, namely the maximum air-gap and the quality of the ride. In addition, there is a constraint on the amount of control voltage that can be applied. Hence, performance indices based on the maximum variation in the air-gap error, e, the maximum variation in control voltage v and the maximum acceleration experienced by the passengers $\ddot{x}_2$ are proposed.

The major disturbance to the system is from variations in the guide-way height. These arise from two main sources, the variations due to gradients in the terrain, and guide-way irregularities resulting from sagging between the guide-way supports, maladjustments in the rail and thermal effects. Thus the guide-way height is given by

$$\eta = \eta_1 + \eta_2 \tag{12.15}$$

where η_1 is the disturbance resulting from the guide-way irregularities, and η_2 is the disturbance due to gradients in the terrain. For a 140 m/s vehicle, Müller (1977) has suggested a bound for η_1 as

$$M_1 = \sup\{|\dot\eta_1(t)| : t \geq 0\} = 30 \text{ mm/s} \tag{12.16}$$

and for η_2 as

$$M_2 = \sup\{|\ddot\eta_1(t)| : t \geq 0\} = 100 \text{ mm/s}^2 \tag{12.17}$$

With the inclusion of a stabilising controller, K, maximum outputs for the linear closed-loop system can be calculated using (12.16), (12.17) and (12.6) for the air-gap error, e, passenger acceleration, $\ddot{x}_2$ and control voltage, v, as

$$\hat{e}(K) = M_1 \int_0^\infty |e(\tau, h, K)|\,d\tau + M_2 \int_0^\infty |e(\tau, r, K)|\,d\tau \tag{12.18}$$

$$\hat{\ddot{x}}_2(K) = M_1 \int_0^\infty |\ddot{x}_2(\tau, h, K)|\,d\tau + M_2 \int_0^\infty |\ddot{x}_2(\tau, r, K)|\,d\tau \tag{12.19}$$

$$\hat{v}(K) = M_1 \int_0^\infty |v(\tau, h, K)|\,d\tau + M_2 \int_0^\infty |v(\tau, r, K)|\,d\tau \tag{12.20}$$

12.6 Design using the MoI

The design problem is posed as required by the method of inequalities (MoI) (Zakian and Al-Naib, 1973). The problem is to find a solution p so that the following is satisfied

$$\phi_i(p) \leq \varepsilon_i, \text{ for } i = 1, \ldots, m \tag{12.21}$$

where ϕ_i, $i = 1, \ldots, m$ are objective functions with corresponding tolerable values given by ε_i, and $p \in \mathcal{P}$ is a vector of design parameters chosen from some design set $\mathcal{P}$.

For the design of the maglev EMS control system, the objective functions are defined as

$$\begin{aligned}\phi_1 &= \hat{e}(K) \\ \phi_2 &= \hat{\ddot{x}}_2(K) \\ \phi_3 &= \hat{v}(K)\end{aligned} \tag{12.22}$$

From Section 12.4, tolerable bounds on the outputs are

$$\begin{aligned}\varepsilon_1 &= 5 \text{ mm} \\ \varepsilon_2 &= 500 \text{ mm/s}^2 \\ \varepsilon_3 &= 90 \text{ V}\end{aligned} \tag{12.23}$$

Satisfaction of inequalities (12.21) means that inequalities (12.7) and hence (12.1) are satisfied, and the critical control system will operate satisfactorily.

Measurements of the air-gap error, e, and the passenger cabin acceleration, $\ddot{x}_2$, are used for feedback. The control voltage v is related to the outputs by

$$v = \begin{bmatrix} K_1 & K_2 \end{bmatrix} \begin{bmatrix} e \\ \ddot{x}_2 \end{bmatrix} \qquad (12.24)$$

The controllers are parameterised by the design vector $p = [p_1, \ldots, p_{11}]$ as

$$K_1 = \frac{p_3 s^2 + p_2 s + p_1}{s(p_4 s + 1)} \qquad (12.25)$$

$$K_2 = \frac{p_7 s^2 + p_6 s + p_5}{p_9 s^2 + p_8 s + p_7} \qquad (12.26)$$

The secondary suspension stiffness, k, and damping factors, b, are included as design parameters p_{10} and p_{11} respectively.

The problem was implemented and solved in MATLAB using the moving boundaries process (Zakian and Al-Naib, 1973). The following objective functions were obtained

$$\phi_1 = 4.805 \text{ mm}$$
$$\phi_2 = 458.3 \text{ mm/s}^2 \qquad (12.27)$$
$$\phi_3 = 73.34 \text{ V}$$

which satisfy the inequalities (12.21).

The designed controllers are

$$K_1 = \frac{(280.5s^2 + 362.8s + 87.4)}{s(3.159s + 1)} \qquad (12.28)$$

$$K_2 = \frac{(19.13s^2 + 199.4s + 203)}{(46.86s^2 + 141.5s + 1)} \qquad (12.29)$$

and the designed secondary suspension stiffness and damping factors are $k = 90.0$ N/mm and $d = 51.48$ Ns/mm.

To test the performance of designed controllers, the responses of the outputs of the system model given by (12.8) – (12.12) to a test input, $\eta_{test}(t)$, are obtained using SIMULINK. The test input, $\eta_{test}(t) = \eta_{test(1)} + \eta_{test(2)}$, is shown in Figure 12.3 and is chosen (see (12.16) and (12.17)) to correspond to a severe guide-way disturbance. $\eta_{test(1)}$ represents alternating track disturbance derivatives of ± 30 mm/s and $\eta_{test(1)}$ represents alternating gradient derivatives of ± 100 mm/s^2. The response of the air-gap error, e, is shown in Figure 12.4 and is within the bounds of ± 5 mm. The response of the passenger cabin acceleration $\ddot{x}_2$ is shown in Figure 12.5, the bounds of ± 500 mm/s^2 are exceeded slightly, which is acceptable as discussed in Section 12.4. The response of the control voltage v is shown in Figure 12.6 and is within the bounds of ± 90 V.

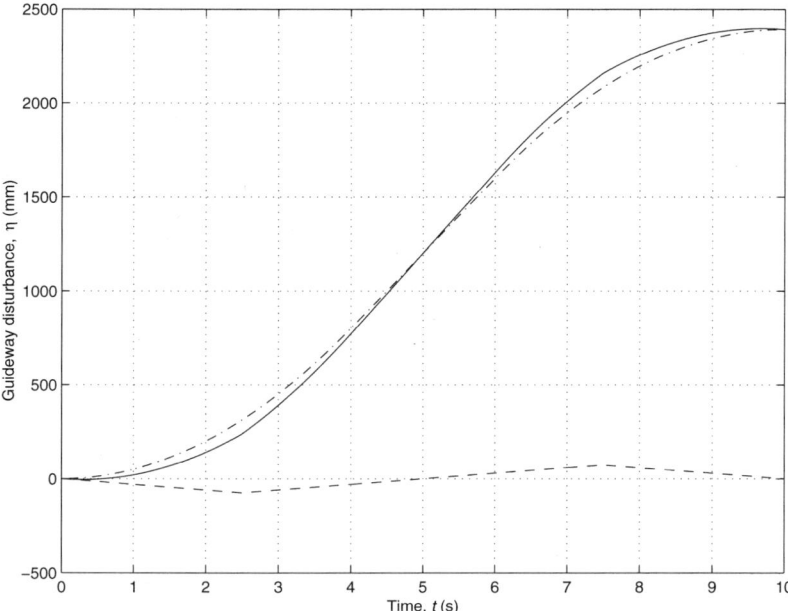

Fig. 12.3. Test input, η_{test}, (——) with disturbance component, $\eta_{test(1)}$, (- - -) and gradient component, $\eta_{test(2)}$, ($\cdot - \cdot -$)

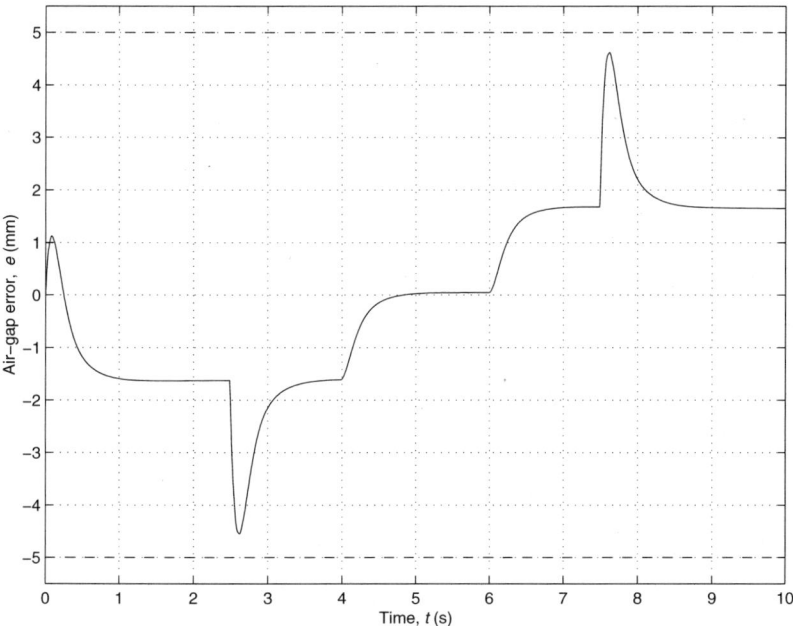

Fig. 12.4. Response of air-gap error, e, to test input

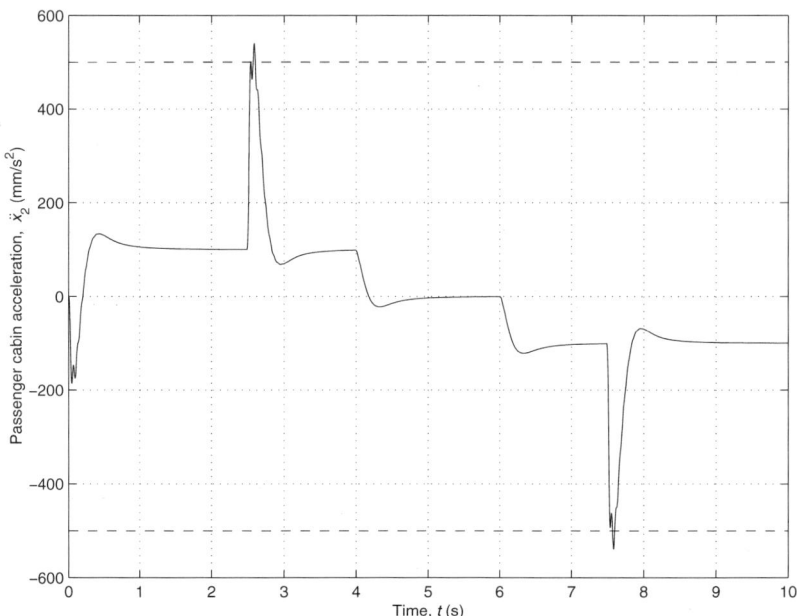

Fig. 12.5. Response of passenger cabin acceleration, $\ddot{x}_2$, to test input

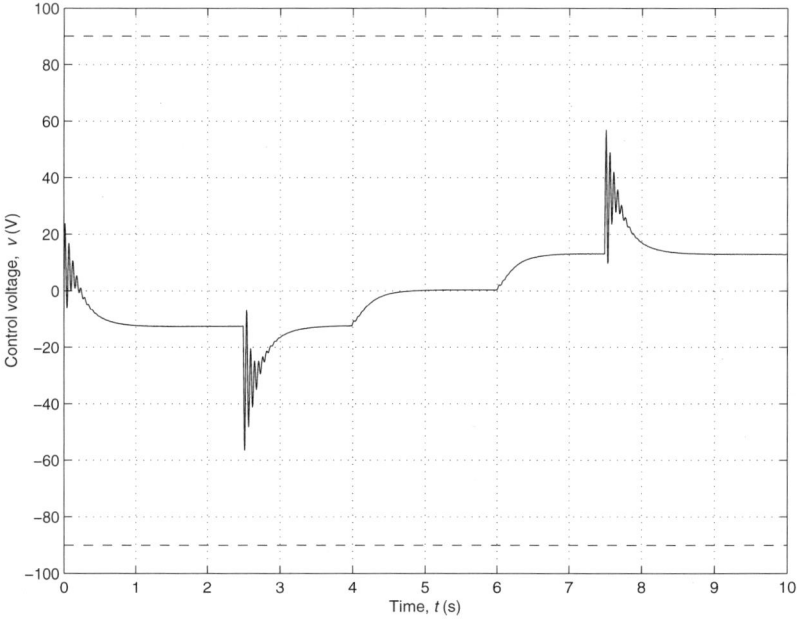

Fig. 12.6. Response of control voltage, v, to test input

12.7 Conclusions

The suspension control of a maglev vehicle is a critical system in that it is necessary to maintain the air-gap. It is also desirable to prevent excessive discomfort to the passengers by restricting the maximum acceleration that they experience. In addition, there is a restriction on the amount of control voltage variation that can be applied.

The main source of disturbance to the system results from variations in the guide-way, due to variations in the terrain and to guide-way irregularities. These variations can be modelled by input spaces characterised by bounds on the derivatives. The method of inequalities was used to design a controller for a maglev suspension system that can maintain the air-gap as well as restricting the maximum acceleration experienced by passengers. The controller was also tested in simulation.

The design is performed with the assumption of linearity. Although the design is tested with a non-linear model, performance can only be guaranteed for the linearised model. Furthermore, no account is taken of uncertainty in the model. Methods for analysing non-linear critical control systems have been developed (de Figueiredo and Chen, 1989; Blanchini, 1995; Lu, 1998). A theoretical basis for accounting for model uncertainty has been developed by Zakian (1984) and extended by Whidborne *et al* (2001)(see also Whidborne *et al*, 2000). The utilisation of these for this problem remains for further work.

References

Armstrong, D.S. (1984), Magnet/rail systems — a critical review of the options, *Proc. IMechE Conf. on Maglev Transport — Now and for the Future*, Solihull, U.K., pp. 59–66.

Blanchini, F. (1995), Non-quadratic Lyapunov functions for robust-control, *Automatica*, 31(3):451–461.

de Figueiredo, R.J.P. and G.R. Chen (1989), Optimal disturbance rejection for nonlinear control systems, *IEEE Trans. Autom. Control*, 34(12):1242–1248.

Holmer, P. (2003), Faster than a speeding bullet train, *IEEE Spectrum*, 40(8):30–34.

Kortüm, W. and A. Utzt (1984), Control law design and dynamic evaluations for a maglev vehicle with a combined lift and guidance suspension system, *ASME J. Dyn. Syst. Meas. & Control*, 106:286–292.

Lu, W.M. (1998), Rejection of persistent $\mathcal{L}_\infty$-bounded disturbances for nonlinear systems, *IEEE Trans. Autom. Control*, 43(12):1692–1702.

Müller, P.C. (1977), Design of optimal state-observers and its application to maglev vehicle suspension control, *Proc. 4th IFAC Symp. Multivariable Technological Systems*, Fredericton, Canada, pp. 175–182.

Ross, D. (ed.) (2003), China first with magnetic levitation, *IEE Review*, 49(2):17.

Sinha, P.K. (1987), *Electromagnetic Suspension: Dynamics and Control*, Peter Peregrinus, London.

Whidborne, J.F. (1993), EMS control system design for a maglev vehicle - A critical system, *Automatica*, 29(5):1345–1349.

Whidborne, J.F. and G.P. Liu (1993), *Critical Control Systems: Theory, Design and Applications*, Research Studies Press, Taunton, U.K.

Whidborne, J.F., V. Zakian, T. Ishihara, and H. Inooka (2000), Practical matching conditions for robust control design, *Proc. 3rd Asian Contr. Conf.*, Shanghai, China, pp. 1463–1468.

Whidborne, J.F., T. Ishihara, H. Inooka, T. Ono and T. Satoh (2001), Some practical matching conditions for robust control system design, King's College London Mechanical Engineering Department Report EM/2001/10, London, U.K.

Zakian, V. (1984), A framework for design: Theory of majorants, Technical Report 604, Control Systems Centre, UMIST, Manchester, U.K.

Zakian, V. (1986), A performance criterion, *Int. J. Control*, 43(3):921–931.

Zakian, V. (1987), Input spaces and output performance, *Int. J. Control*, 46(1):185–191.

Zakian, V. (1989), Critical systems and tolerable inputs, *Int. J. Control*, 49(4):1285–1289.

Zakian, V. and U. Al-Naib (1973), Design of dynamical and control systems by the method of inequalities, *Proc. IEE*, 120(11):1421–1427.

13 Critical Control of Building under Seismic Disturbance

Suchin Arunsawatwong

Abstract. The control of a building subject to a seismic disturbance is critical in the sense that the drift in each interstorey is required to remain strictly within a prescribed bound, despite any earthquake excitation. This is because any violation of the bound can lead to the collapse of the building. This chapter describes the design of such a control system using the principle of matching and the method of inequalities, where all earthquakes supposed to be possible to happen are explicitly taken into account and are modelled as functions such that the two norms of the magnitude and rate of change are uniformly bounded by respective constants. As a result, one can ensure that (provided a design solution is found) the building is tolerant to all the earthquakes that can be modelled in this way. A design is carried out for the case of a six storey laboratory-scaled building. The numerical results demonstrate that the design framework employed here gives a realistic formulation of the design problem and therefore is suitable for designing critical systems in practice.

13.1 Introduction

A critical system (Zakian, 1989; see also Whidborne and Liu, 1993), as explained in Chapter 1, is a system whose outputs are required to remain strictly within prescribed bounds in the presence of external inputs impinging on the system; any violation of these bounds can give rise to unacceptable (and perhaps catastrophic) consequences.

A principal aim in designing such a system is to ensure that the absolute value of any critical scalar output v of the system lies within a prescribed bound for all time and for all *possible inputs* (that is, inputs that can happen or are likely to happen). That is to say,

$$|v(t,f)| \leq \varepsilon \quad \forall t \ \forall f \tag{13.1}$$

where $v(t, f)$ denotes the value of an output v in response to a possible input f at time t, and ε is a prescribed bound on v.

Clearly, the effect of the input f on the system is of great importance. For such a design problem to be realistically formulated, it is therefore important that all the possible f's be considered explicitly and performance measures that are directly related to the criterion (13.1) be used.

Control of a building excited by seismic disturbances is critical, since each interstorey drift is required to remain strictly within a prescribed bound, despite earthquake excitation; any violation of the bound resulting in the possibility of collapse of the building. Although there has been much research on the design of building control systems (see, for example, Tadjbaksh and Rofooei, 1992; Schmitendorf *et al*, 1994; Jabbari *et al*, 1995; Chopra, 1998, and the references therein), none has explicitly taken into account the notion of critical systems. Moreover, the performance measures used in those works (see *ibid.*) are hardly related to the criteria (13.1), which are actually those used by civil engineers to evaluate the performance of the control system. Consequently, designers have had to perform simulations with a number of earthquake waveforms after obtaining a design result, so as to check whether the building is tolerant to all the earthquakes that can happen or are likely to happen. This usually involves repeated redesign and simulation of a performance, a process that can be very time-consuming.

This chapter describes the design of control system for buildings subject to seismic disturbances. The design is based on the principle of matching (Zakian, 1991, 1996) and the method of inequalities (Zakian and Al-Naib, 1973). The design problem is formulated by requiring that:

- each interstorey drift of the building be maintained within the prescribed bound for all possible earthquakes;
- the control force generated by the actuator be restricted within a certain range during operation.

In this connection, the principal design criteria used in this work are expressed explicitly as inequalities of the form

$$\phi_v \leq \varepsilon \tag{13.2}$$

which is equivalent to (13.1), where the performance measure ϕ_v is defined as

$$\phi_v := \sup_{f \in \mathcal{F}} \sup_{t \geq 0} |v(t, f)| \tag{13.3}$$

the bound ε is the largest value of ϕ_v that can be tolerated; v represents either an interstorey drift or the control force; the possible set $\mathcal{F}$ represents a set including all earthquakes that used to happen or can happen or are likely to happen, and is defined as

$$\mathcal{F} := \{f : \|f\|_2 \leq M \text{ and } \|\dot{f}\|_2 \leq D\} \tag{13.4}$$

and the bounds M and D can be easily determined from past records. It may be noted that the approach adopted here was first presented in Arunsawatwong *et al* (2002). From the above, it is evident that this approach facilitates a realistic formulation of the design problem, provided ϕ_v can be calculated in practice.

13 Critical Control of Building under Seismic Disturbance

To illustrate the effectiveness of the design Framework, a numerical design is carried out for the case of a six storey laboratory-scaled building. Notice that the critical control problem formulated in this way contains six critical output variables (see Section 13.4) and other variables, connected with the actuators, that are required to be bounded. It is clear that only the design framework, based on the principle of matching and the principle of inequalities, adopted in this chapter can deal with the problem in a systematic fashion.

Since in reality earthquake waveforms do not vary persistently for all time but are transient and since the possible set $\mathcal{F}$ contains only transient inputs, it is obvious that $\mathcal{F}$ is an appropriate representation of a set of possible earthquake waveforms. It is interesting to note that the earthquake excitation considered by Kelly *et al* (1987) are subject to the restrictions of the form

$$\|f\|_\infty \leq M \text{ and } \|\dot{f}\|_\infty \leq D \tag{13.5}$$

The set of possible earthquakes characterised by (13.5) contains both *persistent inputs* (that is, those that vary persistently for all time) and *transient inputs* (that is, those that do not) and consequently includes many *fictitious inputs* (that is, inputs that cannot happen or are not likely to happen) which can give rise to more conservatism in the design. For further details on transient and persistent inputs, see Zakian (1989, 1996) and Chapter 1.

The organisation of the chapter is as follows. Section 13.2 recapitulates Lane's (1995) method (see Chapter 2) for computing the performance measure ϕ_v. Section 13.3 explains the model of the building used in this work. Section 13.4 explains how the design problem is formulated. In Section 13.5, the numerical results are presented. In Section 13.6, the discussion and conclusions are given.

13.2 Computation of Performance Measure

The design requirement in (13.1) gives rise to a problem of evaluating the associated performance measure ϕ_v. Once ϕ_v is readily computable, the inequality (13.2) becomes a useful design criterion. Obviously, the calculation of ϕ_v is a cumbersome task if effected directly from the definition in (13.3). In the following, Lane's (1995) method, which provides a practical tool for computing ϕ_v, is reviewed. For the details on the theory and the proofs, see Lane (1995) and also Chapter 2.

Consider a linear, time-invariant and non-anticipative system whose input f and output v are related by the convolution integral

$$v(t,f) = \int_{-\infty}^{\infty} h(t-\tau)f(\tau)d\tau \tag{13.6}$$

where h denotes the impulse response of the system whose transfer function is $H(s)$ (see Figure 13.1).

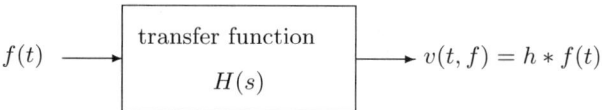

Fig. 13.1. System whose input and output are related by convolution integral

Lane (1995) notices that the infinite-dimensional optimisation problem in (13.3) is convex. Using this fact, he proves that such a problem is equivalent to the following one-dimensional minimisation problem:

$$\phi_v = \min\{\psi(\theta) : 0 \leq \theta \leq \pi/2\} \tag{13.7}$$

where the objective function ψ and its derivative are given by

$$\psi(\theta) = \left(\|g\|_2^2 \cos\theta + \|\dot{g}\|_2^2 \sin\theta\right)^{1/2} \times \left(M^2 \cos\theta + D^2 \sin\theta\right)^{1/2} \tag{13.8}$$

$$\frac{d\psi}{d\theta} = \frac{D^2 \|g\|_2^2 - M^2 \|\dot{g}\|_2^2}{2\psi(\theta)} \tag{13.9}$$

and g is the impulse response of the system whose transfer function $G(s)$ is given by

$$G(s) = \frac{H(s)}{(\sqrt{\cos\theta} + s\sqrt{\sin\theta})^2} \tag{13.10}$$

Further, he points out that the problem (13.7) is quasi convex and therefore can be solved effectively by a straightforward bisection algorithm. See Lane (1995) and also Chapter 2 for the details of the algorithm.

The function ψ can readily be determined once the numbers $\|g\|_2$ and $\|\dot{g}\|_2$ are obtained. For the purpose of this chapter, attention is restricted only to finite-dimensional systems. In this connection, assume that $H(s)$, and hence $G(s)$, are rational transfer functions; and let a state-space realisation of $G(s)$ be given by $\{A_1, b_1, c_1, 0\}$. There are two ways of calculating $\|g\|_2$ and $\|\dot{g}\|_2$.

- **Method I:** Using well-known results (see, for example, Lane, 1995 and the references therein), it can be shown that $\|g\|_2^2$ and $\|\dot{g}\|_2^2$ are readily given by

$$\|g\|_2^2 = c_1 W_{\text{con}} c_1^T, \tag{13.11}$$

$$\|\dot{g}\|_2^2 = (c_1 A_1) W_{\text{con}} (c_1 A_1)^T \tag{13.12}$$

where W_{con} is the controllability gramian of the realisation $\{A_1, b_1, c_1, 0\}$ and satisfies the Lyapunov equation

$$A_1 W_{\text{con}} + W_{\text{con}} A_1^T + b_1 b_1^T = 0 \tag{13.13}$$

Lane (1995) used this method for computing $\|g\|_2$ and $\|\dot{g}\|_2$.

- **Method II:** It can be easily verified that g and $\dot{g}$ are obtained from the following differential-algebraic equations:

$$\begin{aligned}\dot{x}(t) &= A_1 x(t), \quad x(0) = b_1 \\ g(t) &= c_1 x(t) \\ \dot{g}(t) &= c_1 A_1 x(t)\end{aligned} \qquad (13.14)$$

When the system (13.14) is asymptotically stable, $g(t)$ and $\dot{g}(t)$ decay to zero exponentially as $t \to \infty$. In this case, one can therefore approximate $\|g\|_2^2$ and $\|\dot{g}\|_2^2$ using the following truncated integrals:

$$\|g\|_2^2 \approx \int_0^T g^2(t)dt \quad \text{and} \quad \|\dot{g}\|_2^2 \approx \int_0^T \dot{g}^2(t)dt \qquad (13.15)$$

where the number T is chosen to be sufficiently large. In practice, the numerical solution of (13.14) is readily obtainable using, for example, Zakian I_{MN} recursions (see Zakian, 1975, and also Arunsawatwong, 1998, for details), which are efficient and reliable even if the equations are stiff. The integration in (13.15) can readily be performed using basic numerical algorithms. It is also of interest to note that using this approach to computing $\|g\|_2$ and $\|\dot{g}\|_2$, Lane's (1995) method can be extended to the case of systems with time-delays.

If Method I fails to give satisfactory results then Method II may provide an alternative tool in computing $\|g\|_2^2$ and $\|\dot{g}\|_2^2$. When the system matrix A_1 has eigenvalues that are very widely distributed on the complex plane, the gramian W_{con} obtained from solving (13.13) may suffer from round-off errors in the computation. If this is the case, then the computed W_{con} has a large condition number (for example, 10^{39}) and is not positive definite (sometimes causing the computed values of $\|g\|_2^2$ and $\|\dot{g}\|_2^2$ to be negative). Recall that for a controllable and observable system, if $G(s)$ is stable then W_{con} is positive definite. Accordingly, when it is found that W_{con} is not positive definite, Method II is preferable and is employed instead of Method I.

Let p denote a vector of design parameters (or a controller). In solving the inequality

$$\phi_v(p) \leq \varepsilon \qquad (13.16)$$

by numerical methods, it is necessary to obtain a *stability point*—that is, a point p such that

$$\phi_v(p) < \infty. \qquad (13.17)$$

This is because search algorithms, in general, are able to seek a solution of (13.17) only if they start from such a point; see Zakian (1987) and Chapter 1 for the details. In order to formulate the problem of determining a stability point so that it is suitable for solution by numerical methods, it is necessary to replace (13.17) by an equivalent statement that is soluble by numerical methods.

Assume that a state-space realisation of $H(s)$ is $\{A, b, c, d\}$. Then it is easy to show under mild assumptions that $\phi_v < \infty$ if and only if all the eigenvalues of A lie in the open left half of the complex plane, that is to say,

$$\alpha(p) < 0 \qquad (13.18)$$

where the number $\alpha(p)$ is called the *spectral abscissa* of A and defined as

$$\alpha(p) := \max_i \{\operatorname{Re} \lambda_i(A)\} \qquad (13.19)$$

and $\lambda_i(A)$ denotes an eigenvalue of A. Notice that $\alpha(p) < \infty$ for all values of p and that α can be computed economically in practice. Accordingly, the inequality (13.18) is always soluble by numerical methods (Zakian, 1987) and hence provides a useful computational tool for determining a stability point.

13.3 Model of Building

In this work, a control design will be performed for a six storey laboratory-scaled building. To this end, this section describes a linear, lumped-parameter model of the building in planer motion. The model is taken from Kelly *et al* (1987) (see also Schmitendorf *et al*, 1994). For the purpose of demonstrating the usefulness of the present framework to the design of critical systems, the model is modified by setting the spring and damping constants k_0 and c_0 to large values so that the effect of the base isolation system can be neglected; see Table 13.1. Note, in passing, that most of buildings found in practice have no base isolation system.

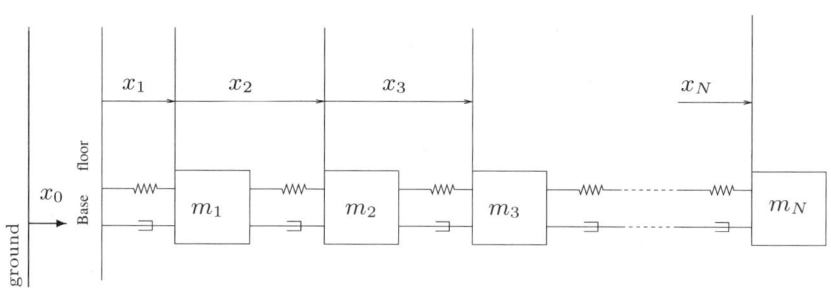

Fig. 13.2. Lumped-parameter model of the building based on interstorey drifts.

Consider the lumped-parameter model of the building shown in Figure 13.2. Assume that the actuator is installed at the sixth floor of the building and the actuator dynamic is neglected. Using Newton's law, one can easily derive the following differential equations.

$$\ddot{x}_1 = -\frac{c_0}{m_1}\dot{x}_1 + \frac{c_1}{m_1}\dot{x}_2 - \frac{k_0}{m_1}x_1 + \frac{k_1}{m_1}x_2 - \ddot{x}_0$$
$$\ddot{x}_2 = \frac{c_0}{m_1}\dot{x}_1 - (\frac{c_1}{m_1} + \frac{c_1}{m_2})\dot{x}_2 + \frac{c_2}{m_2}\dot{x}_3 + \frac{k_0}{m_1}x_1 - (\frac{k_1}{m_1} + \frac{k_1}{m_2})x_2 + \frac{k_2}{m_2}x_3$$
$$\ddot{x}_3 = \frac{c_1}{m_2}\dot{x}_2 - (\frac{c_2}{m_2} + \frac{c_2}{m_3})\dot{x}_3 + \frac{c_3}{m_3}\dot{x}_4 + \frac{k_1}{m_2}x_2 - (\frac{k_2}{m_2} + \frac{k_2}{m_3})x_3 + \frac{k_3}{m_3}x_4 \quad (13.20)$$
$$\ddot{x}_4 = \frac{c_2}{m_3}\dot{x}_3 - (\frac{c_3}{m_3} + \frac{c_3}{m_4})\dot{x}_4 + \frac{c_4}{m_4}\dot{x}_5 + \frac{k_2}{m_3}x_3 - (\frac{k_3}{m_3} + \frac{k_3}{m_4})x_4 + \frac{k_4}{m_4}x_5$$
$$\ddot{x}_5 = \frac{c_3}{m_4}\dot{x}_4 - (\frac{c_4}{m_4} + \frac{c_4}{m_5})\dot{x}_5 + \frac{c_5}{m_5}\dot{x}_6 + \frac{k_3}{m_4}x_4 - (\frac{k_4}{m_4} + \frac{k_4}{m_5})x_5 + \frac{k_5}{m_5}x_6$$
$$\ddot{x}_6 = \frac{c_4}{m_5}\dot{x}_5 - (\frac{c_5}{m_5} + \frac{c_5}{m_6})\dot{x}_6 + \frac{k_4}{m_5}x_5 - (\frac{k_5}{m_5} + \frac{k_5}{m_6})x_6 + \frac{1}{m_6}u$$

where x_i ($i = 1, 2, \ldots, 6$) is the interstorey drift of the ith floor[1]; x_0 is the ground displacement relative to an inertial reference frame; m_i ($i = 1, 2, \ldots, 6$) is the mass of the ith floor; k_i and c_i ($i = 0, 1, \ldots, 5$) denote the spring and damping coefficients, respectively, in connection with the ith floor and the $(i-1)$th floor; and u is the control force generated by the actuator.

Let $f := \ddot{x}_0$ be the ground acceleration caused by earthquake. By defining the state vector

$$x_p := [x_1, x_2, \ldots, x_6, \dot{x}_1, \dot{x}_2, \ldots, \dot{x}_6]^T$$

and by letting all the interstorey drifts x_i be the measured outputs

$$y(t) := [x_1, x_2, x_3, x_4, x_5, x_6]^T$$

it follows immediately from (13.20) that the building dynamics is described by

$$\begin{aligned} \dot{x}_p(t) &= A_p x_p(t) + B_p u(t) + E_p f(t) \\ y(t) &= C_p x_p(t) \end{aligned} \quad (13.21)$$

where the matrices A_p, B_p, C_p, E_p are given by

$$A_p = \begin{bmatrix} 0_{6\times 6} & I_{6\times 6} \\ -A_K & -A_C \end{bmatrix} \quad (I_{6\times 6} = \text{identity matrix}, \; 0_{6\times 6} = \text{zero matrix})$$

---

[1] an interstorey drift of the ith floor is the displacement of the ith floor relative to that of the $(i-1)$th floor.

$$A_K = \begin{bmatrix} \frac{k_0}{m_1} & -\frac{k_1}{m_1} & 0 & 0 & 0 & 0 \\ -\frac{k_0}{m_1} & \frac{k_1}{m_1}+\frac{k_1}{m_2} & -\frac{k_2}{m_2} & 0 & 0 & 0 \\ 0 & -\frac{k_1}{m_2} & \frac{k_2}{m_2}+\frac{k_2}{m_3} & -\frac{k_3}{m_3} & 0 & 0 \\ 0 & 0 & -\frac{k_2}{m_3} & \frac{k_3}{m_3}+\frac{k_3}{m_4} & -\frac{k_4}{m_4} & 0 \\ 0 & 0 & 0 & -\frac{k_3}{m_4} & \frac{k_4}{m_4}+\frac{k_4}{m_5} & -\frac{k_5}{m_5} \\ 0 & 0 & 0 & 0 & -\frac{k_4}{m_5} & \frac{k_5}{m_5}+\frac{k_5}{m_6} \end{bmatrix}$$

$$A_C = \begin{bmatrix} \frac{c_0}{m_1} & -\frac{c_1}{m_1} & 0 & 0 & 0 & 0 \\ -\frac{c_0}{m_1} & \frac{c_1}{m_1}+\frac{c_1}{m_2} & -\frac{c_2}{m_2} & 0 & 0 & 0 \\ 0 & -\frac{c_1}{m_2} & \frac{c_2}{m_2}+\frac{c_2}{m_3} & -\frac{c_3}{m_3} & 0 & 0 \\ 0 & 0 & -\frac{c_2}{m_3} & \frac{c_3}{m_3}+\frac{c_3}{m_4} & -\frac{c_4}{m_4} & 0 \\ 0 & 0 & 0 & -\frac{c_3}{m_4} & \frac{c_4}{m_4}+\frac{c_4}{m_5} & -\frac{c_5}{m_5} \\ 0 & 0 & 0 & 0 & -\frac{c_4}{m_5} & \frac{c_5}{m_5}+\frac{c_5}{m_6} \end{bmatrix}$$

$B_p = [0,\ 0,\ 0,\ 0,\ 0,\ 0,\ 0,\ 0,\ 0,\ 0,\ 0,\ 1/m_6]^T$

$C_p = [I_{6\times 6}\ \ 0_{6\times 6}], \qquad D_p = [0,\ 0,\ 0,\ 0,\ 0,\ 0]^T$

$E_p = [0,\ 0,\ 0,\ 0,\ 0,\ 0,\ -1,\ 0,\ 0,\ 0,\ 0,\ 0]^T$

The mass, spring and damping coefficients (m_i, k_i and c_i) of the building are listed in Table 13.1.

Table 13.1. Model coefficients (Kelly et al, 1987)

Mass (kg)	Spring (kN/m)	Damping (kNs/m)
$m_1 = 6,800$	$k_0 = 600,000,000$	$c_0 = 120,000$
$m_2 = 5,897$	$k_1 = 33,732$	$c_1 = 67$
$m_3 = 5,897$	$k_2 = 29,093$	$c_2 = 58$
$m_4 = 5,897$	$k_3 = 28,621$	$c_3 = 57$
$m_5 = 5,897$	$k_4 = 24,954$	$c_4 = 50$
$m_6 = 5,897$	$k_5 = 19,059$	$c_5 = 38$

Regarding the dynamic model of the building mentioned above, it may be noted that the state variables used by Kelly et al (1987) are the displacement and velocity of each floor relative to the ground. By contrast, in this work and also in Schmitendorf et al (1994), the states are defined as the interstorey drifts and velocities. Such a state-space representation is preferable here because the performance specifications used in the design of building control systems are expressed explicitly in terms of the interstorey drifts (see Section 13.4).

13.4 Design Formulation

13.4.1 Control configuration

In the subsequent design, a dynamic output feedback controller is used for the building control system, where the interstorey drifts x_i of each floor are fedback to the controller. The configuration of the feedback control system is shown in Figure 13.3.

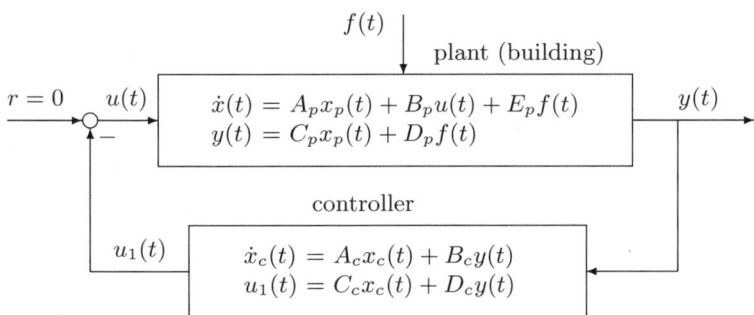

Fig. 13.3. Control system configuration

Suppose that the controller is represented by the following state equations:
$$\dot{x}_c(t) = A_c x_c(t) + B_c y(t)$$
$$u_1(t) = C_c x_c(t) + D_c y(t) \qquad (13.22)$$

where u_1 is the controller output. It is easy to verify that the state equations of the closed-loop control system are given by

$$\begin{bmatrix} \dot{x}_p(t) \\ \dot{x}_c(t) \end{bmatrix} = \begin{bmatrix} A_p - B_p D_c C_p & -B_p C_c \\ B_c C_p & A_c \end{bmatrix} \begin{bmatrix} x_p(t) \\ x_c(t) \end{bmatrix} + \begin{bmatrix} E_p - B_p D_c D_p \\ B_c D_p \end{bmatrix} f(t) \quad (13.23)$$

It is noted (see, for example, Kuhn and Schmidt, 1987, and also the references therein) that not every element of the matrices in equation (13.22) is independent; therefore, the maximum number of independent parameters is less than the total number of the elements of A_c, B_c, C_c and D_c. More specifically, when there is no constraint on the controller, the number n of independent parameters required to describe the input-output behaviour of the controller (13.22) is given by

$$n = n_c(m + r) + mr$$

where n_c is the order of the controller (that is, $n_c = \dim x_c$), m is the number of the measured outputs y_i (that is, $m = \dim y$), and r is the number of the actuators (that is, $r = \dim u$).

Although there are a number of state-space representations that can be used for the controller, the one given by Kuhn and Schmidt (1987) is employed here. Since there is one actuator used in the control system, it follows immediately that $r = 1$ and thus the associated matrices A_c, B_c, C_c, D_c are of the form

$$A_c = \begin{bmatrix} 0 & 0 & \ldots & 0 & p_1 \\ 1 & 0 & \ldots & 0 & p_2 \\ 0 & 1 & \ldots & 0 & p_3 \\ \vdots & \vdots & & \vdots & \vdots \\ 0 & 0 & \ldots & 1 & p_{n_c} \end{bmatrix}, \quad C_c = [0\ 0\ \ldots 0\ 1]$$

$$B_c = \begin{bmatrix} p_{n_c+1} & p_{n_c+2} & \cdots & p_{n_c+m} \\ p_{n_c+m+1} & p_{n_c+m+2} & \cdots & p_{n_c+2m} \\ \vdots & \vdots & & \vdots \\ p_{n_c+m(n_c-1)+1} & p_{n_c+m(n_c-1)+2} & \cdots & p_{n_c+n_cm} \end{bmatrix}$$

$$D_c = \begin{bmatrix} p_{n_c(m+1)+1} & p_{n_c(m+1)+2} & \cdots & p_{n_c(m+1)+m} \end{bmatrix}$$

(13.24)

where $p_1, p_2, \ldots, p_{n_c(m+1)+m}$ are design parameters. For example, when a first-order controller is used (that is, $n_c = 1$), the matrices A_c, B_c, C_c, D_c are

$$\begin{aligned} A_c &= p_1, \quad B_c = [p_2,\ p_3,\ p_4,\ p_5,\ p_6,\ p_7] \\ C_c &= 1, \quad D_c = [p_8,\ p_9,\ p_{10},\ p_{11},\ p_{12},\ p_{13}] \end{aligned}$$

(13.25)

For further details on the controller structure, see Kuhn and Schmidt (1987).

13.4.2 Design specifications

According to the building standard (see Article 1630.10 *Storey Drift Limitation* in Uniform Building Code, 1997), it is noted that the interstorey drifts should not exceed 0.025 times the storey height for structures having a fundamental period of less than 0.7 seconds and should not exceed 0.020 times the storey height for structures having a fundamental period of 0.7 seconds or greater. This standard leads to the principal design specifications for the building control system subject to seismic disturbances.

Suppose that the building described in Section 13.3 is one metre high. Then, from the above standard, one can arrive at the design requirements that each interstorey drift x_i ($i = 1, 2, \ldots, 6$) of the scaled building model should be maintained within ± 25 mm during earthquake excitation; otherwise the building may collapse. Furthermore, because of the limitation of the actuator, it is desirable that the control force u generated by the actuator should be restricted within the range ± 100 kN during operation.

Hence, the principal design objectives for the building control system are expressed as the following inequalities:

$$\phi_{x_i}(p) \leq 25 \text{ mm}, \ i = 1, 2, \ldots, 6$$
$$\phi_u(p) \leq 100 \text{ kN} \qquad (13.26)$$

where p is the vector of design parameters; the performance measures ϕ_{x_i} and ϕ_u are defined, according to (13.3), as

$$\phi_{x_i}(p) := \sup_{f \in \mathcal{F}} \sup_{t \geq 0} |x_i(t, f, p)|$$
$$\phi_u(p) := \sup_{f \in \mathcal{F}} \sup_{t \geq 0} |u(t, f, p)|$$

Notice that ϕ_{x_i} is seen as the greatest magnitude of the interstorey drift x_i for all possible earthquakes and for all time. It should be noted that the design criteria (13.26) are the actual requirements used by structure engineers to assess the performance of the building control system.

From the specifications (13.26), it is obvious that the design problem has several objectives. Consequently, in accordance with the method of inequalities (Zakian and Al-Naib, 1973), the problem is formulated as a set of inequalities. In solving the inequalities (13.26) by numerical methods, it is necessary that search algorithms start from a stability point p, for which $\phi_{x_i}(p) < \infty$ for all i and for which $\phi_u(p) < \infty$. In general, such a point p can be found by solving the inequality

$$\alpha(p) \leq -\varepsilon \qquad (13.27)$$

where $\alpha(p)$ is the spectral abscissa of A_{cl} (see (13.19)); A_{cl} is the system matrix of (13.23), which is given by

$$A_{\text{cl}} = \begin{bmatrix} A_p - B_p D_c C_p & -B_p C_c \\ B_c C_p & A_c \end{bmatrix}$$

and the bound ε is a small positive number. For the case of building control systems, since the building is open-loop stable, one may obtain a stability point easily by hand.

In this work, the set of inequalities (13.26) is solved by a numerical search algorithm called the *moving boundaries process*. For the details of the algorithm, see Zakian and Al-Naib (1973).

13.5 Numerical Results

This section presents the numerical results of the design problem formulated in Section 13.4. At this point, the possible set $\mathcal{F}$ needs to be defined. Accordingly, the bounds M and D are chosen in such a way that the set $\mathcal{F}$ includes some earthquakes that used to happen.

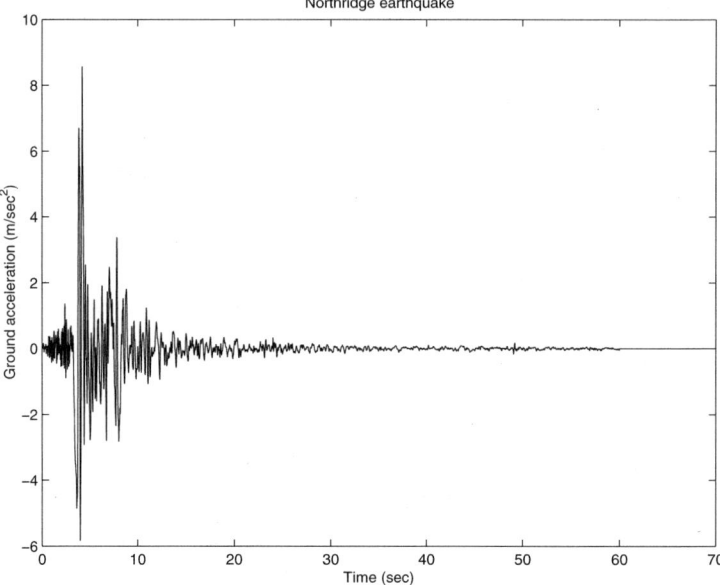

Fig. 13.4. Northridge earthquake waveform (ground acceleration)

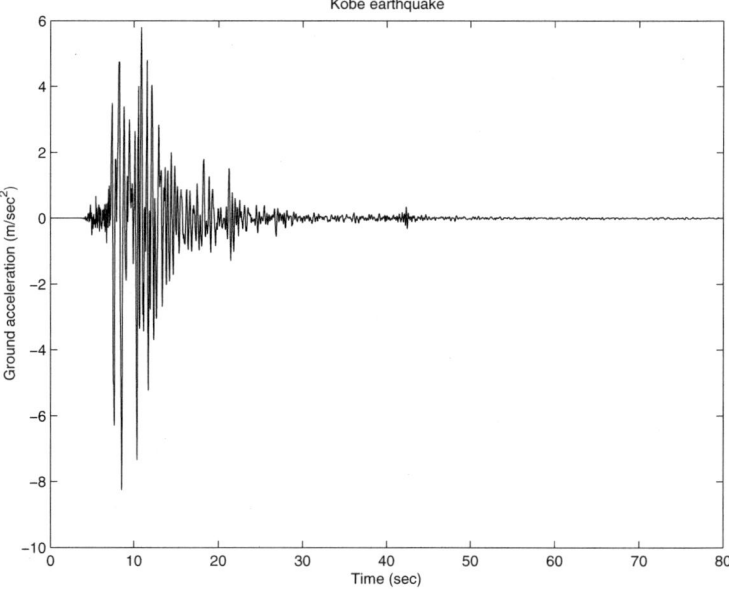

Fig. 13.5. Kobe earthquake waveform (ground acceleration)

Let M and D be given as follows:

$$M = 7.24 \text{ m/s}^2 \text{ and } D = 106.79 \text{ m/s}^3$$

By straightforward calculation, it is easily found that $\mathcal{F}$ includes well-known earthquakes such as the Northridge, the Kobe, the El Centro and the Hachinohe earthquakes. In the following, however, only the Northridge and the Kobe earthquakes will be used for verifying the design results because they are more severe than the others in the sense that they have larger two-norms of both magnitude and rate of change. The waveforms of the Northridge and the Kobe earthquakes are shown in Figures 13.4 and 13.5, respectively.

In obtaining a design solution, one chooses a controller structure, which corresponds to the dimension of the design parameter p, and then solves the inequalities (13.26). At the beginning, controllers of order $n_c = 1, 2, 3$ (see (13.22) and (13.24)) were tried. After a number of iterations, the moving boundaries process was not able to locate a design solution. For this reason and because the restriction on the control force u is not so rigid as those on the interstorey drifts x_i, the problem is then reformulated by relaxing the bound on ϕ_u so that the design objectives become

$$\begin{aligned}\phi_{x_i}(p) &\leq 25 \text{ mm}, i = 1, 2, \ldots, 6 \\ \phi_u(p) &\leq 220 \text{ kN}\end{aligned} \qquad (13.28)$$

It may be also noted that ϕ_u represents the greatest possible magnitude of the control force u and may rarely occur.

By using the moving boundaries process, a second-order controller satisfying (13.28) is found and results in the following controller matrices:

$$A_c = \begin{bmatrix} 0 & 1.2491 \times 10^4 \\ 1 & 3.9084 \times 10^2 \end{bmatrix}, \qquad C_c = \begin{bmatrix} 0 & 1 \end{bmatrix}$$

$$B_c = \begin{bmatrix} -8.9920 \times 10^5 & -3.8142 \times 10^7 \\ -2.8770 \times 10^8 & 4.1409 \times 10^7 \\ 1.2568 \times 10^7 & 1.0362 \times 10^8 \\ 9.6043 \times 10^6 & 1.1179 \times 10^8 \\ 6.2444 \times 10^6 & -1.2329 \times 10^7 \\ 3.9589 \times 10^7 & 6.3476 \times 10^9 \end{bmatrix}^T, \qquad D_c = \begin{bmatrix} -1.5923 \times 10^6 \\ 2.9484 \times 10^7 \\ -9.1729 \times 10^6 \\ -1.1844 \times 10^7 \\ -2.2942 \times 10^7 \\ 4.1263 \times 10^6 \end{bmatrix}^T$$

The corresponding values of the performance measures are listed in Table 13.2.

To verify the design, the building control system is excited with the Northridge and the Kobe earthquakes. The system's responses to both earthquakes are shown in Figures 13.6 and 13.7. Note that the response x_1 is not shown because, as can be seen from Table 13.2, its magnitude is very small compared with the other responses x_i ($i = 2, 3, \ldots, 6$). It can be seen from these figures that all the interstorey drifts x_i are kept within the range ± 25 mm and the magnitude of the control force u is less than 100 kN.

Table 13.2. Performance measures of the control system

performances measures	bounds
$\phi_{x_1} = 0.016$ mm $\leq$	25 mm
$\phi_{x_2} = 25.0$ mm $\leq$	25 mm
$\phi_{x_3} = 24.7$ mm $\leq$	25 mm
$\phi_{x_4} = 19.8$ mm $\leq$	25 mm
$\phi_{x_5} = 16.7$ mm $\leq$	25 mm
$\phi_{x_6} = 14.1$ mm $\leq$	25 mm
$\phi_u = 215$ kN $\leq$	220 kN

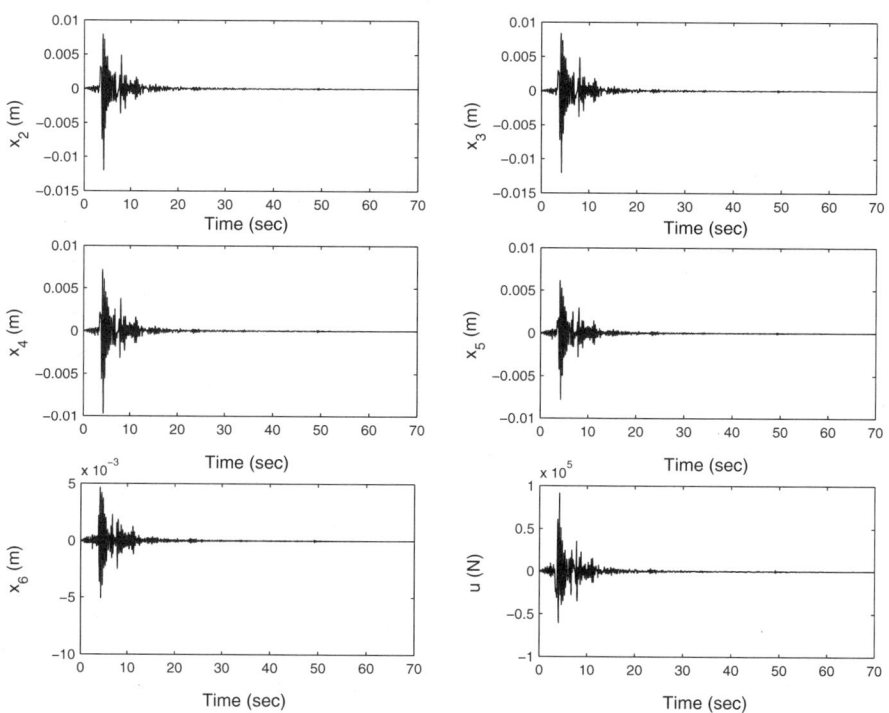

Fig. 13.6. Responses to the Northridge earthquake

13.6 Discussion and Conclusions

This chapter designs a critical system for controlling a building under earthquake excitation by using Zakian's framework, which comprises two main parts: the principle of matching (Zakian, 1991, 1996) and the method of inequalities (Zakian and Al-Naib, 1973). The principle of matching plays an important role in suggesting what kind of inequality should be used so that the control design problem is formulated in a realistic and accurate manner,

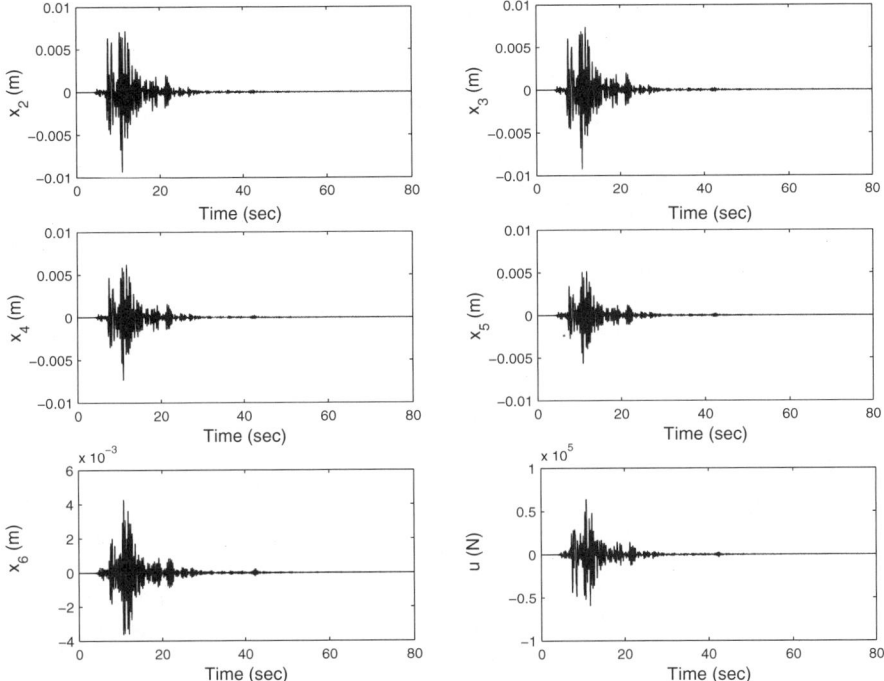

Fig. 13.7. Responses to the Kobe earthquake

whereas the method of inequalities suggests the design problem be cast as a set of inequalities. As an adjunct to the principle of matching, Lane's (1995) theory is used to provide the useful design inequalities (13.26), which are in keeping with the method of inequalities, for the present design problem. The numerical results demonstrate that the framework is suitable for designing critical systems in practice, especially building control systems under earthquake excitation.

Acknowledgements

The author wishes to thank Mr. Yothin Rakvongthai, his former undergraduate student, for his assistance in computing the numerical results.

References

Arunsawatwong, S. (1998). Stability of Zakian I_{MN} recursions for linear delay differential equations. *BIT*, 38(2):219–233 .

Arunsawatwong, S., C. Kuhakarn and T. Pinkaew (2002). Critical system design for control of a building under earthquake excitation. *Proceedings of the 4th Asian Control Conference*, Singapore, pp. 1997–2001.

Chopra, A.K. (editor) (1998). Special issue on benchmark problem. *Earthquake Engineering and Structural Dynamics*, 27:1127–1397.

Jabbari, F., W.E. Schmitendorf and J.N. Yang (1995). H_∞ control for seismic-excited buildings with accelerations feedback. *ASCE Journal of Engineering Mechanics*, 121:994–1002.

Kelly, J.M., G. Leitmann and A.G. Soldatos (1987). Robust control of base-isolated structures under earthquake excitation. *Journal of Optimization Theory and Applications*, 53:159–180.

Kuhn, U. and G. Schmidt (1987). Fresh look into the design and computation of optimal output feedback control using static output feedback. *International Journal of Control*, 46:75–95.

Lane, P.G. (1995). The principle of matching: a necessary and sufficient condition for inputs restricted in magnitude and rate of change. *International Journal of Control*, 62:893–915.

Schmitendorf, W.E., F. Jabbari and J.N. Yang (1994). Robust control techniques for building under earthquake excitation. *Earthquake Engineering and Structural Dynamics*, 23:539–552.

Tadjbaksh, I.G. and F. Rofooei (1992). Optimal hybrid control of structures under earthquake excitation. *Earthquake Engineering and Structural Dynamics*, 21:233–252.

Uniform Building Code (1997). *Chapter 16: Structural Design Requirements*, Volume 1. International Conference of Building Officials, Whittier, California.

Whidborne, J.F. and G. P. Liu (1993). *Critical Control Systems*. Research Studies Press, Taunton.

Zakian, V. (1975). Properties of I_{MN} and J_{MN} approximants and applications to numerical inversion of Laplace transforms and initial-value problems. *Journal of Mathematical Analysis and Applications*, 50(1):191–222.

Zakian, V. (1987). Design formulations. *International Journal of Control*, 46:403–408.

Zakian, V. (1989). Critical systems and tolerable inputs. *International Journal of Control*, 49:1285–1289.

Zakian, V. (1991). Well-matched systems. *IMA Journal of Mathematical Control and Information*, 8:29–38.

Zakian, V. (1996). Perspectives on the principle of matching and the method of inequalities. *International Journal of Control*, 65:147–175.

Zakian, V. and U. Al-Naib (1973). Design of dynamical and control systems by the method of inequalities. *Proceedings of the Institution of Electrical Engineers*, 120:1421–1427.

14 Design of a Hard Disk Drive System

Takahiko Ono

Abstract. The position of the magnetic head of a hard disk drive used in a computer is a critical variable. This chapter considers the design of a hard disk drive system to achieve control of this critical variable. The control system is designed as the servo system required to maintain the position of the head over the desired track with an accuracy of 10% of the track pitch in the presence of the air turbulence caused by the disk rotation and the measurement noise.

14.1 Introduction

A hard disk drive (HDD) is most widely used as a memory storage for saving data in a computer system. The inside view of a typical HDD system is illustrated in Figure 14.1. The basic components are one or more disks to store data, the swing arm to support the magnetic heads for reading and writing data, the voice coil motor (VCM) to actuate the swing arm and the electronic circuits for signal processing and interface. The disk is logically divided into

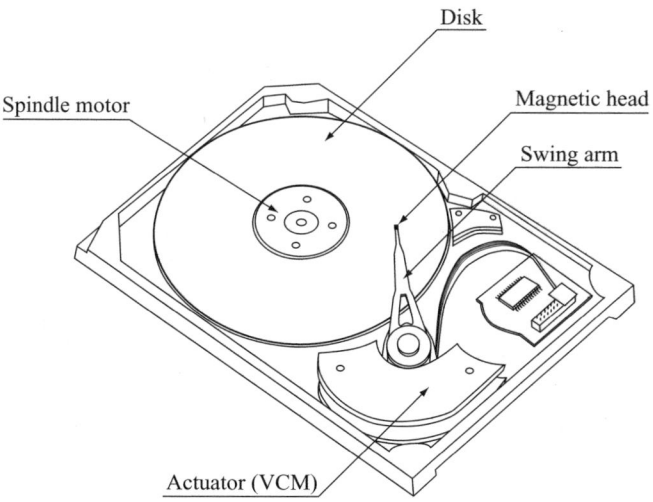

Fig. 14.1. HDD system

concentric circles called tracks and each track is divided into small segments called sectors. The data is divided into pieces according to the sector size and recorded in the sectors on the same track or sometimes different tracks. When the HDD receives a request of reading or writing data from the operating system, the VCM rotates the swing arm and keeps the magnetic head over the desired track. In this time, the required accuracy in positioning the magnetic head at the track is usually less than 10% of the track pitch, the distance between two adjacent tracks. If this requirement is violated, the HDD has to wait, for reading or writing data, until the disk rotates one more time. This results in undesirable operation delay in a computer system. In this sense, the position of the head is a critical variable and the HDD system is a critical system.

The typical controller of an HDD system to control this critical variable is a combination of PI controllers, lead/lag compensators and notch filters, which are designed based on the classical control theory. To achieve better control performance, other control methods have also been proposed and tested on the real systems: LQG/LTR techniques (Hanselmann and Engelke, 1988; Weerasooriya and Phan, 1995), multi-sensing controller design using acceleration feedback (Kobayashi et al, 1997), dual-stage servo control design using neural network (Sasaki et al, 1998), multirate control design (Chen et al, 1998; Hara and Tomizuka, 1999), robust control design based on H_∞ theory (Hirata et al, 1999), perfect tracking control for a step reference (Chen, 2000), track following control using a disturbance observer (White et al, 2000), fixed-structure H_∞ controller design (Ibaraki and Tomizuka, 2001), μ-synthesis (Ohno, Abe and Maruyama, 2001) and real-time parameter optimisation (Hao et al, 2003). However, these methods do not necessarily ensure attainment of control of the critical variable. The purpose of this chapter is to demonstrate the design of HDD systems in terms of critical control, using the framework built upon by the principle of matching and the method of inequalities described in Part I. The simulation results are also presented and the power of the framework is shown.

This chapter uses the following notations. $\mathbb{R}$ and $\mathbb{R}_+$ denote the set of all real numbers and the set of non-negative real numbers, respectively. For a function $f : \mathbb{R}_+ \mapsto \mathbb{R}$, its p-norm is defined as follows

$$\|f\|_p := \begin{cases} \int_0^\infty |f(t)| dt & \text{for} \quad p = 1 \\ \text{ess sup}\{|f(t)| : t \in \mathbb{R}_+\} & \text{for} \quad p = \infty \end{cases}$$

14.2 HDD Systems Design

The operation of positioning a head is divided into two phases: the track seeking and the track following. In the seeking phase, the head is accelerated or decelerated by the VCM and moved to the target track as quickly as

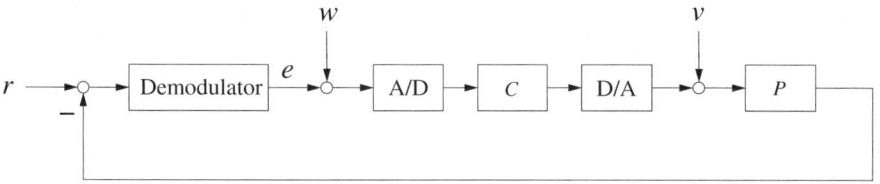

Fig. 14.2. Block diagram of the control system

possible. The control in this phase is called the seek control. In the following phase, the head is maintained over the track by position control until the system finishes reading or writing data. The control in this phase is called the following control. When the HDD system receives a request signal for reading or writing data from the operating system, the following control is performed after the seek control. If the data is fragmented and stored in several different tracks, this control sequence repeats until the data is completely read from the disk. Usually, the system has two exclusive controllers for the seek and following control and switches from one to another according to the phase. This section focuses only on the design of a controller for the following control.

The mechanical part of the system, including the VCM and the swing arm, is a continuous-time system, while the controller is a discrete-time system since it is implemented digitally. In this sense, an HDD system is a sampled-data system. As a method for designing such a system, this chapter takes the digital redesign: First, design the analog controller which attains control of the critical variable in a pure continuous-time framework. Then implement it digitally.

14.2.1 Modelling

Consider an HDD system for a portable computer. The basic hardware specifications are as follows: The diameter of a disk is 2.5 inch, the number of tracks is 9212, the track pitch is 2.76 μm, the rotational speed of a disk is 4200 rpm and the sampling frequency is 5400 Hz. The block diagram of the control system is shown in Figure 14.2. The block P represents the integrated system comprising the power amplifier, the VCM, the swing arm and the head. Its input is the command signal to the amplifier and its output is the position of the head. The block C is the track-following controller. It is designed as a linear time-invariant controller. The signal r is an input concerned with the target track, the signal v is a disturbance and the signal w is a measurement noise. The signal e is the position error of the head from the centre of the track. It is a variable to be controlled and simultaneously a measured output. For convenience, let the magnitude of these signals be normalised by the track pitch.

The swing arm is not rigid but flexible, so essentially the system P has several modes of vibration. The model of P including three dominant modes of vibration is approximately given by

$$P(s) = k \left(\sum_{n=1}^{3} \frac{a_n}{s^2 + 2\zeta_n \omega_n s + \omega_n^2} \right) e^{-\tau s} \tag{14.1}$$

where k is a constant gain, τ is a time delay, and a_n, ζ_n and ω_n are the nth residue, damping factor and resonant frequency, respectively. The nominal values of these parameters are given as follows:

$$k = 3.25 \times 10^8, \quad \tau = 2.30 \times 10^{-5}$$
$$a_1 = 1.00, \quad \zeta_1 = 0.60, \quad \omega_1 = 30 \text{ [Hz]}$$
$$a_2 = -1.40, \quad \zeta_2 = 0.06, \quad \omega_2 = 4700 \text{ [Hz]}$$
$$a_3 = 0.40, \quad \zeta_3 = 0.02, \quad \omega_3 = 6300 \text{ [Hz]}$$

The delay τ is mainly the computational time of a controller. For simplicity, the term of the delay $e^{-\tau s}$ is approximated by the 1st-order Pade approximation, that is,

$$e^{-\tau s} \simeq \frac{1 - 0.5\tau s}{1 + 0.5\tau s} \tag{14.2}$$

The second and third vibration modes of $P(s)$ have higher resonance frequencies than the Nyquist frequency, 2700 Hz. So it is impossible to control them with a digital controller. If they cause the violation of the critical variable, it is necessary to introduce some additional analog compensators, for example, the notch filters so as to attenuate them.

Generally, a track is not purely circular but a little distorted. Defining r as the distance from the average circular trajectory of a track as shown in Figure 14.3, it is viewed as a persistent input comprising the periodic elements with frequencies that are multiples of the disk rotation frequency. For the system considered here, the input r is known only to the extent that it consists of periodic elements with frequencies of less than 150 Hz, mainly 70 Hz and 140 Hz, and has the maximum rate of change of 470. Thus, r can be modelled by

$$r = F_r r_o, \quad r_o \in \mathcal{F}_r \tag{14.3}$$

where F_r is a low-pass filter with cutoff frequency of 150 Hz and $\mathcal{F}_r$ is the set of all piecewise smooth functions with bound 470 on the first derivative and zero initial condition:

$$\mathcal{F}_r := \left\{ f : \mathbb{R}_+ \mapsto \mathbb{R} : \begin{array}{l} f : \text{piecewise smooth} \\ \|\dot{f}\|_\infty \leq 470, \ f(0) = 0 \end{array} \right\} \tag{14.4}$$

The input v is a persistent and random low-frequency disturbance caused by

14 Design of a Hard Disk Drive System 359

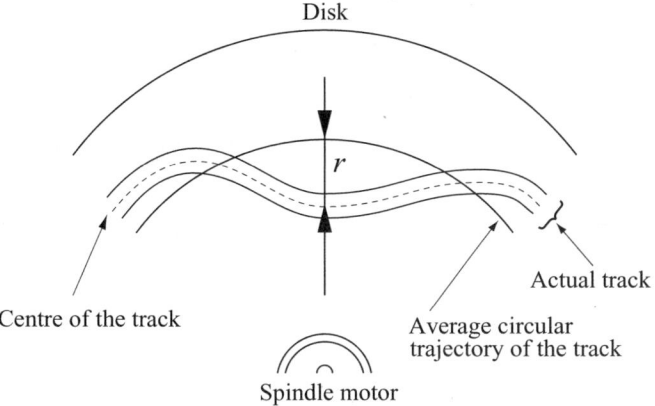

Fig. 14.3. Input r

the air turbulence generated by the disk rotation. For the system considered, the input v is known only to the extent that it consists of random elements with frequencies of less than 50 Hz and has the maximum magnitude of 0.0005. So v can be modelled by

$$v = F_v v_o, \quad v_o \in \mathcal{F}_v \tag{14.5}$$

where F_v is a low-pass filter with cutoff frequency of 50 Hz and $\mathcal{F}_v$ is the set of all piecewise continuous functions with bound 0.0005 on the magnitude:

$$\mathcal{F}_v := \left\{ f : \mathbb{R}_+ \mapsto \mathbb{R} : \begin{array}{l} f : \text{piecewise continuous} \\ \|f\|_\infty \leq 0.0005 \end{array} \right\} \tag{14.6}$$

The measurement noise w works as a persistent and random high-frequency input. It is known only to the extent that it consists of random elements with frequencies of more than 50 Hz and has the maximum magnitude of 0.005. So it can be modelled by

$$w = F_w w_o, \quad w_o \in \mathcal{F}_w \tag{14.7}$$

where F_w is a high-pass filter with cutoff frequency of 50 Hz and $\mathcal{F}_w$ is the set of all piecewise continuous functions with bound 0.005 on the magnitude:

$$\mathcal{F}_w := \left\{ f : \mathbb{R}_+ \mapsto \mathbb{R} : \begin{array}{l} f : \text{piecewise continuous} \\ \|f\|_\infty \leq 0.005 \end{array} \right\} \tag{14.8}$$

The filters F_r, F_v and F_w work as the shaping filters to remove the fictitious inputs not happening in the real environment from the sources $\mathcal{F}_r$, $\mathcal{F}_v$ and $\mathcal{F}_w$. They can also be interpreted as the weights for exogenous inputs. Therefore

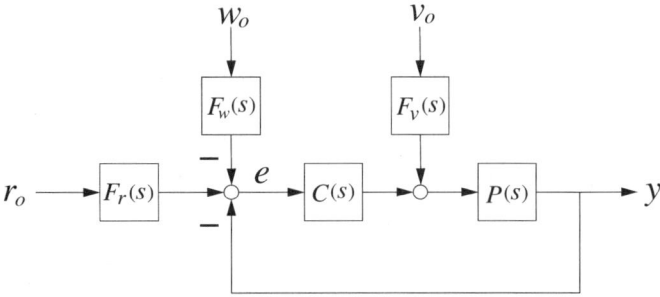

Fig. 14.4. Block diagram for the design of a track-following controller

such filters are convenient especially when the frequency characteristic of exogenous inputs is incorporated into design.

Under the modelling above, the system diagram for the design of a controller can be drawn as shown in Figure 14.4. In the design, F_r, F_v and F_w are regarded as parts of the system, and r_o, v_o and w_o are treated as exogenous inputs instead of r, v and w.

14.2.2 Specifications

The role of the track-following controller is to maintain the position of the head over the target track with an accuracy of 10% of the track pitch while reducing the effect of the disturbance and the measurement noise. In this sense, the tracking error is viewed as a critical variable. Meanwhile, the parameters of $P(s)$ change according to the inside temperature and they differ slightly for every HDD system. Since the parameter variation affects the regulating performance of the critical variable, the system should be robust to it. The problem of the robustness of the system can be handled in terms of the gain and phase margin. Accordingly the design specifications are given as follows:

(a) The primary objective is to maintain the critical variable e within 0.1 for all $r_o \in \mathcal{F}_r$, $v_o \in \mathcal{F}_v$ and $w_o \in \mathcal{F}_w$ for the block diagram of Figure 14.4. This is stated in the form of the inequality

$$\sup\{\|e(r,v,w)\|_\infty : r_o \in \mathcal{F}_r,\ v_o \in \mathcal{F}_v,\ w_o \in \mathcal{F}_w\} \leq 0.1 \qquad (14.9)$$

Let $z_1(t,h)$, $z_2(t,\delta)$ and $z_3(t,\delta)$ denote the step response of the transfer function from r_o to e, the impulse response of the transfer function from v_o to e and the impulse response of the transfer function from w_o to e, respectively. Given the filters F_r, F_v and F_w as linear time-invariant filters. Then, according to Part I, the inequality (14.9) can be rewritten as

$$\mu_1\|z_1(h)\|_1 + \mu_2\|z_2(\delta)\|_1 + \mu_3\|z_3(\delta)\|_1 \leq 0.1 \tag{14.10}$$

where $\mu_1 = 470$, $\mu_2 = 0.0005$ and $\mu_3 = 0.005$.

(b) The secondary objective is to obtain the gain margin g_M and the phase margin θ_M for the open-loop transfer function $P(s)C(s)$ such that

$$\begin{aligned} g_M &\geq 10\,[\text{dB}] \\ \theta_M &\geq 30\,[\text{deg}] \end{aligned} \tag{14.11}$$

The specifications above are stated in the form of inequalities. Using the method of inequalities, the controller which fulfills these specifications is designed.

14.2.3 Design using the Method of Inequalities

The method of inequalities is a numerical method to find a p to satisfy the inequalities

$$\phi_i(p) \leq \varepsilon_i \quad (i = 1, 2, \ldots) \tag{14.12}$$

where p is a vector characterising a controller, $\phi_i(p)$ is an objective function and ε_i is the tolerance of $\phi_i(p)$. Such a p is known as an admissible solution. For the design of the track-following controller,

$$\begin{aligned} \phi_1(p) &= \mu_1\|z_1(h,p)\|_1 + \mu_2\|z_2(\delta,p)\|_1 + \mu_3\|z_3(\delta,p)\|_1 \\ \phi_2(p) &= 1/g_M(p) \\ \phi_3(p) &= 1/\theta_M(p) \end{aligned} \tag{14.13}$$

and

$$\begin{aligned} \varepsilon_1 &= 0.1 \\ \varepsilon_2 &= 1/10 \\ \varepsilon_3 &= 1/30 \end{aligned} \tag{14.14}$$

For the objective function $\phi_1(p)$ to be finite, $z_1(t,h,p)$ must converge to zero as t increases. This means that the controller must have at least one integrator. This condition is also necessary to avoid a steady state error to the constant disturbance modelled by (14.5). Generally, the nth-order linear time-invariant controller including an integrator is parameterised by a $2n$-dimensional real vector. For this reason, set $p = [p_1, p_2, \ldots, p_{2n}] \in \mathbb{R}^{2n}$ and parameterise the controller as

$$C(s) = \frac{p_1(s^n + p_2 s^{n-1} + \cdots + p_n s + p_{n+1})}{s(s^{n-1} + p_{n+2} s^{n-2} + \cdots + p_{2n})} \tag{14.15}$$

The inequality solver called the moving boundaries process (Zakian and Al-Naib, 1973) is used to find the admissible solutions to the inequalities defined by (14.13) and (14.14). The filters F_r, F_v and F_v are chosen so that they cut off the fictitious inputs not reflecting the actual frequency characteristics. As such filters, the 4th-order Butterworth filters are used. The cutoff frequencies of F_r, F_v and F_w are set to 150 Hz, 50 Hz and 50 Hz, respectively. These filters are given by the function **butter** in the MATLAB SIGNAL PROCESSING TOOLBOX. In this case, the exogenous inputs modelled by (14.3), (14.5) and (14.7) have the bounds $\|\dot{r}\|_\infty = 610$, $\|v\|_\infty = 0.0006$ and $\|w\|_\infty = 0.006$. So they well reflect the characteristics of the actual inputs.

First, a parameter search was tried for $p \in \mathbb{R}^2$. But any admissible solution could not be found. Next, a search was performed for $p \in \mathbb{R}^4$ and $p \in \mathbb{R}^6$. As a result, the admissible solutions could be found. The solution for $p \in \mathbb{R}^4$ gives

$$C(s) = \frac{0.196(s + 736.7)(s + 701.5)}{s(s + 13753.3)} \tag{14.16}$$

which attains

$$\phi_1 = 0.0985$$
$$\phi_2 = 1/11.5 \tag{14.17}$$
$$\phi_3 = 1/50.3$$

This controller has the same structure as a PI controller and a lead compensator. Meanwhile, the solution for $p \in \mathbb{R}^6$ gives

$$C(s) = \frac{0.117(s + 2900)(s + 2544.7)(s + 90.6)}{s(s + 13765.7)(s + 2500)} \tag{14.18}$$

which attains

$$\phi_1 = 0.0988$$
$$\phi_2 = 1/15.9 \tag{14.19}$$
$$\phi_3 = 1/33.2$$

In the next section, the control performance attained by these controllers is examined by simulation.

14.3 Performance Evaluation

To evaluate the control performance of the designed controllers, the frequency and time responses of the system are examined. For the purpose of reference, the response for the reference controller is also given. It is the combination of a PI controller and a lead compensator, which was used by the HDD maker.

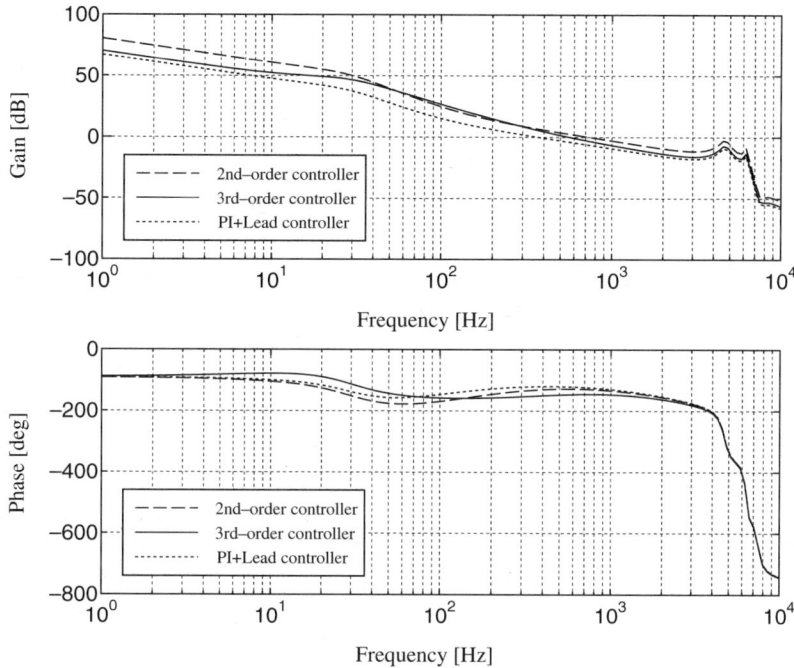

Fig. 14.5. Bode diagram for the open-loop transfer function

But, note that it was designed for the same HDD, but not under the same specifications as considered in this chapter.

First, observe the frequency response. The Bode diagram for the open-loop transfer function $P(s)C(s)$ is illustrated in Figure 14.5. The broken line shows the response for the 2nd-order controller (14.16), the solid line shows the one for the 3rd-order controller (14.18) and the dotted line shows the one for the reference controller. The typical performance measures are listed in Table 14.1. From this table, it is found that all controllers have enough gain and phase margins. However, the gain crossover frequency for the designed controllers is higher. This means that they have larger servo bandwidth than the reference controller. Figure 14.6 shows the Bode magnitude of the closed-loop system from r to e. The closed-loop system from w to e has the same frequency response. Figure 14.7 is the Bode magnitude of the closed-loop system from v to e. From these figures, it can be seen that the designed controllers have better tracking and disturbance rejection performances than the reference controller. Instead, they are more sensitive to the measurement noise at over 800 Hz.

Next, observe the control performance in time domain. The controllers are discretised at the sampling frequency of 5400 Hz. Generally, the specifi-

Table 14.1. Performance measures for the frequency response

	2nd-order controller	3rd-order controller	Reference controller (PI+Lead)
Gain margin [dB]	11.5	15.9	17.4
Phase margin [deg]	50.3	33.2	59.5
Gain crossover frequency [Hz]	737.0	563.8	370.4
Phase crossover frequency [Hz]	3182.5	2956.3	3243.9

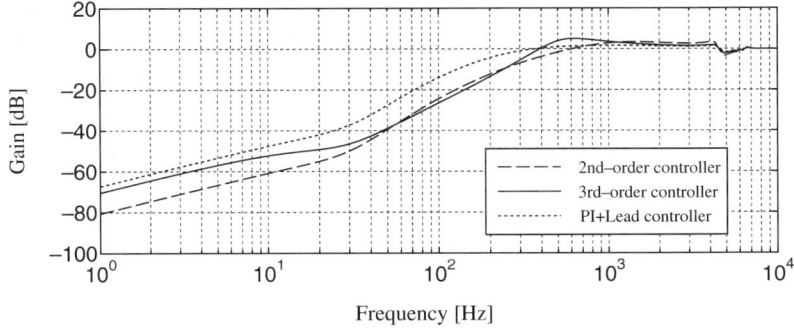

Fig. 14.6. Bode magnitude of the closed-loop transfer function from r to e

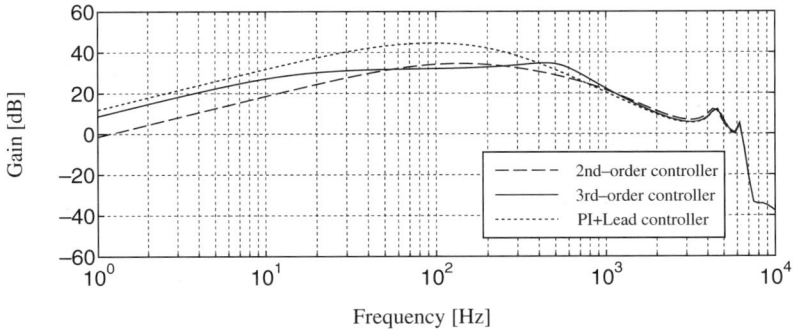

Fig. 14.7. Bode magnitude of the closed-loop transfer function from v to e

cations, particularly the inequality (14.9), are not necessarily guaranteed by the discretised controller. By simulating the time response to the test inputs, let us examine whether the critical variable can really be regulated or not. For the block diagram of Figure 14.2, the time response is simulated using MATLAB SIMULINK. In the simulation, the exact time delay of 23 μs is included in the system instead of the Pade approximation (14.2). Assuming

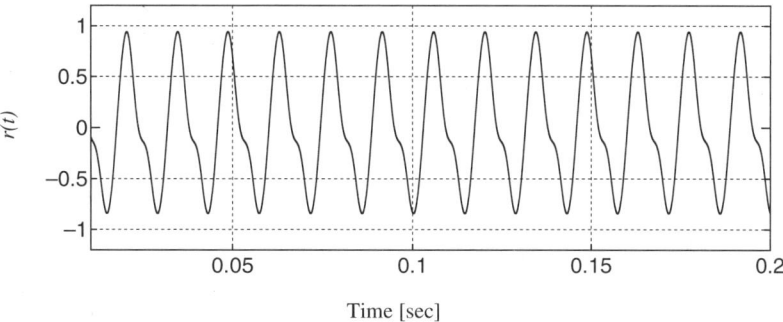

Fig. 14.8. Reference $r(t)$

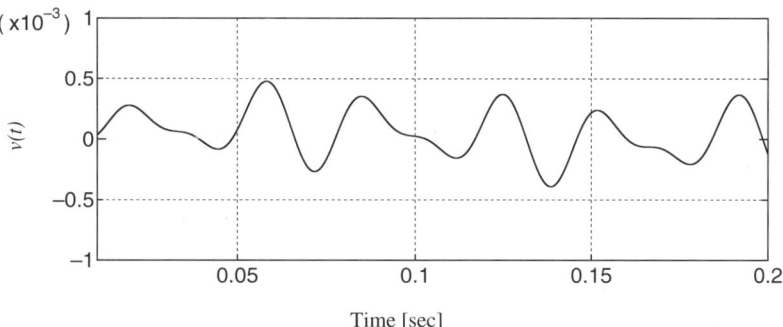

Fig. 14.9. Disturbance $v(t)$

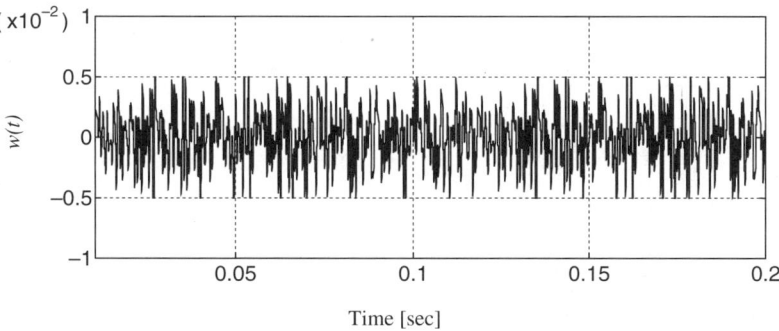

Fig. 14.10. Measurement noise $w(t)$

that the plant P is being perturbed, the 6th-order linear model, which has the following parameters, are used instead of the nominal model.

k : 2 dB smaller than the nominal value
$\zeta_1, \zeta_2, \zeta_3$: 50 % smaller than the nominal value
$\omega_1, \omega_2, \omega_3$: 5 % smaller than the nominal value

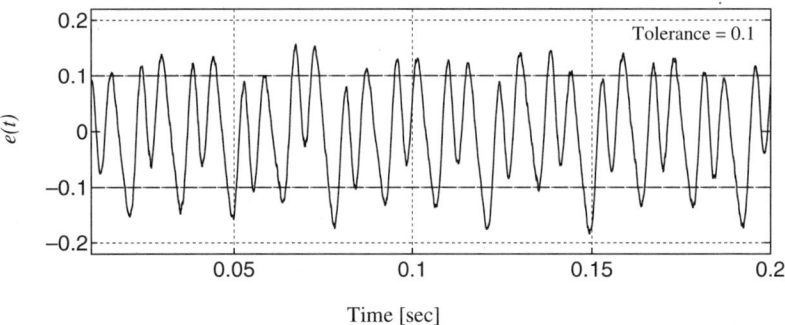

Fig. 14.11. Tracking error to the test inputs for the conventional controller

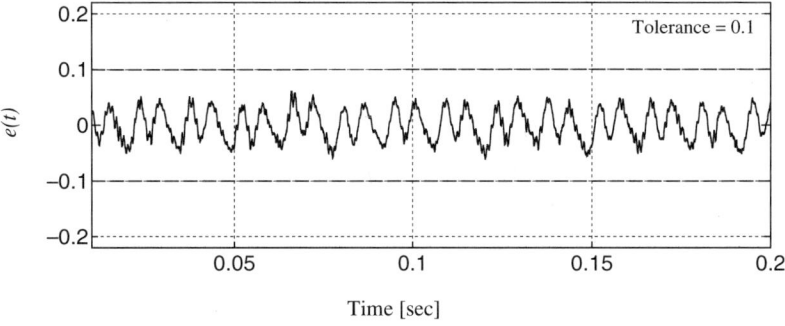

Fig. 14.12. Tracking error to the test inputs for the 2nd-order controller

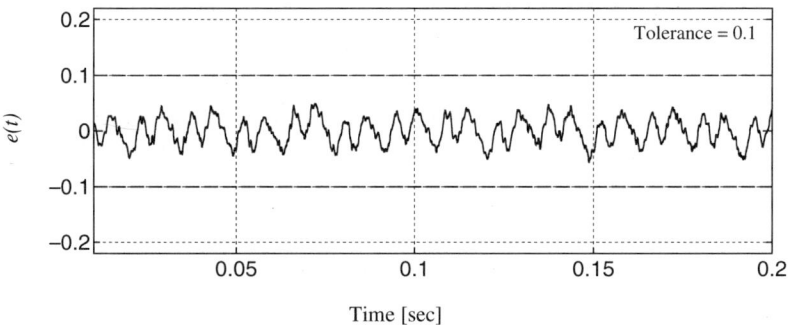

Fig. 14.13. Tracking error to the test inputs for the 3rd-order controller

Figures 14.8 to 14.10 show the applied test inputs. Figures 14.11 to 14.13 show the tracking errors to these inputs for the reference controller and the designed controllers. It is noted that all the signals are normalised by the track pitch. The broken line shows the boundary of the tolerance of the tracking error. As these figures indicate, the reference controller cannot maintain the tracking error within 0.1, while the designed controllers attain control of the critical variable. Compared with the reference controller, they reduce the tracking error to about 1/3. The 2nd-order controller is more sensitive to the noise than the 3rd-order controller. This is because the servo bandwidth of the former is much higher than that of the latter.

14.4 Conclusions

This chapter demonstrated the design of an HDD system based on the framework built upon by the principle of matching and the method of inequalities. The HDD system is the critical system which requires the position of the head to be maintained over the target track with an accuracy of 10% of the track pitch. Three factors interfering with the position control were considered: the distortion of the trajectory of a track, the disturbance caused by the air turbulence generated by the disk rotation and the measurement noise. They are modelled by the function spaces comprising the inputs restricted in magnitude or rate of change. The design specifications are stated as a conjunction of the inequalities on the bounds of the tracking error and the gain and phase margin. The problem of finding the controller satisfying the inequalities was formulated into an admissibility problem and the controller was successfully obtained by the method of inequalities. In the simulation, the control performance was evaluated in frequency and time domain. The power of the framework was illustrated by comparing with the conventional design based on the classical control theory.

This chapter achieves the robustness to the parameter perturbation of a plant by obtaining enough gain and phase margin. However, this approach does not necessarily guarantee the bound of a critical variable by nature. To attain control of critical variables for perturbed systems, it is necessary to develop the methods for robust matching (Zakian, 1996; Whidborne et al, 2001).

References

Chen, R., G. Guo, T. Huang and T.S. Low, (1998) Optimal multirate control design for hard disk drive servo systems. *IEEE Trans. Magnetics*, 34:1898–1900.

Chen, B.M, (2000) *Robust and H_∞ control*, Springer.

Hanselmann, H., A. Engelke, (1988) LQG-control of a highly resonant disk drive head positioning actuator. *IEEE Trans. Industrial Electronics*, 35:100–104.

Hao, Q., R. Chen, G. Guo, S. Chen and T.S. Low, (2003) A gradient-based track-following controller optimization for hard disk drive. *IEEE Trans. Industrial Electronics*, 50:108–115.

Hara, T., and M. Tomizuka, (1999) Performance enhancement of multi-rate controller for hard disk drives. *IEEE Trans. Magnetics*, 35:898–903.

Hirata, M., T. Atsumi, A. Murase and K. Nonami, (1999) Following control of a hard disk drive by using sampled-data H_∞ control. *Proc. of the 1999 IEEE International Conference on Control Applications*, pp. 182–186.

Ibaraki, S., M. Tomizuka, (2001) Tuning of a hard disk drive servo controller using fixed-structure H_∞ controller optimization. *ASME Journal of Dynamic Systems, Measurement, and Control*, 123:544–549.

Kobayashi, M., T. Yamaguchi, T. Yoshida, H. Hirai, (1997) Carriage acceleration feedback multi-sensing controller for sector serve systems. *International Conference on Micromechatronics for Information and Precision Equipment*.

Ohno, K., Y. Abe, T. Maruyama, (2001) Robust following control design for hard disk drives. *Proc. of the 10th IEEE International Conference on Control Applications*.

Sasaki, M., T. Suzuki, E. Ida, F. Fujisawa, M. Kobayashi, H. Hirai, (1998) Track-following control of a dual-stage hard disk drive using a neuro-control system. *Engineering Applications of Artificial Intelligence*, 11:707–716.

Weerasooriya, S., D.T. Phan, (1995) Discrete-time LQG/LTR design and modeling of a disk drive actuator tracking servo system. *IEEE Trans. Industrial Electronics*, 42:240–247.

Whidborne, J.F., T. Ishihara, H. Inooka, T. Ono and T. Satoh, (2001) Some practical matching conditions for robust control system design. *King's College London Mechanical Engineering Department Report EM/2001/10*, London, U.K.

White, M. T., M. Tomizuka and C. Smith, (2000) Improved track following in magnetic disk drives using a disturbance observer. *IEEE/ASME Trans. Mechatronics*, 5:3-11.

Zakian, V., U. Al-Naib, (1973) Design of dynamical and control systems by the method of inequalities. *Proc. IEE Control & Science*, 120:1421–1427.

Zakian, V., (1996) Perspectives on the principle of matching and the method of inequalities. *Int. J. Contr.*, 65:147–175.

15 Two Studies of Robust Matching

Oluwafemi Taiwo

Abstract. Two case studies of robust matching are reviewed in this chapter. One relates to systems with recycle and the other to the brake control of a heavy-duty truck. For the recycle system, a combination of design criteria for vague infinite-dimensional plants together with a special compensator for plants with recycle, is exploited to design a controller for a plant with recycle when the parameters of the process in the recycle path are subject to variations. The method gives a control system with guaranteed performance for large variations in process parameters. On the other hand, the problems encountered while designing a robust controller for the braking system of a heavy-duty truck include vagueness in the input function, variations in the plant parameters and the presence of dead time. Moreover, the system is critical. Appropriate design criteria are used to guarantee desired performance despite these difficulties.

15.1 Introduction

In this chapter, two applications of Zakian's (1984, 1996) design framework to vague systems are reviewed. In previous work, delayed and retarded control systems (Bada 1987 and Taiwo 1986), were designed to be robustly matched to their respective environments. These works are reviewed here. Plant vagueness in both cases is caused by variations in dead times and plant gains that are not known precisely but are known only to be in a certain range.

15.1.1 Systems with Recycle

Recycle operation is commonplace in the process industries and often leads to considerable deterioration of plant dynamics (Denn and Lavie 1982, Taiwo 1985, 1986, 1996, Luyben 1994, Wu *et al* 2002). By using a recycle compensator, Taiwo (1985, 1986, 1996) has shown that the dynamics of such processes can be considerably improved and simplified. Among those who have investigated the efficacy of Taiwo's compensator are Scali and Ferrari (1997, 1998, 1999). However, the exact theoretical performance of the system incorporating a recycle compensator can only be realised in practice when the model of the process in the recycle path is accurately known. Even for this situation, the desired performance of the control system is not guaranteed by conventional methods if the operating point changes by way of, for instance, the recycle flow rate changing.

This chapter describes a design method that guarantees robust matching for plants with recycle, whenever the model uncertainty or plant variations satisfy certain conditions.

15.1.2 Brake Control

Reliable and effective braking systems are an essential part of the design of heavy-duty trucks for material transport in the mining industry. Although conventional braking systems are generally adequate in the case of uphill running in open–pit mines, downhill running in mountain type mines gives rise to severe braking problems. Taking into account safety and financial considerations, the need exists for a robust braking control system to ensure strict regulation of truck speed over the wide range of operating conditions. Such strict regulation is the essential characteristic of a critical control system (Zakian, 1989).

Difficulties that arise in designing such a system are three-fold. First, the braking system dynamics include an amount of dead time, which is of the same order of magnitude as the plant time constants. Second, the parameters of the braking system transfer function depend on the load carried by the truck, which is subject to non-negligible variations. These variations must be taken into account in ensuring the required performance of the braking system. Finally, the system is subjected to disturbances that are not precisely known *a priori*; this leads to the consideration of a set of possible inputs instead of just one specific input function.

The presence of dead time complicates control system design in many ways. It has a destabilising effect on the closed-loop control system. For certain operating conditions, it can lead to a severe deterioration in system performance, verging on closed-loop instability (Bada 1984, Martin *et al* 1977, Ross 1977).

It also invalidates the use of conventional design techniques developed for rational transfer functions. The conventional tuning methods for delayed systems have been shown to be inadequate because they often lead to unstable designs (Bada 1984). An alternative approach, which consists of replacing the non-rational plant transfer function $G(s)$ by a rational approximant $G^*(s)$ and proceeding with conventional design techniques that cope with rational transfer functions, can fail to give satisfactory results if $G^*(s)$ is not sufficiently close to $G(s)$. Moreover, it is not possible to predict the performance of the system, having $G(s)$ for its plant model, from that of its approximant with $G^*(s)$ as nominal plant model (Bada 1985).

When the pertinent parameters of the plant, such as the dead time, the gain or the time constants, change during operation, the controller parameters that give adequate performance for one set of conditions can cause sluggish or even oscillatory responses for another set of conditions (Chiang and Durbin 1981). If the changes are considerable and the stability region limited, then critical tolerances are exceeded and even instability can result.

This means that strict regulation is not maintained for all operating conditions. Hence the need to consider parameter changes in order to guarantee required performance.

15.2 Robust Matching and Vague Systems

The design framework developed by Zakian (1984, 1996; see Chapter 1) can be used to design control systems for non-linear, infinite-dimensional, time-varying and vague systems. This framework, which is a natural extension of his previous work (Zakian 1978, 1979a, 1983), is summarised below for the case of vague infinite-dimensional systems such as those given in Sections 15.3 and 15.4 below.

15.2.1 Input Spaces

A system with input $f : \mathbb{R} \to \mathbb{R}$ and output $v_i(f) : \mathbb{R} \to \mathbb{R}$ maps an input space into an output space. As usual, $f^{(1)}(t)$ is the derivative of f, $df(t)/dt$, $v_i(t, f)$ is the value of the output at time t, so that $v_i(f) : t \to v(t, f)$, the exponential function is denoted by $t \to \exp \sigma(\cdot)$. If X is a space of piecewise continuous function $x : [0, \infty) \to \mathbb{R}$ then define

$$\| x \|_n = \left\{ \int_0^\infty |x(t)|^n dt \right\}^{1/n}, \quad 1 \le n < \infty \tag{15.1}$$

$$\| x \|_\infty = \sup \{|x(t)| : t \ge 0\} \tag{15.2}$$

The functions $\| \cdot \|_n$ and $\| \cdot \|_\infty$ map X into the extended half-line $[0, \infty]$.

Let n, δ, D, M denote real parameters such that $1 < n \le \infty$, $0 \le \delta < \infty$, $0 < D < \infty$, $0 < M < \infty$. The input space $F(n, \delta, D, M)$ comprises all $f : \mathbb{R} \to \mathbb{R}$ such that $f(t) = 0$ for all $t \le 0$, $f^{(1)}(t)$ is piecewise continuous with

$$\|f\|_\infty \le M \tag{15.3}$$

and

$$\beta(t) \le D, \quad t \in \mathbb{R} \tag{15.4}$$

where

$$\beta(t) = \begin{cases} \left\{ \int_0^t \left|\exp(-\delta\lambda)f^{(1)}(t-\lambda)\right|^n d\lambda \right\}^{1/n} & \text{if } 1 < n < \infty \\ \sup \left\{ \left|f^{(1)}(t-\lambda)\exp(-\delta\lambda)\right| : \lambda \ge 0 \right\} & \text{if } n = \infty \\ \left\|f^{(1)}(t)\right\|_\infty, & \text{if } n = \infty \text{ and } \delta = 0 \end{cases} \tag{15.5}$$

For $1 < n < \infty$ the functions $t \to \beta^n(t)$ can be thought of as the output of a filter with transfer function $(s+n\delta)^{-1}$ and input $|f^{(1)}(t)|^n$. If $1 < n < \infty$

and $\delta = 0$ then $\beta = |f^{(1)}(t)|^n$. $\beta(t)$ is a measure of the rate of change of f, thus, the condition (15.4)) clearly constitutes a restriction on the rate of change of input.

The input space F can be regarded either as a set of test functions f with respect to which the control system has to behave uniformly well or as a mathematical model of the environment in which the system has to operate.

15.2.2 Robust Performance

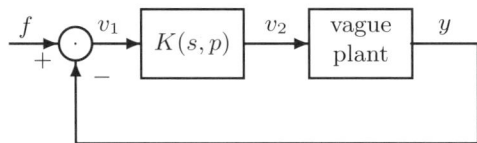

Fig. 15.1. Control system with vague plant

Suppose a vague plant in its domain of operation is characterised by a set $\mathcal{G}$ of impulse response functions g (Figure 15.1), so that for every input $v_2 \in \mathcal{V}_2$, there is a $g \in \mathcal{G}$ such that the plant output y is given by $\mathcal{L}(y) = \mathcal{L}(g)\mathcal{L}(v_2)$, where $\mathcal{L}(\cdot)$ denotes the Laplace transformation. Assume that the plant input v_2 is restricted to the set $\mathcal{V}_2$ defined by

$$\mathcal{V}_2 = \{v_2 : \|v_2\|_\infty \leq \varepsilon_2\} \tag{15.6}$$

For the system in Figure 15.1, the input f is said to be tolerable if

$$\|v_i(f)\|_\infty \leq \varepsilon_i, \quad i = 1, 2, \text{ for all } g \in \mathcal{G} \tag{15.7}$$

Then, with the notation

$$\hat{v}_i \triangleq \sup\{\|v_i(f)\|_\infty : g \in \mathcal{G}\} \tag{15.8}$$

the tolerable set T_m is defined by

$$T_m = \left\{ f : \bigwedge_{i \in \hat{m}} (v_i(f) \leq \varepsilon_i) \right\}, \quad \hat{m} \triangleq \{1, 2, \ldots, m\} \tag{15.9}$$

where $\wedge$ is the symbol for conjunction.

Consider the feedback system in Figure 15.1, where the variables we wish to control are v_1 and v_2 and the vague plant has a set of transfer functions $\mathcal{L}(g) = G(s)$. For a given environment characterised by the possible set F, the system of Figure 15.1 and the environment are said to be robustly matched if $F \subseteq T_2$.

In order to find practical conditions for a robust match between the system of Figure 15.1 and its environment, consider another system, which is identical to that of Figure 15.1, except that the plant has a transfer function $G^*(s)$ called the nominal plant.

Define the notation

$$\phi(p) = \sup\{\hat{v}_i(f) : f \in F\} \tag{15.10}$$

The design problem is solved when the controller parameters p of $K(s,p)$ are computed which satisfy $A_2(p)$ where

$$A_2(p) = \{\phi_1(p) \leq \varepsilon_1\} \wedge \{\phi_2(p) \leq \varepsilon_2\} \tag{15.11}$$

The first step in solving the inequalities (15.11) is that of determining a point p that makes $\phi_i(p)$ finite. Thereafter, a difficulty arises because $\phi_i(p)$ is computationally expensive to evaluate directly from the defining expression (15.10). Zakian (1984, 1996) therefore proposes to replace $\phi_i(p)$ by a majorant $\hat{\phi}_i(p)$ such that

$$\hat{\phi}(p) \geq \phi(p), \quad i = 1, 2 \tag{15.12}$$

where $\hat{\phi}_i(p)$ is given by

$$\hat{\phi}_i(p) = \frac{\sigma_i^*(p)M + \gamma_i^*(p)D}{1 - \theta(p)}, \quad \theta(p) < 1 \tag{15.13}$$

and

$$\theta(p) = \|\hat{v}_2^* - \hat{v}_2^*(\infty)\|_1 A + \sigma_2^*(p)B + \gamma_2^*(p)C \tag{15.14}$$

$$A = \sup\{|z(0)| : g \in \mathcal{G}\} \tag{15.15}$$

$$B = \sup\{\|z\|_1 : g \in \mathcal{G}\} \tag{15.16}$$

$$C = \sup\left\{\left\|z^{(1)}\right\|_1 : g \in \mathcal{G}\right\} \tag{15.17}$$

$$z = g - g^* \tag{15.18}$$

$$\sigma_i^*(p) = |\hat{v}_i^*(\infty)|, \quad i = 1, 2 \tag{15.19}$$

$$\gamma_i^*(p) = \begin{cases} \left\{\int_0^\infty |\exp(\delta\lambda)(\hat{v}_i^*(\lambda) - \hat{v}_i^*(\infty))|^m \, d\lambda\right\}^{1/m}, & i = 1, 2, \\ & 1 < m < \infty, \, \delta > 0 \\ \|v_i^*(\lambda) - \hat{v}_i^*(\infty)\|_1, \quad i = 1, 2, & m = 1, \, \delta = 0 \end{cases} \tag{15.20}$$

$$\|x\|_1 = \int_0^\infty |x(\lambda)| d\lambda \tag{15.21}$$

The positive numbers m and n are chosen such that

$$m^{-1} + n^{-1} = 1 \tag{15.22}$$

The $\hat{v}_i^*$, $i = 1, 2$, which depend on p, are the unit responses defined by $\hat{v}_i^* = v_i^*$ if $f = 1$. $\hat{v}_i^*$ is the response of the control system in Figure 15.1 when the transfer function $G(s)$ of the vague plant is replaced by a strictly proper rational transfer function $G^*(s)$, and we recall that if $G = G^*$ under normal operating conditions, then $G^*(s)$ minimises A, B and C.

Hence the design problem is solved in practice by searching for a p that satisfies $\hat{A}_2(p)$, where

$$\hat{A}_2(p) = \left\{ \hat{\phi}_1(p) \leq \varepsilon_1 \right\} \wedge \left\{ \hat{\phi}_2(p) \leq \varepsilon_2 \right\} \wedge \left\{ \theta(p) < 1 \right\} \tag{15.23}$$

By using (15.23) a robust match is ensured, because of the obvious relation

$$\hat{A}_2(p) \Rightarrow A_2(p) \Leftrightarrow F \subseteq T_2 \tag{15.24}$$

It is also obvious that $\hat{A}_2(p)$ is a practical sufficient condition for a match ($F \subseteq T_2$).

Search methods (see Chapter 1, Section 1.7 and Part III) can be used to solve the inequalities (15.23). The computational steps employed here are now outlined.

Starting from an arbitrary point $p^0 \in \mathbb{R}^n$, the moving boundaries process (Zakian and Al-Naib 1973) is used to locate a point that satisfies the inequality

$$l(p) < 0 \tag{15.25}$$

where $l(p)$ is the abscissa of stability of the nominal system (see Chapter 1).

Since $\theta(p)$ increases with growing difference between nominal and other operating points, the numbers $\hat{\phi}_i(p)$ become smaller as the actual plant transfer function G and the transfer function of the nominal plant G^* become closer for a given controller. I_{MN} approximants (Zakian 1975; Arunsawatwong 1998) are used to compute time responses from which the numbers $\hat{\phi}_i(p)$ appearing in inequalities (15.23) are calculated.

15.3 Robust Matching for Plants with Recycle

The design criteria for vague infinite-dimensional plants and the recycle compensator for plants with recycle, are used to design a controller for a plant with recycle. The method guarantees robust matching for large variations in the process parameters (Taiwo 1986).

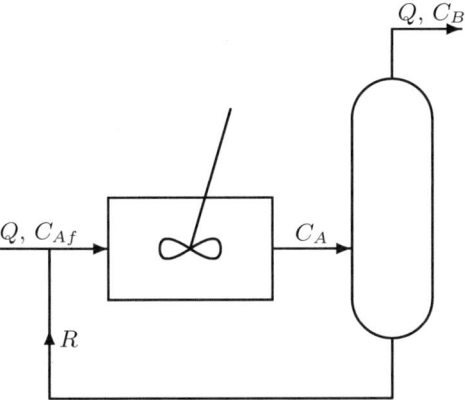

Fig. 15.2. Schematic diagram of a continuous stirred-tank reactor with recycle

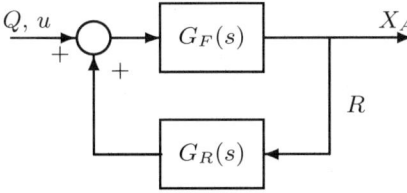

Fig. 15.3. Block diagram of a continuous stirred-tank reactor with recycle

15.3.1 Dynamic Model of a Recycle Process and the Specification of a Recycle Compensator

Consider an isothermal continuous stirred tank reactor in which a first-order irreversible reaction (involving incompressible compounds) A→B takes place and whose effluent is separated in a distillation column downstream (Figure 15.2). It is assumed that the products of the distillation column are almost pure and B is taken from the top, while A, which is the bottom product, is fed with fresh feed to the reactor. We assume that the column top product is controlled by manipulating the reflux rate, while the bottom quality is controlled using steam. To avoid undue interaction, only the top loop is under tight control, while the bottom loop is relatively loosely controlled. In order to maintain constant-quality feed to the column, we manipulate the fresh-feed concentration to the reactor. The main aim of this section is to investigate the design of a dynamically desirable control system for this loop, whose simplified block diagram is given in Figure 15.3, from which it is easily seen (see also Taiwo 1984) that

$$G_A(s) = \frac{X_A(s)}{U(s)} = \frac{G_F(s)/(1+r)}{1 - [r/(r+1)]G_F(s)G_R(s)} \qquad (15.26)$$

where X_A is the deviation in reactor effluent concentration C_A, U is the deviation in feed concentration C_{Af} and r is the recycle ratio R/Q. R is the recycle flow rate, while Q is the feed flow rate. G_F and G_R are defined in Figure 15.3.

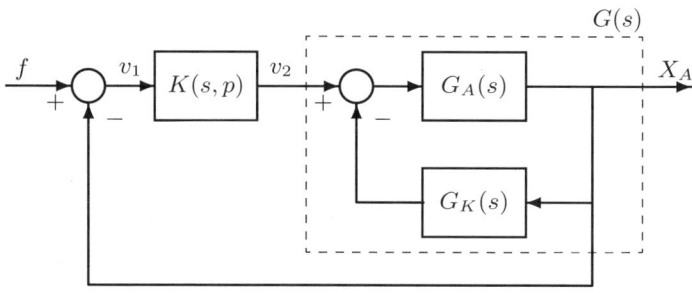

Fig. 15.4. Feedback system with recycle compensator

Taiwo (1984) has shown that the response time of $G_A(s)$ can be several thousand times that of $G_F(s)$. However, by using a recycle compensator, the dynamic response of $G_A(s)$ can be considerably improved. Figure 15.4 shows the basic ideas in the design of a feedback controller for the plant. $K(s,p)$ is the controller, $G_K(s)$ is the recycle compensator, while $G(s)$ is the transfer function of the inner loop. By prescribing $G_K(s)$ as

$$G_k(s) = rG_R(s) \qquad (15.27)$$

it is easy to prove that $G(s)$ is given by

$$G(s) = G^*(s) = \frac{X_A(s)}{V_2(s)} = \frac{G_F(s)}{1+r} \qquad (15.28)$$

Note that $G^*(s)$, except for a reduced gain, has the same structure as $G_F(s)$, and its dynamics (the response time and stability) are a great improvement on those of the uncompensated plant given by (15.26).

The effective transfer function given by (15.28) will only apply if $G_R(s)$ remains constant and would not be so if, for example, the operating point of the plant were to change. In order to take account of the effects of the parameter variations on the dynamic behaviour of the plant, we let the transfer functions in (15.26)–(15.28) denote the description of the plant at its normal operating point, which we assume is accurately known. Thus (15.28) holds only at the normal operating point. For other operating points, the transfer function of the compensated plant, that is the inner loop, is given by

$$G(s) = \frac{G'_F(s)(1+r')}{1 - [r'/(r'+1)]G'_R(s)G'_F(s) + [r/(r'+1)]G_R(s)G'_F(s)} \quad (15.29)$$

where current plant variables are primed. From (15.29), note that a class of transfer functions is defined where each element of the class corresponds to a different operating point.

For our purposes we adopt the particular plant considered by Denn and Lavie (1982) and Taiwo (1984). At its normal operating point

$$G_F(s) = \frac{1}{0.25s + 1} \quad (15.30)$$

$$r = 1 \quad (15.31)$$

$$G_R(s) = \frac{1.95 \exp(-8s)}{8s + 1} \quad (15.32)$$

$$G^*(s) = \frac{0.5}{0.25s + 1} \quad (15.33)$$

In this work we assume that variations in the plant occur only in the time delay and the steady-state gain of the recycle process. The transfer function of the recycle process can therefore be written as

$$G_R(s) = \frac{k_R \exp(-Ts)}{8s + 1} \quad (15.34)$$

In practice, variations in k_R and T can be caused by changes in purity of column bottom product and the recycle flow rate respectively. Although changes in the recycle flow rate will normally lead to changes in g_F and r in practice, we shall, for simplicity, assume that these are constant. Hence, with $G_F(s)$ in (15.30), r in (15.31) and $G_R(s)$ in (15.34), the transfer function of the compensated plant, which is the inner loop in Figure 15.4, is given by

$$G(s) = \frac{4s + 0.5}{2s^2 + 8.25s + 1 - 0.5k_R \exp(-Ts) + 0.975 \exp(-8s)} \quad (15.35)$$

Note the non-rational nature of this compensated plant transfer function, which therefore gives rise to a control system with infinite-dimensional state-space. Note also that, under normal operating conditions, $G = G^*$, where unlike G, G^* is a strictly proper rational function.

If we assume that both positive and negative changes in the parameters k_R and T can occur and that the maximum change is 40% of the normal operating values in both these parameters, then the class G of transfer function

$G(s)$ that the compensated plant can have is given by

$$\mathcal{G} = \left\{ g : G(s) = \frac{4s + 0.5}{2s^2 + 8.25s + 1 - 0.5k_R \exp(-Ts) + 0.975 \exp(-8s)}, \right.$$
$$\left. 1.17 \leq k_R \leq 2.73, \; 4.8 \leq T \leq 11.2 \right\} \quad (15.36)$$

Our aim is to design a controller that guarantees robust matching for all the plants in the G given in (15.36).

15.3.2 Design of a Controller for a Plant with Recycle

In this section we use the technique described above to design a controller for the class of plants given by (15.36). The design problem can therefore be stated thus. Find a controller that gives a closed-loop system with guaranteed robust matching for this vague system.

In order to solve this problem, the compensated control system is represented by its transfer function corresponding to the nominal plant at the normal operating point, $G^*(s)$. Then in order to evaluate $\theta(p)$, A, B and C defined in (15.15)–(15.17) are first computed. Since k_R and T vary independently, the computation of B and C ($A = 0$ for this example) involves a numerical search in $\mathbb{R}^2$. This search revealed that B and C occurred when $k_R = 2.73$ and $T = 11.2$. Note that $A = B = C = 0$ at the normal operating conditions, indicating that $G^*(s)$ in fact minimises A, B, and C.

It now remains to specify the input space and the performance inequalities. Following Taiwo (1986), $n = \infty$, $m = 1$ and $\delta = 0$ in (15.5) and (15.20), furthermore, it is assumed that the input space has the following characteristics:

$$M = 1, \quad D = 1 \quad (15.37)$$

while the numbers ε_i are given by

$$\varepsilon_1 = 0.25, \quad \varepsilon_2 = 15 \quad (15.38)$$

Upon specifying a proportional-plus-integral (PI) controller of the form

$$K(s,p) = p_1 + p_2/s \quad (15.39)$$

The moving boundaries process located the point

$$p = (19.2, 54.1) \quad (15.40)$$

guaranteeing robust matching. The closed-loop system has the following characteristics:

$$\hat{\phi}_1(p) = 0.22, \quad \hat{\phi}_2(p) = 14.96, \quad \theta(p) = 0.84 \quad (15.41)$$

The goodness of the present design is shown by the fact that the largest possible error $v_i(t)$ in the system is about one fifth of the magnitude of the largest possible input $f(t)$, while the largest possible controller output $v_2(t)$ is under just fifteen times the magnitude of the largest possible input $f(t)$.

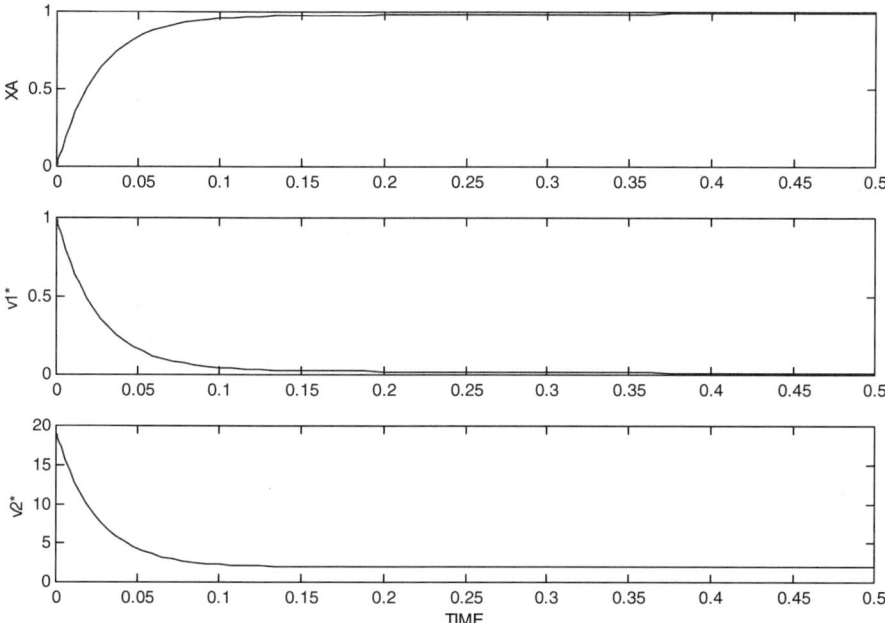

Fig. 15.5. Time responses of the closed-loop system to unit step input

The unit step responses of the closed-loop system at the normal operating point are shown in Figure 15.5. Similar responses were obtained for $x_A(t)$ and $v(t)$ at other operating points in G. However, $v_2(\infty)$ increases from 1.22 to 2.78 as k_R decreases from 2.73 to 1.17.

15.3.3 Conclusions and Discussion

It has been demonstrated that, by using Zakian's design framework and the Taiwo's recycle compensator, a robustly matched system can be designed when the plant is subject to large but known parameter variations. Although the technique has been demonstrated here for a recycle process with two variable parameters, it is clear that the same technique can be applied to plants with more variable parameters. Furthermore, Taiwo (1996) not only derived a generalised recycle compensator applicable to the multivariable situation but also demonstrated the favourable performance of systems with the compensator in the design of robust multivariable controllers.

In its more general setting, Zakian's framework is applicable to non-linear and time varying plants in addition to vague infinite-dimensional plants. Applications of the framework to plants with time delay without parameter variations have been undertaken by Bada (1984, 1985).

Most plants with recycle are often very lightly damped (Denn and Lavie 1982, Taiwo 1985, 1986, 1996, Luyben 1994, Scali and Ferrari 1997, 1998, 1999 and Wu et al 2002), and as a consequence of the exponential terms in the denominator of the transfer functions of such plants, their frequency responses are rather peculiar, being 'horn-shaped' as well as possessing several 'spikes' (Denn and Lavie 1982, Taiwo 1985). This means that the usual rules of thumb for specifying the parameters of a controller when designing in the frequency domain are not applicable. However, by employing a recycle compensator, a rational transfer function $G^*(s)$ is obtained for the normal operating point, and this can be easily used for designing feedback controllers using Zakian's framework. When designed this way, the control system exhibits robust matching, despite severe parameter variations in the plant.

15.4 Robust Matching for the Brake Control of a Heavy-duty Truck – a Critical System

Reliable and effective braking systems are an essential part of the design of heavy-duty trucks for material transport in the mining industry. Although conventional braking systems are generally adequate in the case of uphill running in open–pit mines, downhill running in mountain type mines gives rise to severe braking problems. These problems arise because the truck braking system is subject to variations in dead time, changes in the input function and variations in its other parameters. If these changes are considerable and the stability region limited, then critical tolerances can be exceeded and even instability may result if conventional control system design methods are used. This means that strict regulation must be maintained for all operating conditions in order to avoid catastrophic situations caused by inadequacy or failure of the brake system. Taking into account safety and financial considerations, the need exists for a robust braking control system to ensure strict regulation of truck speed over the wide range of operating conditions. Such strict regulation is the essential characteristic of a critical control system (Zakian, 1989). Zakian's framework is used to guarantee robust matching despite the difficulties.

15.4.1 Truck Braking Control System Model

Consider the truck model shown schematically in Figure 15.6. The block diagram of the braking control system is shown in Figure 15.7. Its main components are a microprocessor that acts as a controller, an air-servo valve that converts the electrical signal from the controller to air pressure, a power changer that converts the air pressure into hydraulic pressure, a hydraulic brake which generates the required braking force, a speed sensor in the feedback loop and a safety circuit that warns against excessive hydraulic brake temperature. Associated with the controller is a pulse counter. This counts the number of pulses (from the speed sensor) in a period equal to the sample period $T = 0.25$s. Assuming that the counter reads the truck speed at $T/2$, it is seen that there is a dead time of 0.125s associated with the control system. The safety circuit is ignored for control-system design purposes as this is usually very fast.

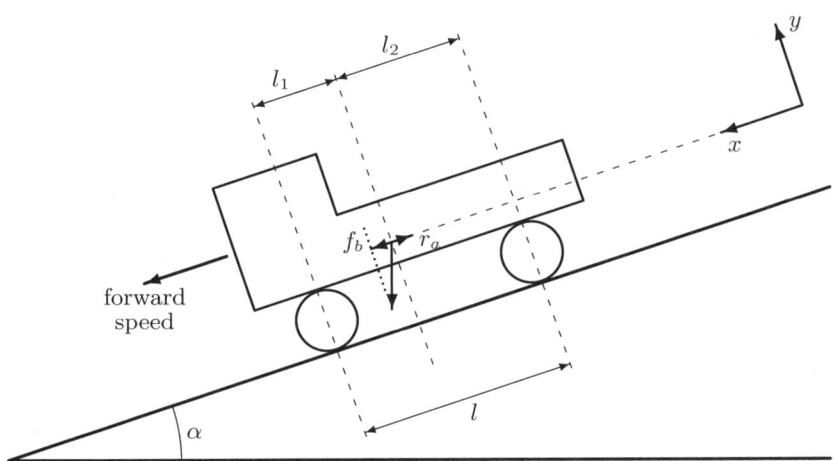

Fig. 15.6. Schematic truck model

After solving the equations arising from the application of the principles of rigid body motion and making adequate assumptions (see Bada 1987), the equivalent simplified model of the truck braking system, which is now shown in 15.8, is obtained. The transfer function $G_1(s)$ includes the air-servo valve, the power changer and the braking force generator. It is given (Bada 1987) by

$$G_1(s) = \frac{a\exp(-\tau s)}{(1+bs)(1+cs)} \tag{15.42}$$

where the normal values of the gain a, the time constant of the servo valve b, the time constant of the power changer c and the dead time τ as given by Bada (1987) will be used in the sequel.

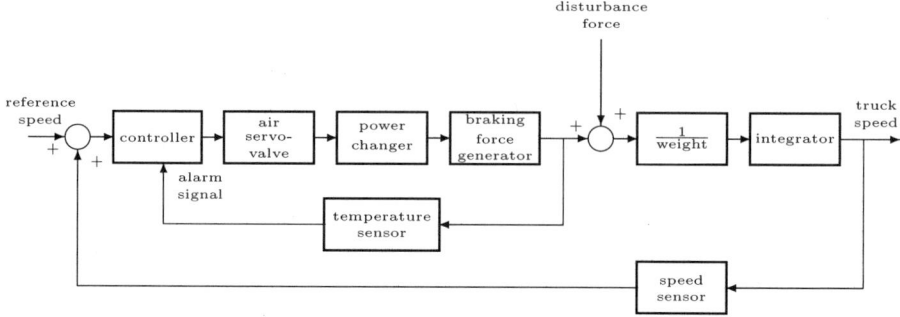

Fig. 15.7. Truck braking control loop

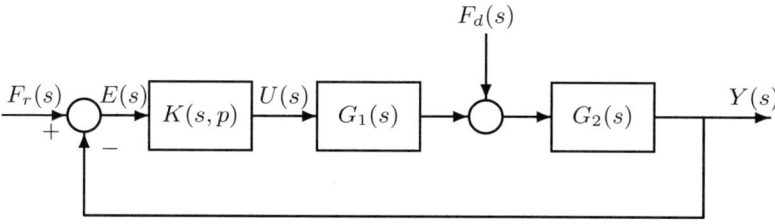

Fig. 15.8. Simplified truck braking control loop

The normal value of τ was obtained on the assumption of linear changes in the truck speed. However, as this is not always the case in road conditions such as those prevailing in mines, variations in the nominal value of τ should be taken into consideration in the design.

The transfer function $G_2(s)$ is given by

$$G_2(s) = \frac{1}{ds} \tag{15.43}$$

where d is the total weight of the truck and its contents. Another source of plant vagueness arises from the operating conditions. As the maximum load of material carried depends on the mine design and the safety regulations enforced and because the truck is to be designed for a wide range of mine conditions, variations in the parameter d should also be taken into account in the design.

Although it is straightforward designing a system exhibiting robust matching for the case where F_d is non-zero, the case with F_d set equal to zero is considered here. Hence the final block diagram of the control system is as shown in Figure 15.1, where the transfer function of the vague plant is

$$G(s) = \frac{(a/d)\exp(-\tau s)}{s(1+0.4s)(1+0.1s)} \tag{15.44}$$

and the class of plants is given by

$$\mathcal{G} = \left\{ g : G(s) = \frac{(a/d)\exp(-\tau s)}{s(1+0.4s)(1+0.1s)}, \quad 0.01 \leq \tau \leq 0.15, \right.$$
$$\left. 208.05 \leq (a/d) \leq 346.75 \right\} \tag{15.45}$$

The analysis here is similar to that given in Section 15.2, except that here, $n = 1.25$, $m = 5.00$, $\delta = 0.001$, $M = 11$ and $D = 0.76$, in (15.3), (15.4), (15.5) and (15.20), see Bada (1987).

15.4.2 Robust Controller Design

At the nominal operating point, the transfer function $G^0(s)$ of the plant braking system is given by

$$G^0(s) = \frac{277.86\exp(-0.125s)}{s(1+0.4s)(1+0.1s)} \tag{15.46}$$

An accurate low-order approximant for $G^0(s)$ is given by

$$G^*(s) = \frac{N(s)}{sD(s)} \tag{15.47}$$

where:

$$N(s) = 0.198s^2 - 13.406s + 277.86 \tag{15.48}$$

$$D(s) = 0.156 \times 10^{-5} s^5 + 0.119 \times 10^{-3} s^4 + 0.436 \times 10^{-2} s^3$$
$$+ 0.081 s^2 + 0.567 s + 1.0 \tag{15.49}$$

By comparing the step and impulse responses of $G^*(s)$ with those of $G^0(s)$ it can be shown that the approximant is sufficiently accurate.

The manufacturer's design specifications call for a very simple controller form. By choosing the controller to be of proportional-plus-integral form (15.39), the controller is to be found satisfying (15.23) with $\varepsilon_i = 2.52$ and 10.00 for $i = 1, 2$ respectively.

In order to calculate $\theta(p)$, it should be noted that $A = 0$ for the approximant (15.46). Moreover, $\sigma_2^*(p) = 0$, thus making $B = 0$, (because the approximant has a pole at the origin). Therefore, only the value of C is needed

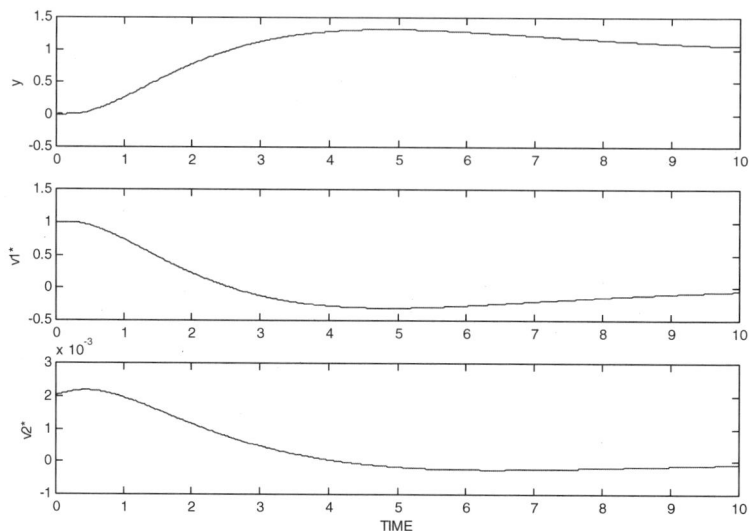

Fig. 15.9. Unit step responses of the closed-loop system

to compute $\theta(p)$. This was determined from the defining equation (15.17). Thereafter, the moving boundaries process was used to solve the problem (15.23).

A satisfactory solution is provided by the PI controller (15.39) with $p = (0.00203, 0.0005)$, for which

$$\begin{aligned}\sigma_1^*(p) &= 0.00 \\ \gamma_1^*(p) &= 0.96 \\ \sigma_2^*(p) &= 0.00 \\ \gamma_2^*(p) &= 0.22 \times 10^{-2} \\ \theta(p) &= 0.71\end{aligned} \qquad (15.50)$$

giving

$$\begin{aligned}\hat{\phi}_1(p) &= 2.52 \\ \hat{\phi}_2(p) &= 0.58 \times 10^{-2}\end{aligned} \qquad (15.51)$$

which satisfy the design criteria. The step responses at the nominal operating point are shown in Figure 15.9. Similar responses were observed at other operating points.

15.4.3 Conclusion

The example of the braking system for a heavy-duty truck shows that Zakian's design framework, can be effectively used to solve the problems which arise from the presence of dead time in the control loop and variations in the plant parameters. The procedure leads to a controller of simple form, which guarantees robust matching for the critical brake control of a heavy duty truck.

Three situations are considered in Bada (1987), *viz.*, variations in the dead time τ alone, variations in a/d only and the situation of simultaneous variations in both τ and a/d. Only the case of simultaneous variations in both τ and a/d is reported here for brevity and also because it is the most difficult.

References

Arunsawatwong, S., (1998) Stability of Zakian I_{MN} recursions for linear delay differential equations, BIT, 38:219-233.

Bada, A.T., (1984) Design of delayed control systems with Zakian's method, *Int. J. Control*, 40:297-316.

Bada, A.T., (1985) Design of delayed control systems using Zakian's framework, *IEE Proc. D*, 132(6):251-256.

Bada, A.T., (1987) Robust brake control for a heavy-duty truck, *IEE Proc. D*, 134:1-8.

Chiang, H.S., and Durbin, L.H., (1981) Variable gain dead-time compensation for the second-order time lag case, *ISA Trans.*, 20(3):1-12.

Denn, M.M., and Lavie, R., (1982) Dynamics of plants with recycle, *Chem. Engry J.*, 24,:55-59.

Luyben, W.L., (1994) Snowball effects in reactor-separator processes with recycle, *Industrial & Eng. Chem. Res.*, 33:299-305.

Martin, J.Jr., Coropio, A.B., and Smith, C.L., (1977) Comparison of tuning methods for temperature control, *ISA Trans.*, 16:53-57.

Ross, C.W., (1977) Evaluation of processes for dead-time processes, *ISA Trans.*, 16:25-34.

Scali, C. and Ferrari, F., (1999) Performance of control systems based on recycle compensators in integrated plants, *J. Proc. Contr.*, 9:425-437.

Scali, C., Antonelli, R. and Ferrari, F., (1998) Design of control systems for chemical plants with recycle streams, *Trends in Chem. Eng.*, 4:33-43.

Scali, C. and Ferrari, F., (1997) Control of systems with recycle by means of compensators, *Comp. Chem. Eng.*, 20:S267-S272.

Taiwo, O., (1986) The design of robust control systems for plants with recycle, *Int. J. Control*, 43:671-678.

Taiwo, O. and Krebs, V., (1996) Robust control systems design for plants with recycle, *Chem. Eng. J. & Biochem. Eng. J.*, 61:1-6.

Taiwo, O., (1985) The dynamics and control of plants with recycle, *J. Nig. Soc. Chem. Engr. J.*, 4:96-107.

Wu, K.L., Yu, C.C., Luyben, W.L. and Skogestad, S., (2002) Reactor/separator processes with recycle – 2. Design for composition control, *Comp. Chem. Eng.*, 27:401-421.

Zakian, V., (1984) A framework for design: theory of majorants. Control Systems Centre Report 604, UMIST, Manchester, England.

Zakian, V., (1978) The performance and sensitivity of classical control systems, *Int. J. Syst. Sci.*, 9:343-355.

Zakian, V., (1979a) New formulations for the method of inequalities, *Proc. IEE*, 126:579-584.

Zakian, V., (1983) A criterion of approximation for the method of inequalities, *Int. J. Control*, 37:1103-1112.

Zakian, V., (1986) A performance criterion, *Int. J. Control*, 43:921-931.

Zakian, V., (1996) Perspectives on the principle of matching and the method of inequalities, *Int. J. Control*, 65:147-175.

Zakian, V., (1979b) Computation of abscissa of stability by repeated use of Routh test, *IEEE Trans. Autom. Control*, 24(4):604-607.

Zakian, V., (1975) Properties of I and J approximants and applications to numerical inversion of Laplace transforms and initial value problems, *J. Math. Anal & Appl.*, 50:191-222.

Zakian, V., and Al-Naib, U., (1973) Design of Dynamical and control Systems by the method of inequalities, *Proc. IEE*, 120:1421-1427.

Index

Abbosh, F.G., 266
Abe, Y., 356
abscissa of stability, 11, 62, 67, 69, 227, 228, 279, 280
actuator saturation, 260
adaptive control, 266
admissibility problem, 232
– first, 69, 70
– second, 69
admissible, 69
– design, 22, 29
admissible set, 22, 29, 62, 69
– empty, 90
– size, 86
advanced turbofan engine, 254
affine function, 100
Ahson, S.I., 254, 277, 278, 280
Al-Janabi, T.H., 212, 253
Al-Naib, U., 13, 19, 61, 73, 79, 143, 211, 219, 226, 229, 231, 233, 253, 287, 312, 320, 323, 329, 333, 334, 340, 349, 352, 374
alpha condition
– definition, 38
α-level set, 103
Aly, A.W., 212
analytical method, *see* optimal control method
Anderson, B.D.O., 166
approximately worst input, 98, 103, 113, 115
approximately worst inputs, 107
Argus 308, 256
Armstrong, D.S., 332
Arunsawatwong, S., 67, 92, 340, 343, 374
Åström, K.J., 252
Atherton, D.P., 289

Atsumi, T., 356
augmented possible set, *see* augmented set
augmented set, 15, 32, 34
– definition, 36
auxiliary vector index, 232, 234, 236, 247
average crossover
– function, 300
– operator, 300

Bada, A.T., 212, 369, 370, 380, 381, 383, 385
Balachandran, R., 253, 254, 269
Balas, G.J., 311
Bamieh, B.A., 166
banana function, 196
bandwidth, 289
Barratt, C.H., 10, 68, 70, 98, 99, 101, 125, 128, 129, 135
Belletutti, J.J., 254, 268
Berry, M.W., 252, 253, 267–269
biggest log-modulus method, *see* biggest log-modulus tuning method
biggest log-modulus tuning method, 253, 254, 269
Birch, B.J., 58, 121
bisection algorithm, 100, 104, 115, 342
Blanchini, F., 337
BLT method, *see* biggest log-modulus tuning method
Bode, H.W., 12
Bollinger, K.E., 212, 253
Bongiorno, Jr., J.J, 122
bound, 232
– constant, 69
– of possible set, 44

388 Index

Boyd, S.P., 10, 68, 70, 98, 99, 101, 125, 128, 129, 135
braking system, 370
Burrows, C.R., 212

camel function, 196, 197
Cartesian product, 33
– set, 60
Cartesian space, 305
Chan, W.S., 67
Chang, S.S.L., 122
characteristic polynomial, 5
characteristics locus design method, 268
Chen, B. M., 356
Chen, D., 253
Chen, G.R., 337
Chen, J., 212, 288
Chen, R., 356
Chen, S., 356
Chen, T., 166
Chiang, H.S., 370
Chiang, R.Y., 311, 318
Chidambaram, M., 212, 253, 254, 269
Chilali, M., 156
Chipperfield, A.J., 312
Cho, W.H., 253
Chopra, A.K., 340
chromosomal representation, 300
closure, 128
Coelho, C.A.D., 212, 253
Coelho, L.D.S., 325
Coello, C.A., 231
complementary sensitivity function, 298
complex conjugate, 111, 113
complex function, 49
complex input, 46
condition for robust match, 374
condition for stability
– Maxwell's, 7, 62
– Nyquist's, 7, 11, 253
conditionally linear, 17
– system, 13, 17, 23
conflict
– between interaction and damping, 273, 276
– between stability margin and interaction, 279, 280

constraint
– eigenvalue, 299
– function, 131
– functional, 130
– hard, 17
– soft, 17
control
– advanced turbofan engine, 269–276
– conventional definition, 8, 251
– decentralised, *see* non-centralised
– definition, 18, 61
– distillation column, 255–269, 320–325
– informal definition, 6, 18
– maglev vehicle, 327–337
– non-centralised, 64, 251, 252
– non-interacting, 252, 253, 264, 267, 269
– robust, 311–325
– turbo-alternator, 277–280
controllability grammian, 114
controllable, 5
controller
– decentralised, *see* non-centralised
– efficiency and structure, 26
– integral, 54
– non-centralised, 64, 254, 255
– parametrisation, 65
conventional foundation for design, 3–11
conventional versus inequalities formulation, 26
convergence space, 71
– size, 86
convex optimisation, 58, 68, 122
convex optimisation problem, 98
– finite-dimensional, 100, 115, 126
– infinite-dimensional, 115, 125, 342
– sensitivity theory, 110
convolution, 99
– integral, 10, 32, 44, 97, 124
convolution integral, 341
Cook, P.A., 212, 253
Coonick, A.H., 255
coprime factorisation, 315
Coropio, A.B., 370
crisis in control, 11–12
criterion for stability, *see* condition for stability

critical control, 288
- multi-objective, 302–308
- robust, 304
critical control system, *see* critical system
critical design, 106
critical output port, 15
critical system, 13, 15, 16, 26, 165, 303, 339, 340, 344, 352, 380
Crossley, T.R., 212, 254
crossover, 300
crossover frequency, 289, 291
- normalised, 290
crowding, 239
cutting-plane algorithm, 129

Dahshan, A.M., 212, 254
Dakev, N.V., 312
Daley, S., 288–290
damping, 289
Davison, E.J., 251, 253, 277
De Hoff, R.L., 270
dead time, 370, 380, 381
Deb, K., 231, 240
decentralised PI control, *see* non-centralised proportional-plus-integral control
decoupling, 252
degree of input-output stability, 8
delay, *see* time delay
delta function, 43
Denn, M.M., 369, 377, 380
dense, 128
design
- extreme, 27
- extremely good, 31
- framework, 19, 341
- freedom, 24
- point, 232
design space, 20, 61, 69
- constrained, 20
- system, 11
Desoer, C.A., 67
de Figueiredo, R.J.P., 337
diagonally dominant, 253
digital redesign, 166
direct design, 166
direct transmission matrix, 5, 8, 18
discrete design, 166

distillation column, 251, 253, 254, 301, 375
Dixon, L.C.W., 196
Dixon, R., 289
Doyle, J.C., 114, 311, 320
DPI controller, *see* diagonal proportional-plus-integral controller
dual problem, 100, 115
Dumont, G.A., 231, 243
Durbin, L.H., 370

Edgar, T.F., 252, 253, 267, 268, 289
Efrati, T., 252
eigenstructure assignment, 288, 294–297
- multi-objective, 297–299
- multi-objective robust, 294
- toolbox, 302
eigenvalue, 261, 274–276, 278
- assignment, 294
eigenvector matrix
- left, 296
- right, 296
El Dahshan, A., 212
El Singaby, M., 212
elitist strategy, 300
ellipsoid algorithm, 101, 103, 111, 129, 135
- constraint iteration, 112
ellipsoid method, *see* ellipsoid algorithm
Engelke, A., 356
environment, 3
- appropriately restricted, 17
environment-system couple, 6, 18, 19
- least conservative, 35
- matched, 14, 34
- well-designed, 15, 34, 41
- well-matched, 14, 34
- wind-gust/wind-turbine, 98, 106
- wind-turbine/wind-gust, 115
environment-system model, 48, 49
error, 4
- vector, 259
exhaustive search, 71
- totally, 71
exponential function, 371
external input space, *see* input space
extreme tolerance to disturbance, 15
extremely tolerant system, 19, 34

390 Index

– definition, 38

Fagervik, K., 252
fast-discretisation, 166, 174
Fawzy, A.S., 212
Ferrari, F., 369, 380
Filipic, B., 231
filter-system combination, 4, 8
finiteness
– function, 69
– point, 69
– search space, 71
– space, 69
first-order plant with dead time, 289
fitness, 300
– assignment, 239
five shifted camel functions
– with four pairs of global minima, 198
– with one pair of global minima, 204–208
five shifted Rosenbrock functions
– with five equal global minima, 198
– with one global minimum, 200–203
Fleming, P.J., 231, 232, 239, 240, 312, 325
Fonseca, C.M., 231, 232, 239, 240
FOPDT, see first-order plant with dead time
Foss, A.S., 252
Fourier coefficients, 126
Fourier series expansion, 126
Fourier transform, 111, 113
– inverse, 113
FPI controller, see full proportional-plus-integral controller
Francis, B.A., 114, 166, 320
frequency response method, 254
frequency-domain identification, 291, 293
Fujisawa, F., 356
full matrix, 254
fully controllable, 53
fully designable, 53

GA, see genetic algorithm
Gahinet, P., 156
gain margin, 7, 9, 253, 289, 291
– normalised, 290
Gan, O.P., 253

Gelatt, C.D., 219
genetic algorithm, 88, 231, 295, 299
– multi-objective, 231, 238, 240, 247
– multiple objective, 239
– non-dominated sorting, 240
– vector evaluated, 239–241
genetic drift, 238
genetic inequalities solver, 238–243
Gershgorin band, 253
Giesy, D.P., 288
global convergence, 83
global search methods, 219–228
Glover, K., 311, 312, 314, 316, 319, 320
Goldberg, D.E., 231, 238–240
Gould, L.A., 62
Grace, A., 291
gradient, 104, 105, 112
Gram-Schmidt procedure, 214
Gray, J.O., 212, 253
Green, M., 314
Gtefenstette, J.J., 231, 242
Gu, D.-W., 22, 68, 212, 220–222, 225, 226, 232, 288, 312, 320, 325
Guo, G., 356

Hackney, R.D., 270, 272
Hagglund, T.J., 252
Halevi, Y., 252
Hall, Jr., W.E., 270
Hang, C.C., 252
Hanselmann, H., 356
Hao, Q., 356
hard disk drive, 355
Hardy space, 304
HDD, see hard disk drive
$\mathcal{H}^\infty$
– approach, 68
– optimisation problem, 304
– optimisation technique, 307
– theory, 311–325
Hirai, H., 356
Hirata, M., 356
Ho, W.K., 253
holder, 168
Hölder's inequality, 45, 57, 59
Holland, J.H., 231
Holmer, P., 328
Horowitz, I.M., 122, 325
Hougen, J.O., 252, 253, 267, 268

Hovd, M., 253, 254
Hoyle, D.J., 314, 325
$\mathcal{H}^2$ optimisation problem, 304
Hu, Y.A., 253
Huang, T., 356
Hyde, R.A., 314, 325
hydraulic actuator, 135
hypercube, 194

IAE, see Integral Absolute Error
Ibaraki, S., 356
Ibrahim, A., 253
Ida, E., 356
ideal match
− definition, 35
ill-posed problem, 234
Imai, Y., 212
I_{MN} approximants, 92, 262, 374
I_{MN} recursion, 343, see I_{MN}
  approximants
impulse response, 43, 45–47, 49, 51, 53, 92, 97, 111, 126, 127, 143, 152, 153, 360, 383
− function, 44, 372
incremental array, 76
inequalities approach, 22
inferior couple, 39
initial population, 300
inner approximation, 128
Inooka, H., 88, 171, 174, 179, 180, 212, 232, 233, 337, 367
input, 3
− fictitious, 341
− persistent, 60, 121, 124, 135, 169, 341
− transient, 47, 60, 97, 124, 149, 166, 169, 341
input space, 303, 304, 329, 371
input-output sensitivity, 8
− minimal, 8
instantaneous transmission, 51
Integral Absolute Error, 10, 53, 57
Integral Square Error, 10, 53, 59
internal uncertainty, 303
interstorey drift, 340, 345
Inverse Nyquist Array method, 65, 254, 277
ISE, see Integral Square Error
Ishihara, T., 88, 171, 174, 179, 180, 193, 212, 232, 233, 337, 367

isothermal continuous stirred tank
  reactor, 375
iterative continuous cycling method, 253
iterative numerical method, 66

Jabbari, F., 340
Jackson, R., 58, 121
Jamshidi, M., 325
Janabi, T.H., 212
Johnson, K.G., 288
Jones, A.H., 231
Jones, D.F., 221

Kadirkamanathan, V., 288
Kaiser, J.F., 62
Kasenally, E.M., 312, 314, 317–320, 322, 325
Katebi, M.R., 212
Katebi, S.D., 212
Kautsky, J., 294, 301
Keller, J.P., 166
Kelly, J.M., 341, 344, 346
Keogh, P.S., 212
Kessler, C., 289
Khachiyan, L.G., 135
Khargonekar, P.P., 288, 320
Kimura, H., 295
King, S.J., 319
Kirkpatrick, S., 219
Kobayashi, M., 356
Kobe earthquake, 351
Kortüm, W., 330–332
Kreisselmeier, G., 254
Krishnakumar, K., 231
Kristinsson, K., 231, 243
Krohling, R.A., 325
Kronecker delta, 259
Kuhn, T.S., 11, 16
Kuhn, U., 347, 348
Kung, H.T., 242
Kunugita, E., 252
Kwon, B.H., 296
Kwong, S., 325

l^1/L^1 optimisation problem, 304
labour, 71
− end putative, 86
− putative, 84

Lagrange multiplier, 100
Lagrangian, 111, 115
– dual, 99
– duality, 98
– function, 100
Lamont, G.B., 231
Landau, I.D., 289
Lane, P.G., 46, 58, 60, 98, 100, 101, 111, 115, 212, 341–343, 353
Laplace transform, 47, 51, 59, 97, 372
Laplace transformation, *see* Laplace transform
Lavie, R., 369, 377, 380
lead compensator, 362
Lee, J.T., 253
Lee, T.H., 253
lifting, 166
Limebeer, D.J.N., 312, 314, 317–320, 322, 325
limit of design, 31, 38
linear couple, 15, 40
– definition, 40
linear environment-system couple, *see* linear couple
linear input-output transformation, 46
linear matrix inequalities, 68, 145, 147
linear quadratic Gaussian
– control, 288
– optimisation problem, 304
Liu, G.P., 166, 212, 288–290, 294, 295, 299, 302–307, 327, 339
Liu, T.K., 88, 193, 232, 233, 312
LMI, *see* linear matrix inequalities
local minimum, 90
local search, 71
– globally convergent, 72
– method, 71
– stuck, 84
– universally globally convergent, 73
local trap, 89
Loh, A.P., 252
Louie, S.G., 220–224
Low, T.S., 356
LQG optimisation problem, *see* linear quadratic Gaussian optimisation problem
Lu, W.M., 337
Luccio, F., 242

Luyben, M.L., 253, 254, 267
Luyben, W.L., 252–254, 267, 268, 369, 380
Lyapunov equation, 114

Maarleveld, A., 251
MacFarlane, A.G.J., 254, 268
Maciejowski, J.M., 64, 65, 81, 254
majorant, 59, 373
Man, K.F., 325
margin of stability, 8, 67
Martin, Jr., J., 370
Maruyama, T., 356
matching
– condition, 98, 107, 108
– definition, 32
– inverse problem, 14, 41
MATLAB, 113, 114, 243, 311, 323, 334
– Control System Toolbox, 113, 114
– LMI Control Toolbox, 162
– Optimization Toolbox, 291
– Robust Control Toolbox, 163, 318
– Signal Processing Toolbox, 362
– Simulink, 364
Mattson, S.E., 115
maximal input, 122
maximal interaction, 262
maximum controller output, 261
Maxwell, J.C., 3, 6, 62
Mayne, D.Q., 83
MBP, *see* moving boundaries process
MBP(R), *see* moving boundaries process with Rosenbrock trial generator
McFarlane, D.C., 312, 314, 316, 319
McGinnis, R.G., 254
measures of sensitivity, 9
Megginson, R.E., 128
Mellichamp, D.A., 289
Melsa, J.L., 254, 270–272
method of inequalities, 13, 15, 19, 24, 46, 61–68, 231, 232, 240, 241, 246, 251, 253–255, 266–268, 270, 276, 277, 280, 307, 311, 312, 319–320, 325, 333, 339, 353
Metropolis algorithm, 220
Metropolis, N., 220
MGA, *see* multi-objective genetic algorithm

Miller R.J., 270, 272
MIMO system, *see* multi-input multi-output system
minimal admissible set, 28
minimal realisation, 262, 271, 273, 278
minimax
- optimisation, 291
- process, 81
Mirrazavi, S.K., 221
modal analysis, 254
mode, 5
- uncontrollable, 7
model
- non-parametric, 291
- paramtric transfer function, 289
MOGA, *see* multiple objective genetic algorithm
MoI, *see* method of inequalities
Monica, T.J., 253
Moore, B.C., 294
moving boundaries process, 72, 73, 79, 81, 89, 143, 193, 211, 233, 245, 261–264, 273, 320, 323, 334, 349, 351, 374, 378
- with Rosenbrock trial generator, 89, 213, 227
multi-input multi-output system, 294
multi-modal, 196
multi-objective
- control, 287, 288
- minimisation, 221
- optimisation, 81, 246
- performance function, 299
- problem, 299
multivariable control system, 312, 313
Munro, N., 227, 253
Murad, G., 212, 288, 312, 320, 325
Murase, A., 356
mutation operator, 300
Müller, P.C., 333

NA method, *see* node array method
NA&MBP(R), *see* node array method employing MBP(R)
Nakanishi, E., 252
Nakayama, H., 233
negotiated inequalities, 26
Nelder-Mead dynamic minimax method, 323

Nemirovskii, A.S., 135
Nemirovsky, A., 162
Nesterov, Y., 162
Newton, G.C., 62
Ng, W.Y., 212, 323
NIC control, *see* non-interacting control
Nichols, N.B., 289
Nichols, N.K., 294, 301
Nicholson, H., 254, 277, 278, 280
Niederlinski, A., 251, 252
Niksefat, N., 135
node, 73, 75
- searched, 74
- sequence, 76, 87
node array, 71, 74, 75
node array method, 68–92, 193
- employing MBP(R), 89, 193
non-centralised proportional-plus-integral control, 253
non-divergence rule, 79
non-divergent, 72
non-dominated ranking, *see* Pareto ranking
non-dominated set, 239
Nonami, K., 356
norm
- 1-, 46, 123
- 2-, 46
- $\mathcal{H}^\infty$-, 10, 287, 298, 305, 313
- $\mathcal{H}^2$-, 288, 304
- ∞-, 123
- $\mathcal{L}^1$-, 288
- operator, 10
- p-, 8, 44
- q-, 9
- singular value matrix, 298
Northridge earthquake, 351
NSGA, *see* non-dominated sorting genetic algorithm
Nyquist diagram, 7, 9
Nyquist's method, 62
Nyquist, H., 3

objective, 69
- functional, 130
- vector, 69
objective function, 20, 69, 131, 232
- dual, 100
- performance, 20

- shape, 86
official paradigm, 16
Ohno, K., 356
Ono, T., 68, 171, 174, 179, 180, 337, 367
optimal control method, 10, 27, 64, 66, 254, 277
optimal control techniques, *see* optimal control method
optimal design, 31
optimal-tuning PID control scheme, 291
ordering relation, 31, 77, 233
- conventional, 80
- distinct multi-objective, 79, 81
- inequalities, 79
- multi-objective, *see* distinct multi-objective
- Pareto, 79, 233
- Pareto and inequalities, 80
- scalar, 82
output, 4
- performance, 303
- space, 371
over-specified problem, 27
overshoot, 262, 273, 274

Packard, A., 311
Palmor, Z.J., 252
Pangalos, P., 319
Papoulis, A., 58
Pareto minimiser, 20, 21, 26, 29, 75, 78, 82, 231, 233
- and admissible point, 82
- and minimal admissible set, 30
- strict, 233
Pareto objective, 91
Pareto ranking, 239, 240, 247
Pareto set, 234, 239, 247
- optimal, 238
- strict, 234
Parseval's theorem, 111, 112
Patton, R.J., 212, 288, 294, 295, 299, 302, 303, 305, 306
peak output, 14, 20, 33, 97, 99, 100, 102, 105, 107–110, 115, 121–125, 135, 143
- function, 38
- functional, 14

Pearson, J.B., 166
Peczkowski, J.L., 254, 270–272
perfect match, 15, 34, 35
- definition, 36
Perkins, J.D., 312, 314, 317–320, 322, 325
persistent disturbance, *see* persistent input
persistent function, 48, 121
Pessen, D.W., 289
Phan, D.T., 356
phase margin, 7, 9, 253, 289, 291
- normalised, 290
PI controller, *see* proportional-plus-integral controller
PI7 controller, *see* proportional-plus-integral controller with seven parameters
PID control, *see* proportional, integral and derivative control
PID controller, *see* proportional, integral and derivative controller
PID tuning rule
- integral of absolute error, 289
- integral of squared time weighted error, 289
- modified Ziegler-Nichols, 253
- no-overshoot, 289
- some-overshoot, 289
- symmetric optimum, 289
- Ziegler-Nichols, 253, 289
piecewise continuous function, 43
Pike, D.H., 253
PoI, *see* principle of inequalities
pointwise supremum, 100
pole, 9, 50, 97, 99
- placement, 9
PoM, *see* principle of matching
port, 4
- critical output, 13
- error, 4
- input, 4
- output, 4
Porter, B., 231
possible inputs, 339
possible set, 14, 18, 32, 97, 98, 123

Postlethwaite, I., 22, 68, 212, 220–222, 225, 226, 232, 251, 253, 254, 288, 311–313, 318–320, 325
Power System Stabiliser, 255
Prabhu, E.S., 212
Preparata, F.P., 242
primal problem, 99
principle
– of biological evolution, 231
– of inequalities, 14, 19–31, 34, 61, 79, 219, 231, 233, 235, 254, 287, 288, 295, 299, 311, 312
– of matching, 15, 16, 19, 32–43, 121, 123
– – definition, 32
– of uniform stability, 11, 61, 67
principle of inequalities, 341
principle of matching, 339, 341, 352
probing, 5
process of negotiation, 25, 35
proportional, integral and derivative control, 288
– decentralised, see non-centralised
– multi-objective, 289–291
– non-centralised, 253
– rule-based, 288–289
proportional, integral and derivative controller, 115
– decentralised, see non-centralised
– non-centralised, 252
proportional-plus-integral controller, 63, 262, 280, 362, 378, 384
– diagonal, 251, 253, 264, 266–269
– full, 253, 262, 266–269
– multivariable, 273
– with seven parameters, 263

quasi-convex problem, 100, 115, 342
Queck, C.K., 252

Rademaker, O., 251
Raimirez, W.F., 253
rate of progress, 85
ratio control, 252
rational approximant, 370
Rau, N.S., 277
real space, 61
recycle system, 369
regulator

– multivariable, 278
– problem, 260
Reinelt, W., 122
relay feedback method, 252
response, 5
Ridjnsdorp, J.E., 251, 253
Ritz approximation, 99, 125
robust design procedure, 319
robust matching
– for plants with recycle, 374–380
– for truck braking control, 380–385
robust performance, 316–318
robustness, 298
Rockafellar, R.T., 98, 129
Rofooei, F., 340
root locus, 9
Rosenbluth, A., 220
Rosenbluth, M., 220
Rosenbrock function, see Rosenbrock test function
Rosenbrock test function, 196, 197
Rosenbrock, H.H., 78, 196, 213, 251, 254, 258, 277
Ross, C.W., 370
Ross, D., 328
rotary hydraulic test rig, 292
Rotea, M.A., 288
Routh's test, 62
Royden, H.L., 46
Rutland, N.K., 43, 46, 55, 212

Safonov, M.G., 311, 318
Sahba, M., 83
Sain, M.K., 254, 270–272
SAIS, see simulated annealing inequalities solver
sampled-data system, 5, 165
sampler, 168
Sandhu, P.S., 212, 253
Sasaki, M., 356
Satoh, T., 58, 76, 79, 84, 85, 87–89, 99, 212, 220, 224, 228, 312, 337, 367
saturated set
– definition, 37
saturation type non-linearity, 17
Sawaragi, Y., 233
Scali, C., 369, 380
Schaffer, J.D., 231, 239
Scherer, C., 156

Schmidt, G., 347, 348
Schmitendorf, W.E., 340, 344, 346
Schwanke, C.O., 252, 253, 267, 268
Schwarz' inequality, 99
Schy, A.A., 288
search space, 70
– continuous, 73
– discretised, 71, 73, 74
– non-discretised, see continuous
Seborg, D.E., 253, 289
sensitivity, 7, 9
– function, 298
– individual eigenvalue, 297
– local, 110
– matrix, 40
– of peak output, 98, 108
– overall eigenvalue, 297
Sepehri, N., 135
sequential loop closure and tuning, 252, 253
– improved, 254
settling time, 9, 260, 262, 273
sharing, 239, 240
Shinskey, F.G., 251
Shor, N.Z., 135
sign function, 131
signal
– periodic multi-sine excitation, 293
– persistent bounded casual, 304
– square-integrable, 304
simulated annealing, 87, 219–228
– acceptance criterion, 223
– cooling, 222
– energy function, 223
– termination criterion, 224
simulated annealing inequalities solver, 219–228, 245
singular value, 304
– technique, 298
Sinha, P.K., 330
SISO transfer function, see single-input single-output transfer function
Skira, C. A., 270
Skogestad, S., 251, 253, 254, 311, 313, 318, 319, 369, 380
small gain theorem, 304, 313, 316
Small, L.L., 270, 272
Smith, C., 356

Smith, C.L., 370
Smith, R., 311
smoothing filter, 55
sorting
– brute force, 241
– non-dominated, 242
specifically bounded output, 13
– peak, 14
spectral abscissa, 11, 344, see abscissa of stability
speed-of-response specification, 289
Srinivas, N., 231, 240
Srinivasa Prabhu, E., 254
stability, 6, 298
– BIBO, 47
– external, 7
– input-output, 7, 18, 46
– internal, 10, 67, 315
– p-NBO, 47
– robust, 303, 313, 315
stability margin, 289
stability point, 343
Stainthorp, F.P., 255
state, 5
state-space
– equations, 5, 67
– form, 5, 66
– infinite-dimensional, 377
– model, 181
– nominal model, 331
– realisation, 318
state-space realisation, 114
steady-state error, 260, 289, 291
– normalised, 290
Steinhauser, R., 254
step length, 194
step response, 53, 58, 59, 92, 122, 145, 154, 165, 166, 171, 181–183, 187, 189, 226–228, 245, 246, 260–262, 321, 324, 330, 360, 383
– functional, 229, 321
stochastic control theory, 304
stopping criterion, 84, 98, 101, 102, 105, 115, 210, 211
stopping rule, see stopping criterion
subgradient, 101, 103, 105, 129
subsidiary local search, 79, 82, 90

subsidiary search, *see* subsidiary local search
Sugimoto, K., 212
Suzuki, T., 356
switching function, 122
system, 4
− computer-controlled, 165
− infinite-dimensional, 371
− lightly damped, 275
− non-linear, 371
− time-varying, 371
− vague, 5, 369, 371, 378
Szego, G.P., 196

Tabak, T., 288
Tadjbaksh, I.G., 340
Taiwo, O., 212, 252–255, 261, 262, 264, 267, 270, 272, 273, 277, 280, 369, 374, 376–380
Tamiz, M., 221
Tan, K.K., 252
Tang, K.S., 325
Tanino, T., 233
Tannenbaum, A.R., 114
Teller, A., 220
Teller, E., 220
test problems, 195–215, 224–227, 243–245
Thomas, M.E., 253
time constant, 258
time delay, 5, 92, 258
tolerable input, 32
− definition, 33
tolerable set, 14, 32, 33, 107, 115
tolerance, 13, 22, 69, 232
− flexible, 23
− strict, 23
Tomizuka, M., 356
transfer function, 7–11, 20, 32, 43, 46, 47, 50, 54, 55, 63, 65, 66, 70, 92, 99, 113, 140, 141, 154, 161, 162, 186, 287–289, 298, 304, 360, 371, 372, 374, 376–378, 380–383
− closed-loop, 161, 288, 314, 318, 321
− matrix, 40, 62, 64, 65, 182, 252, 262, 264, 270, 271, 277–279, 315
− non-rational, 370
− open-loop, 363
− proper rational, 6, 97, 115

− rational, 5, 50, 370, 380
− single-input single-output, 311
− stable, 7, 10, 64
− stable unknown, 315
− strictly proper, 99
− strictly proper rational, 6, 43, 46, 92, 374
− unstable, 10
− variable, 304
transient function, 48
trial generator, 78, 84, 222
− Rosenbrock, 78, 84, 89, 193
− simulated annealing, 222–223
trial point, 78, 212
trial point generator, *see* trial generator
turbo-alternator, 277

Unbehauen, H., 303, 305, 306
undershoot, 9
universal set, 69
Urbancic, T., 231
utopian, 233, 234
− non-unique, 237
− unique, 236
Utzt, A., 330–332

vague plant, 372, 374, 383
− infinite-dimensional, 374, 380
Van Dooren, P., 294, 301
Vanderbilt, D., 220–224
Varsek, A., 231
Vasnani, V.U., 252
Vecchi, M.P., 219
VEGA, *see* vector evaluated genetic algorithm
Veldhuizen, D.A., 231
Venn
− diagram, 24, 80, 89
− formulation, 24
Vinante, C.D., 252, 253, 267, 268
Voda, A., 289

Waller, K.V.T., 252, 264
Wang, Q.-G., 252
Weerasooriya, S., 356
well-designed couple
− definition, 39
Whidborne, J.F., 22, 68, 87, 161, 166, 174, 186, 193, 212, 220–222, 225, 226,

232, 288, 303, 307, 312, 319, 320, 325, 327, 329, 337, 339, 367
white Gaussian process, 304
White, M.T., 356
wind turbine and controller system, 114–115
Wonham, W.M., 294
Wood, R.K., 252–254, 267–269
Wu, K.L., 369, 380

Yamaguchi, T., 356
Yang, J.B., 288
Yaniv, O., 325
Yasuoka, H., 252
Yoshida, T., 356
Youla parametrisation, 122
Youn, M.J., 296
Yu, C.C., 252, 253, 369, 380

Yudin, D.B., 135

Zakian, V., 11, 13–15, 17, 19, 29, 31, 34, 35, 41–43, 46–48, 50, 51, 53, 57, 58, 60–62, 66, 67, 70, 73, 79, 80, 82, 91, 92, 97, 98, 106, 121, 123, 143, 145, 150, 165, 193, 194, 209–211, 219, 226, 227, 229, 231, 233, 235, 253, 259, 262, 272, 287, 303, 312, 320, 323, 327, 329, 333, 334, 337, 339–341, 343, 344, 349, 352, 362, 367, 369–371, 373, 374, 380
Zeidler, E., 128
Zhang, P., 255
Zhou, K., 311
Zhuang, M., 289
Ziegler, J.G., 289
Zweiri, Y.H., 319